Introduction to Shape Optimization

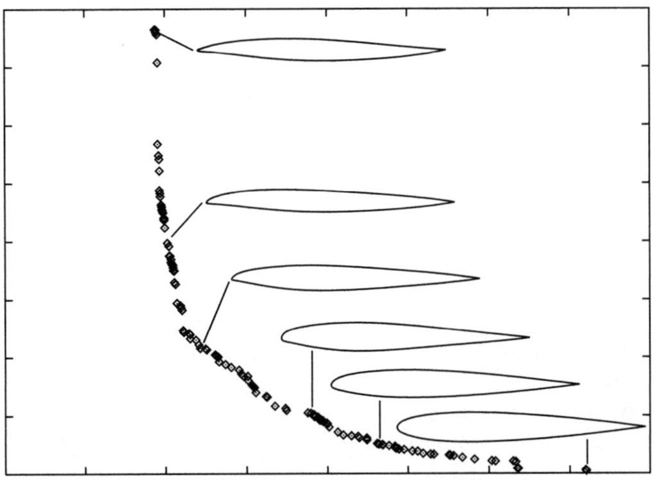

Advances in Design and Control

SIAM's Advances in Design and Control series consists of texts and monographs dealing with all areas of design and control and their applications. Topics of interest include shape optimization, multidisciplinary design, trajectory optimization, feedback, and optimal control. The series focuses on the mathematical and computational aspects of engineering design and control that are usable in a wide variety of scientific and engineering disciplines.

Series Volumes

Haslinger, J. and Mäkinen, R. A. E., *Introduction to Shape Optimization: Theory, Approximation, and Computation*

Antoulas, A. C., *Lectures on the Approximation of Linear Dynamical Systems*

Gunzburger, Max D., *Perspectives in Flow Control and Optimization*

Delfour, M. C. and Zolésio, J.-P., *Shapes and Geometries: Analysis, Differential Calculus, and Optimization*

Betts, John T., *Practical Methods for Optimal Control Using Nonlinear Programming*

El Ghaoui, Laurent and Niculescu, Silviu-Iulian, eds., *Advances in Linear Matrix Inequality Methods in Control*

Helton, J. William and James, Matthew R., *Extending H^∞ Control to Nonlinear Systems: Control of Nonlinear Systems to Achieve Performance Objectives*

Introduction to Shape Optimization

Theory, Approximation, and Computation

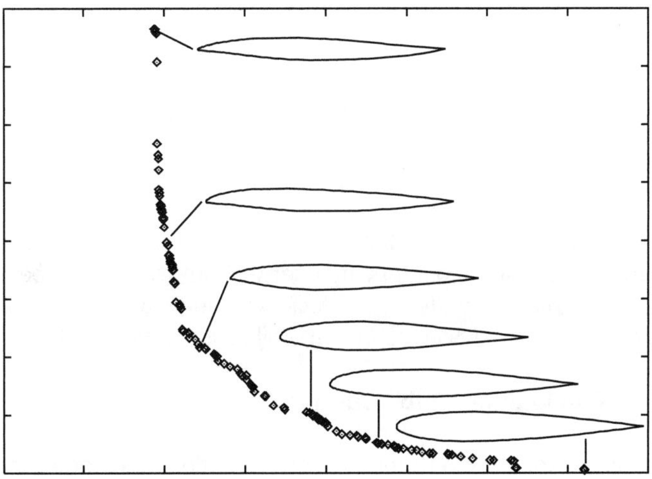

J. Haslinger
Charles University
Prague, Czech Republic

R. A. E. Mäkinen
University of Jyväskylä
Jyväskylä, Finland

Society for Industrial and Applied Mathematics
Philadelphia

10 9 8 7 6 5 4 3 2 1

Library of Congress Cataloging-in-Publication Data
Haslinger, J.
Introduction to shape optimization : theory, approximation, and computation / J. Haslinger, R.A.E. Mäkinen.
p. cm. – (Advances in design and control)
Includes bibliographical references and index.
ISBN 0-89871-536-9
1. Structural optimization–Mathematics. I. Mäkinen, R.A.E. II. Title. III. Series.

TA658.8.H35 2003
624.1'7713–dc21 2003041589

Contents

Preface

Before we explain our motivation for writing this book, let us place its subject in a more general context. Shape optimization can be viewed as a part of the important branch of computational mechanics called *structural optimization*. In structural optimization problems one tries to set up some data of the mathematical model that describe the behavior of a structure in order to find a situation in which the structure exhibits a priori given properties. In other words, some of the data are considered to be parameters (control variables) by means of which one fine tunes the structure until optimal (desired) properties are achieved. The nature of these parameters can vary. They may reflect material properties of the structure. In this case, the control variables enter into coefficients of differential equations. If one optimizes a distribution of loads applied to the structure, then the control variables appear on the right-hand side of equations. In shape optimization, as the term indicates, optimization of the geometry is of primary interest.

From our daily experience we know that the efficiency and reliability of manufactured products depend on geometrical aspects, among others. Therefore, it is not surprising that optimal shape design problems have attracted the interest of many applied mathematicians and engineers. Nowadays shape optimization represents a vast scientific discipline involving all problems in which the geometry (in a broad sense) is subject to optimization. For a finer classification, we distinguish the following three branches of shape optimization:

(i) *sizing optimization*: a typical size of a structure is optimized (for example, a thickness distribution of a beam or a plate);

(ii) *shape optimization* itself: the shape of a structure is optimized without changing the topology;

(iii) *topology optimization*: the topology of a structure, as well as the shape, is optimized by, for example, creating holes.

To keep the book self-contained we focus on (i) and (ii). Topology optimization needs deeper mathematical tools, which are beyond the scope of basic courses in mathematics, to be presented rigorously.

One important feature of shape optimization is its *interdisciplinary character*. First, the problem has to be well posed from the mechanical point of view, requiring a good understanding of the physical background. Then one has to find an appropriate mathematical model that can be used for the numerical realization. In this stage no less than three mathematical disciplines interfere: the theory of partial differential equations (PDEs), approximation of PDEs (usually by finite element methods), and the theory of nonlinear mathematical programming. The complex character of optimal shape design problems makes the presentation of the topic in some respects difficult.

Nowadays, there exist quite a lot of books of different levels on shape optimization. But what is common to all of them is the fact that they are usually focused on some of the above-mentioned aspects, while other aspects are completely omitted. Thus one can find books dealing solely with sensitivity analysis (see [HCK86], [SZ92]) but omitting approximation and computational aspects. We can find excellent textbooks for graduate students ([Aro89], [HGK90], [HA79]) in which great attention is paid to the presentation of basic numerical minimization methods and their applications to simple sizing problems for trusses, for example. But, on the other hand, practically no problems from (ii) (as above) are discussed there. Finally, one can find books devoted only to approximation theory in optimal shape design problems [HN96].

This book is directed at students of applied mathematics, scientific computing, and engineering (civil, structural, mechanical, aeronautical, and electrical). It was our aim to write a self-contained book, including both mathematical and computational aspects of sizing and shape optimization (SSO), enabling the reader to enter the field rapidly, giving more complex information than can be found in other books on the subject. Part of the material is suitable for senior undergraduate work, while most of it is intended to be used in postgraduate work. It is assumed that the reader has some preliminary knowledge of PDEs and their numerical solution, although some review of these topics is provided in Appendix A. Moreover, knowledge of modern programming languages, such as C++ or Fortran 90,[1] is needed to understand some of the technical sections in Chapter 5.

The book has three parts. Part I presents an elementary mathematical introduction to SSO problems. Topics such as the existence of solutions, appropriate discretizations of problems, and convergence properties of discrete models are studied. Results are presented in an abstract, unified way permitting their application not only in problems of solid mechanics (standard in existing books) but also in other areas of mathematical physics (fluid mechanics, electromagnetism, etc.). Part II deals with modern computational aspects in shape optimization. The reader can find results on sensitivity analysis and on gradient, evolutionary, and stochastic type minimization methods, including methods of multiobjective optimization. Special chapters are devoted to new trends, such as automatic differentiation of computer programs and the use of fictitious domain techniques in shape optimization. All these results are then used in Part III, where nontrivial applications in various areas of industry, such as contact stress minimization for elasto-plastic bodies, multidisciplinary optimization of an airfoil, and shape optimization of a dividing tube, are presented.

Acknowledgments. The authors wish to express their gratitude to the following people for their help during the writing of this book: Tomas Kozubek from the Technical University of Ostrava, who computed some of examples in Chapters 1 and 6 and produced most of the figures; Ladislav Luksan from the Czech Academy of Sciences in Prague and Kaisa Miettinen from the University of Jyväskylä for their valuable comments concerning Chapter 4; and Jari Toivanen from the University of Jyväskylä, who computed the numerical example in Section 8.2.

[1]The code in this book can be used with Fortran 90 and its later variants.

This work was supported by grant IAA1075005 of the Czech Academy of Science, grant 101/01/0538 of the Grant Agency of the Czech Republic, grants 44568 and 48464 of the Academy of Finland, and by the Department of Mathematical Information Technology, University of Jyväskylä.

The motto of our effort when writing this book could be formulated as follows: *An understanding of the problem in all its complexity makes practical realization easier and obtained results more reliable.* We believe that our book will assist its readers to accomplish this goal.

J.H. and R.A.E.M.

Notation

Banach and Hilbert spaces

$C^k(A, B)$	functions defined in A, taking values in B, continuously differentiable in the Fréchet sense up to order $k \in \{0\} \cup \mathbb{N}$ ($C^k(A) := C^k(A, \mathbb{R})$);
$C^k(\overline{\Omega})$	functions whose derivatives up to order k are continuous in $\overline{\Omega}$, $k \in \{0\} \cup \mathbb{N} \cup \{\infty\}$ ($C(\overline{\Omega}) := C^0(\overline{\Omega})$);
$C_0^k(\overline{\Omega})$	functions from $C^k(\overline{\Omega})$ vanishing in the vicinity of $\partial\Omega$;
$C^{0,1}(\overline{\Omega})$	Lipschitz continuous functions in $\overline{\Omega}$;
$L^p(\Omega)$	Lebesgue integrable functions in Ω, $p \in [1, \infty[$;
$L^\infty(\Omega)$	bounded, measurable functions in Ω;
$H^k(\Omega)$	functions whose generalized derivatives up to order $k \in \{0\} \cup \mathbb{N}$ are square integrable in Ω ($H^0(\Omega) := L^2(\Omega)$);
$H_0^k(\Omega)$	functions from $H^k(\Omega)$, $k \in \mathbb{N}$, whose derivatives up to order $(k-1)$ in the sense of traces are equal to zero on $\partial\Omega$;
$H^{k,\infty}(\Omega)$	functions from $L^\infty(\Omega)$ whose derivatives up to order $k \in \{0\} \cup \mathbb{N}$ belong to $L^\infty(\Omega)$ ($H^{0,\infty}(\Omega) := L^\infty(\Omega)$);
$V(\Omega)$	subspace of $H^k(\Omega)$;
$\mathbb{V}(\Omega)$	Cartesian product of $V(\Omega)$;
$H^{-k}(\Omega)$	dual space of $H^k(\Omega)$, $k \in \mathbb{N}$;
$H^{1/2}(\Gamma)$	traces on $\Gamma \subset \partial\Omega$ of functions from $H^1(\Omega)$;
$H^{-1/2}(\Gamma)$	dual space of $H^{1/2}(\Gamma)$;

Convergences

$\to$ in X	convergence in the norm of a normed space X (strong convergence);
$\rightharpoonup$ in X	weak convergence in a normed space X;
$\rightrightarrows$ in Q	uniform convergence of a sequence of continuous functions in Q;

Differential calculus

$\dot{f}$, f'	material and shape derivatives, respectively, of f;

$\dfrac{\partial f(x)}{\partial x_i}, \dfrac{\partial^2 f(x)}{\partial x_i \partial x_j}$	first and second order generalized derivatives, respectively, of $f : \mathbb{R}^n \to \mathbb{R}, n \geq 2$;
$D^\alpha f = \dfrac{\partial^{\lvert\alpha\rvert} f}{\partial x_1^{\alpha_1} \cdots \partial x_n^{\alpha_n}}$	αth generalized derivative of $f : \mathbb{R}^n \to \mathbb{R}$, $\lvert\alpha\rvert = \sum_{i=1}^n \alpha_i$, $\alpha_i \in \{0\} \cup \mathbb{N}$;
$f'(\boldsymbol{\alpha}; \boldsymbol{\beta})$	directional derivative of f at a point $\boldsymbol{\alpha}$ and in a direction $\boldsymbol{\beta}$;
∇f	gradient of f;
$\nabla_x f$	partial gradient of f with respect to x;
Df (also $\boldsymbol{J}$)	Jacobian of f;
Δf	Laplacian of f;
$\partial f/\partial \nu$, $\partial f/\partial s$	normal and tangential derivatives, respectively, of f on $\Gamma \subset \partial\Omega$;
curl f	rotation of a function $f : \mathbb{R}^2 \to \mathbb{R}$;
div f	divergence of f;

Domains and related notions

$\mathbb{R}$	set of all real numbers;
$\mathbb{C}$	set of all complex numbers;
$\mathbb{R}^n$	Euclidean space of dimension n;
$\mathbb{N}$	set of all positive integers;
Ω	bounded domain in $\mathbb{R}^n$;
$\overline{\Omega}$	closure of Ω;
$\operatorname{int}\Omega$, $\operatorname{ext}\Omega$	interior and exterior of Ω, respectively;
$\partial\Omega$	boundary of Ω;
Γ	part of the boundary $\partial\Omega$;
$B_\delta(Q)$	δ-neighborhood of a set $Q \subset \mathbb{R}^n$;
conv Q	closed convex hull of Q;

Finite elements

T	triangle;
R	convex quadrilateral;
$P_k(T)$	polynomials of degree $\leq k$ in T;
$Q_1(R)$	four-noded isoparametric element in R;
h	norm of a partition of $\overline{\Omega}$ into finite elements;
$\widehat{\mathcal{T}}_h$, $\widehat{\mathcal{R}}_h$	uniform triangulation and rectangulation of $\overline{\widehat{\Omega}}$, respectively, whose norm is h;
$\mathcal{T}(h, s_\varkappa)$	triangulation of $\overline{\Omega(s_\varkappa)}$, $s_\varkappa \in U_\varkappa^{ad}$, whose norm is h;
$\mathcal{R}(h, s_\varkappa)$	partition of $\overline{\Omega(s_\varkappa)}$, $s_\varkappa \in U_\varkappa^{ad}$, into convex quadrilaterals whose norm is h;
$\Omega_h(s_\varkappa)$	domain $\Omega(s_\varkappa)$ with a given partition $\mathcal{T}(h, s_\varkappa)$, $\mathcal{R}(h, s_\varkappa)$, $s_\varkappa \in U_\varkappa^{ad}$;

$V_h(s_\varkappa)$	finite element space in $\Omega_h(s_\varkappa)$, $s_\varkappa \in U_\varkappa^{ad}$;
$\mathbb{V}_h(s_\varkappa)$	Cartesian product of $V_h(s_\varkappa)$;
u_h	finite element solution;

Fluid mechanics

τ	stress tensor (components $\tau_{ij},\ i, j = 1, \dots, d;\ d = 2, 3$);
ε	strain rate tensor (components $\varepsilon_{ij},\ i, j = 1, \dots, d;\ d = 2, 3$);
p	static pressure;
μ	viscosity of a fluid;
ϱ	density of a fluid;

Linear algebra

$\boldsymbol{x}, \boldsymbol{y}, \boldsymbol{\alpha}$	column vectors in $\mathbb{R}^n$;
$\boldsymbol{x}^{\mathrm{T}}$	transpose of $\boldsymbol{x}$;
$\boldsymbol{A}, \boldsymbol{B}$	matrices $\boldsymbol{A}, \boldsymbol{B}$;
$\boldsymbol{A}^{-1}$	inverse of $\boldsymbol{A}$;
$\boldsymbol{A}^{\mathrm{T}}$	transpose of $\boldsymbol{A}$;
$\lvert\boldsymbol{A}\rvert$	determinant of $\boldsymbol{A}$;
$\operatorname{tr}\boldsymbol{A}$	trace of $\boldsymbol{A}$;
$\boldsymbol{I}$	identity matrix;

Linear elasticity

τ	stress tensor (components $\tau_{ij},\ i, j = 1, \dots, d; d = 2, 3$);
ε	linearized strain tensor (components $\varepsilon_{ij},\ i, j = 1, \dots, d; d = 2, 3$);
c_{ijkl}	elasticity coefficients defining a Hooke's law;
κ, μ	bulk and shear moduli, respectively;
f	density of body forces;
P	density of surface tractions;

Mappings

$A : X \to Y$	A maps X into Y;
A^{-1}	inverse of A;
$\mathcal{L}(X, Y)$	space of all linear continuous mappings from X into Y;
$f \circ g$ (also $f(g)$)	composite function;

Miscellaneous

δ_{ij}	Kronecker symbol;
ν	unit outward normal vector to $\partial\Omega$;
c	generic positive constant;
χ	characteristic function of a set;

Norms and scalar products

$\|\cdot\|$ (also $\|\cdot\|_X$)	norm in a normed space X;
$\lvert\cdot\rvert$ (also $\lvert\cdot\rvert_X$)	seminorm in a normed space X;
$(\cdot,\cdot)$ (also $(\cdot,\cdot)_X$)	scalar product in X;

Norms and scalar products in particular spaces

$\|\boldsymbol{x}\|_p$	p-norm in $\mathbb{R}^n$; i.e., $\|\boldsymbol{x}\|_p := \sqrt[p]{\sum_{i=1}^n \lvert x_i\rvert^p}$, $p \in [1,\infty[$, $\|\boldsymbol{x}\|_\infty := \max_{i=1,\dots,n} \lvert x_i\rvert$;
$\boldsymbol{x}^{\mathrm{T}}\boldsymbol{y}$	scalar product of $\boldsymbol{x}, \boldsymbol{y} \in \mathbb{R}^n$;
$f \cdot g$	scalar product of two vector-valued functions $f, g : \Omega \to \mathbb{R}^n$; i.e., $(f\cdot g)(x) = \sum_{i=1}^n f_i(x)g_i(x)$;
$(f,g)_{k,\Omega}$	scalar product of f, g in $H^k(\Omega)$, $k \in \{0\} \cup \mathbb{N}$;
$\|v\|_{k,\Omega}$, $\lvert v\rvert_{k,\Omega}$	norm and seminorm, respectively, of v in the Sobolev space $H^k(\Omega)$, $k \in \{0\} \cup \mathbb{N}$;
$\|v\|_{k,\infty,\Omega}$, $\lvert v\rvert_{k,\infty,\Omega}$	norm amd seminorm, respectively, of v in the Sobolev space $H^{k,\infty}(\Omega)$, $k \in \{0\} \cup \mathbb{N}$;
$\|v\|_{C^k(\overline{\Omega})}$	norm of v in the space $C^k(\overline{\Omega})$, $k \in \{0\} \cup \mathbb{N}$.

Introduction

As we have already mentioned in the preface, this book consists of three parts and two appendices.

Part I is devoted to mathematical aspects of sizing and shape optimization (SSO). Our aim is to convince the reader that thorough mathematical analysis is an important part of the solution process. In Chapter 1 we present simple optimization problems that at first glance seem to be quite standard. A closer look, however, reveals defects such as the nonexistence of solutions in the classical sense or the nondifferentiability of the solution to a state problem with respect to design variables. These circumstances certainly affect the numerical realization. Further, we present one simple example whose exact solution can be found by hand. We shall see that this is an exceptional situation. Unfortunately, in real-life problems, which are more complex, such a situation occurs very rarely and an appropriate discretization is necessary. By a discretization we mean a transformation of the original problem into a new one, characterized by a finite number of degrees of freedom, which can be realized using tools of numerical mathematics. Then a natural question immediately arises: Is the discretized problem a good approximation of the original one? If yes, in what sense? And is it possible to establish some rules on how to define a good discretization? All these questions are studied in the next chapters.

Chapter 2 starts with two simple SSO problems. We show in detail how to prove the existence of solutions and analyze convergence properties of the discretized problems (Sections 2.1, 2.2). The reason we start with the particular problems is very simple: the main ideas of the proofs remain the same in all SSO problems. This approach facilitates reading the next two sections, in which the existence and convergence results are established in an abstract framework. These results are then applied to particular shape optimization problems governed by various state problems (Section 2.5) of solid and fluid mechanics. To keep our presentation as elementary as possible we confine ourselves to 2D shape optimization problems in which a part of the boundary to be determined is described by the graph of a function. This certainly makes mathematical analysis and numerical realization much easier. Let us mention, however, that the same ideas can be used in the existence analysis of problems in 3D, but the proof will be more technical. Since most industrial applications deal with domains with Lipschitz boundaries, we restrict ourselves to domains satisfying the so-called uniform cone property. For readers who would like to be familiar with the use of more general classes of domains we refer to the recent monograph by Delfour and Zolesio [DZ01] published in the same SIAM series as our book.

Part II is devoted to computational aspects in SSO. Chapter 3 deals with sensitivity analysis or how to differentiate functionals with respect to design variables. Sensitivity analysis is an integral part of any optimization process. Gradient information is important from the theoretical as well as the practical point of view, especially when gradient type minimization methods are used.

We start with sensitivity analysis in algebraic formulations of discretized problems, which is based on the use of the classical implicit function theorem. Then the material derivative approach, a very useful tool in shape derivative calculus, will be introduced. Special attention is paid to sensitivity analysis in problems governed by variational inequalities.

In Chapter 4 we briefly recall both gradient type (local) and gradient free (global) algorithms for the numerical minimization of functions. Gradient type methods include Newton's method, the quasi-Newton method, and the sequential quadratic programming method. Gradient free methods are represented by genetic and random search algorithms. In addition, some methods of multiobjective optimization are presented.

In Chapter 5 we discuss a rather new technique for calculating derivatives, namely automatic differentiation (AD) of computer programs. AD enables us to get accurate derivatives up to the machine precision without any human intervention. We describe in detail how to apply this technique in shape optimization.

Chapter 6 presents a new approach to the realization of optimal shape design problems, based on the use of fictitious domain solvers at the inner optimization level. The classic boundary variation technique requires a lot of computational effort: after any change in the shape, one has to remesh the new configuration, then recompute all data, such as stiffness matrices and load vectors. Fictitious domain methods make it possible to efficiently solve state problems on a uniform grid that does not change during the optimization process. Just this fact considerably increases the efficiency of the inner optimization level. This approach is used for the numerical realization of a class of free boundary value problems.

Part III is devoted to industrial applications. In Chapter 7 we present two problems of SSO of stressed structures, namely multiobjective optimal sizing of a beam under multiple load cases and contact stress minimization of an elasto-plastic body in contact with a rigid foundation. In Chapter 8 we first solve a problem arising in the paper machine industry: to find the shape of the header of a paper machine in order to get an appropriate distribution of a fiber suspension. Next a multidisciplinary and multiobjective problem is solved: we want to optimize an airfoil taking into account aerodynamics and electromagnetics aspects.

For the convenience of the reader two appendices complete the text. In Appendix A some elementary results of the theory of linear elliptic equations, Sobolev spaces, and finite element approximations are collected. Appendix B deals with 2D parametrization of shapes. A good shape parametrization is a key point in a successful and efficient optimization process. Basic properties of Bézier curves are revisited.

Part I

Mathematical Aspects of Sizing and Shape Optimization

Chapter 1

Why the Mathematical Analysis Is Important

In this chapter we illustrate by using simple model examples which types of problems are solved in sizing and shape optimization (SSO). Further, we present some difficulties one may meet in their practical realization. Finally we try to convince the reader of the helpfulness of a thorough mathematical analysis of the problems to be solved.

We start this chapter with a simple sizing optimization problem whose exact solution can be found by hand. Let us consider a simply supported beam of variable thickness e represented by the interval $I = [0, 1]$. The beam is under a uniform vertical load f_0. One wants to find a thickness distribution to maximize the stiffness of the beam. The deflection $u := u(e)$ of the beam solves the following fourth order boundary value problem:

$$\begin{cases} (\beta e^3 u''(x))'' = f_0 \quad \text{in } [0, 1], \\ u(0) = u(1) = (\beta e^3 u'')(0) = (\beta e^3 u'')(1) = 0, \end{cases} \tag{$\mathcal{P}'(e)$}$$

where β is a given positive constant. The stiffness of the beam is characterized by the compliance functional J defined by

$$J(u(e)) = \int_0^1 f_0 u(e)\, dx,$$

where $u(e)$ solves $(\mathcal{P}'(e))$. The stiffer the construction is, the lower the value J attains. Therefore the stiffness maximization is equivalent to the compliance minimization. We formulate the following sizing optimization problem:

$$\begin{cases} \text{Find } e^* \in U^{ad} \quad \text{such that} \\ J(u(e^*)) \leq J(u(e)) \quad \forall e \in U^{ad}, \end{cases} \tag{$\mathbb{P}_1$}$$

where U^{ad} is the set of admissible thicknesses defined as follows:

$$U^{ad} = \left\{ e : [0, 1] \to \mathbb{R}_+ \mid \int_0^1 e(x)\, dx = \gamma \right\}, \quad \gamma > 0 \text{ given.}$$

The integral constraint appearing in the definition of U^{ad} says that the volume of the beam is preserved. Next we show how to find a solution to $(\mathbb{P}_1)$.

For the sake of simplicity we set $\beta = f_0 = 1$ in $[0, 1]$. Instead of the classical formulation $(\mathcal{P}'(e))$ we use its weak form:

$$\begin{cases} \text{Find } u := u(e) \in V \quad \text{such that} \\ \displaystyle\int_0^1 e^3 u'' v'' \, dx = \int_0^1 v \, dx \quad \forall v \in V, \end{cases} \tag{$\mathcal{P}(e)$}$$

where

$$V = \{v \in H^2(I) \mid v(0) = v(1) = 0\}$$

is the Sobolev space of functions vanishing at the endpoints of I (for the definition of Sobolev spaces we refer to Appendix A). In order to release the constraints in $(\mathbb{P}_1)$ that are represented by the state problem $(\mathcal{P}(e))$ and the constant volume constraint we introduce the following Lagrangian:

$$\begin{aligned} \mathcal{L}(e, u, p, \lambda) = & \int_0^1 u \, dx + \int_0^1 e^3 u'' p'' \, dx - \int_0^1 p \, dx \\ & + \lambda \left(\int_0^1 e \, dx - \gamma \right), \quad e \in U^{ad}, \ (u, p) \in V \times V, \ \lambda \in \mathbb{R}. \end{aligned}$$

We now seek stationary points of $\mathcal{L}$, i.e., all points (e, u, p, λ) satisfying

$$\delta\mathcal{L}(e, u, p, \lambda) = 0, \tag{1.1}$$

where the symbol δ stands for the variation of $\mathcal{L}$ with respect to all mutually independent variables without any subsidiary constraints. From the definition of $\mathcal{L}$ it easily follows that

$$\begin{aligned} \delta\mathcal{L}(e, u, p, \lambda) = & \int_0^1 (3e^2 u'' p'' + \lambda) \delta e \, dx + \int_0^1 \delta u \, dx + \int_0^1 e^3 (\delta u)'' p'' \, dx \\ & + \int_0^1 e^3 u'' (\delta p)'' \, dx - \int_0^1 \delta p \, dx + \delta\lambda \left(\int_0^1 e \, dx - \gamma \right), \end{aligned} \tag{1.2}$$

where δe, δu, δp, and $\delta\lambda$ denote the variation of the respective independent variable. From (1.1) and (1.2) we obtain the following optimality conditions satisfied by any stationary point (e, u, p, λ) of $\mathcal{L}$:

$$3e^2 u'' p'' + \lambda = 0 \quad \text{in }]0, 1[; \tag{1.3}$$

$$(e^3 u'')'' = 1 \quad \text{in }]0, 1[, \quad u(0) = u(1) = (e^3 u'')(0) = (e^3 u'')(1) = 0; \tag{1.4}$$

$$(e^3 p'')'' = -1 \quad \text{in }]0, 1[, \quad p(0) = p(1) = (e^3 p'')(0) = (e^3 p'')(1) = 0; \tag{1.5}$$

$$\int_0^1 e \, dx = \gamma. \tag{1.6}$$

Equation (1.4) is the classical form of the state equation while (1.5) represents the so-called adjoint state equation. Comparing (1.4) with (1.5) we see that $p = -u$ so that (1.3) becomes

$$-3e^2 (u'')^2 + \lambda = 0 \quad \text{in }]0, 1[. \tag{1.7}$$

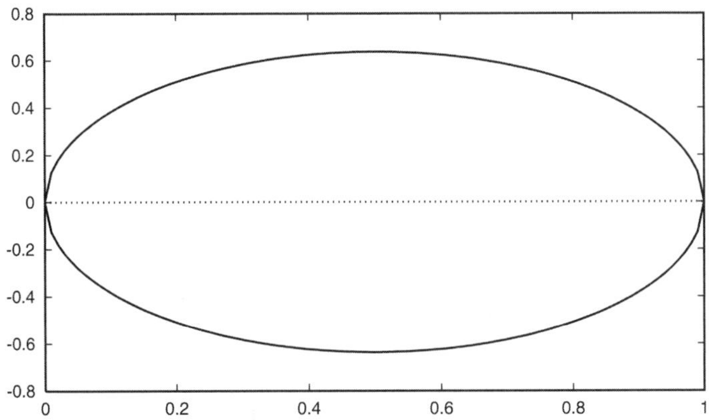

Figure 1.1. *Optimal thickness distribution.*

From this it follows that

$$e|u''| = c = \text{const.} > 0. \tag{1.8}$$

Convention: *Here and in what follows the letter c stands for a generic positive constant attaining different values at different places.*

The bending moment $M := M(x) = (e^3u'')(x)$ satisfies the following boundary value problem:

$$\begin{cases} M''(x) = 1 \quad \text{in }]0, 1[, \\ M(0) = M(1) = 0, \end{cases}$$

whose solution is $M(x) = \frac{1}{2}x(x-1)$. Therefore

$$e(x)u''(x) = \frac{M(x)}{e^2(x)} \leq 0$$

implying together with (1.8) that

$$e(x) = \sqrt{-\frac{M(x)}{c}} = c\sqrt{x(1-x)}, \quad x \in]0, 1[. \tag{1.9}$$

The value of the constant c on the right side of (1.9) can be fixed from the volume constraint. If $\gamma = 1$, then $e(x) = (8/\pi)\sqrt{x(1-x)}$ (see Figure 1.1).

Remark 1.1. It is worth noticing that the stiffest simply supported beam has zero thickness at the endpoints $x = 0, 1$ and, besides the constant volume constraint, *no other assumptions* on admissible thicknesses are present. As we shall see later on this is an exceptional situation. Usually some additional constraints are needed to get a *physically relevant* solution.

The fact that we were able to find the exact solution of the previous optimization problem is exceptional and due to the simplicity of the state problem and the particular form of the cost functional (compliance). Unfortunately most optimization problems we meet in practice are far from being so simple and an approximation is necessary. We proceed as follows: first we decide on an appropriate discretization of the state problem (by using finite elements, for example) and of design parameters (instead of complicated shapes we use spline approximations, for example). Computed solutions to the discrete state problems are now considered to be a function of a finite number of *discrete design variables* $\boldsymbol{\alpha} = (\alpha_1, \dots, \alpha_d)$ fully characterizing discretized shapes or thicknesses. After being inserted into cost functionals, these become functions of $\boldsymbol{\alpha}$. In this way we arrive at a constrained optimization problem in $\mathbb{R}^d$ whose solutions we shall look for.

There are two ways to realize this step. The first is based on the numerical solution of the respective optimality conditions, as in the previous example. This approach, however, has serious drawbacks. First of all one has to derive the optimality conditions. This is usually a difficult task especially in problems whose state relations are given by variational inequalities or even by more complicated mathematical objects. But even if the optimality conditions are at our disposal they are usually so complex that their numerical treatment is not easy at all. For this reason the structural optimization community prefers a more universal approach based on the numerical minimization of functions by means of *mathematical programming methods.* But also in this case one should pay attention to a thorough mathematical analysis in order to obtain additional information on the problem that can be useful in computations. A typical feature is that minimized functions are usually *nonconvex.* This gives rise to some difficulties: nonconvex functions may have *several* local minima; further, the minimization method used may be *divergent* or the result obtained *may depend* on the choice of the initial guess. Below we present very simple SSO problems in one dimension (i.e., $d = 1$) with states described by equations and inequalities involving ordinary differential operators. The dependence of cost functions on the design parameter will be illustrated by the graphs.

Example 1.1. Let

$$U^{ad} = \{e \in \mathbb{R} \mid e_{\min} \le e \le e_{\max}\},$$

where $0 < e_{\min} < e_{\max}$ are given, and consider the optimal sizing problem

$$\min_{e \in U^{ad}} J(e) = \int_0^1 (u_e - 1)^2 \, dx,$$

where u_e solves the following boundary value problem:

$$\begin{cases} -eu_e'' = 2 \quad \text{in }]0,1[, \ e \in U^{ad}, \\ u_e(0) = u_e(1) = 0. \end{cases} \tag{1.10}$$

It is readily seen that $u_e(x) = (x - x^2)/e$ and

$$J(e) = 1 + \frac{1}{30e^2} - \frac{1}{3e}, \qquad e \in U^{ad}.$$

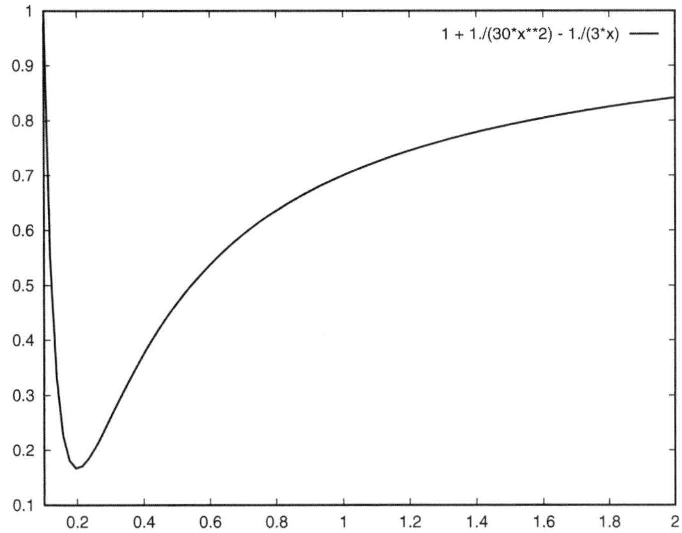

Figure 1.2. *Graph of the cost functional.*

The graph of J is plotted in Figure 1.2. We see that J is unimodal (i.e., for any choice of $e_{\min}, e_{\max}$ it has only one local minimum) but *not convex*. If one tries to minimize J numerically using a simple quadratic approximation of J, the method will not converge provided the initial guess e_0 is large enough, say $e_0 \geq 1/2$.

Example 1.2. (See [Céa81].) Let us consider the following simple prototype of the shape optimization problem:

$$\min_{\alpha \in U^{ad}} J(\alpha) = \int_0^\alpha (u_\alpha - 1)^2 \, dx,$$

where

$$U^{ad} = \{\alpha \in \mathbb{R} \mid 0 \leq \alpha_{\min} \leq \alpha \leq \alpha_{\max}\},$$

$\alpha_{\min} < \alpha_{\max}$ is given, and u_α solves the boundary value problem

$$\begin{cases} -u_\alpha'' = 2 \quad \text{in }]0, \alpha[, \\ u_\alpha'(0) = u_\alpha(\alpha) = 0. \end{cases} \tag{1.11}$$

In contrast to the previous two sizing optimization problems the differential equation here is posed on an *unknown* domain that is to be determined. The solution of this problem is $u_\alpha(x) = \alpha^2 - x^2$ and

$$J(\alpha) = \frac{8}{15}\alpha^5 - \frac{4}{3}\alpha^3 + \alpha, \quad \alpha \in U^{ad}.$$

The graph of J is shown in Figure 1.3. We see that J is nonconvex and it may have one or two local minima depending on the choice of $\alpha_{\min}, \alpha_{\max}$.

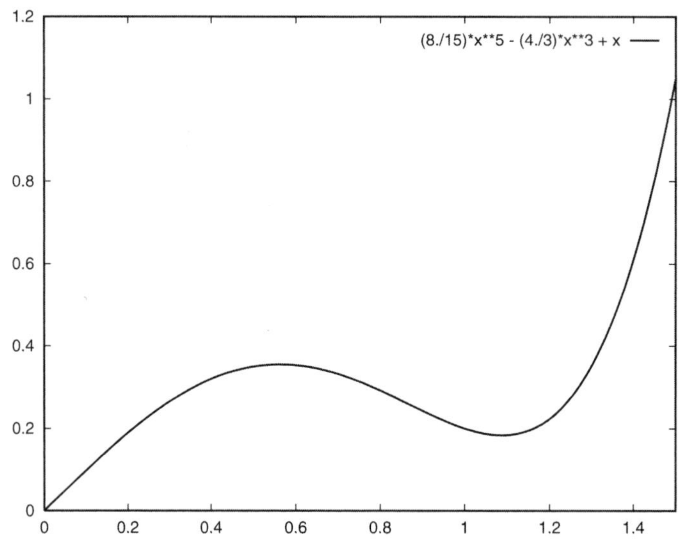

Figure 1.3. *Graph of the cost functional.*

In the previous two examples the resulting functions are *continuously* differentiable. But this is not always so. With an inequality instead of a state equation the situation may be completely different, as will be seen from the following example.

Example 1.3. *(State variational inequality.)* Consider the optimal shape design problem with the same J and U^{ad} as in Example 1.2 but with u_α solving the following variational inequality:

$$u_\alpha \in K(\alpha) : \quad \int_0^\alpha u'_\alpha(v' - u'_\alpha)\,dx \geq \int_0^\alpha 2(v - u_\alpha)\,dx \quad \forall v \in K(\alpha), \tag{1.12}$$

where

$$K(\alpha) = \{v \in H^1(]0, \alpha[) \mid v(0) = 0,\ v(\alpha) \leq 1\}, \quad \alpha \in U^{ad}.$$

It is well known that the solution u_α to (1.12) exists and is unique as follows from Lemma A.1 in Appendix A. Integrating by parts on the left-hand side of (1.12) it is easy to show that u_α satisfies the following set of conditions:

$$\begin{cases} -u''_\alpha(x) = 2 \quad \forall x \in]0, \alpha[, \\ u_\alpha(0) = 0,\ u_\alpha(\alpha) \leq 1,\ u'_\alpha(\alpha) \leq 0,\ (u_\alpha(\alpha) - 1)u'_\alpha(\alpha) = 0. \end{cases} \tag{1.13}$$

It is readily seen that the function

$$u_\alpha(x) = \begin{cases} -x^2 + 2\alpha x, & \alpha < 1, \\ -x^2 + \dfrac{\alpha^2 + 1}{\alpha} x, & \alpha \geq 1, \end{cases} \tag{1.14}$$

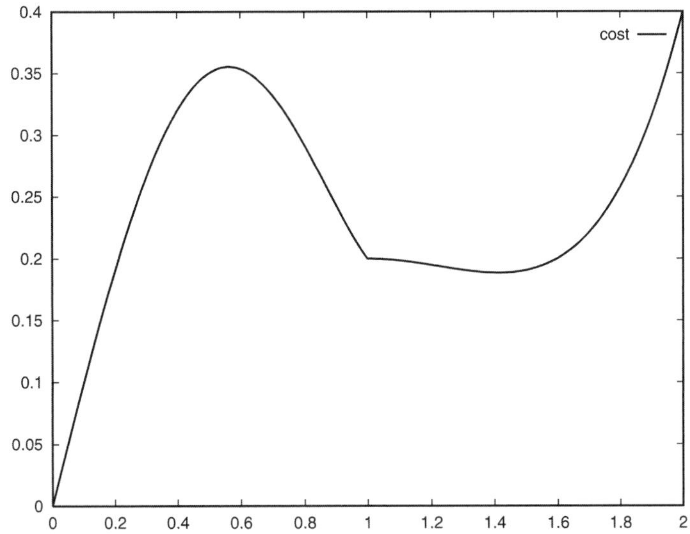

Figure 1.4. *Graph of the cost functional.*

satisfies (1.13) and its substitution into J yields the following analytic expression of J as a function of the design parameter α:

$$J(\alpha) = \begin{cases} \dfrac{8}{15}\alpha^5 - \dfrac{4}{3}\alpha^3 + \alpha, & \alpha < 1, \\ \dfrac{1}{30}\alpha^5 - \dfrac{1}{6}\alpha^3 + \dfrac{1}{3}\alpha, & \alpha \geq 1. \end{cases}$$

A direct computation shows that $J'_-(1) = -1/3$ and $J'_+(1) = 0$; i.e., J *is not differentiable* at $\alpha = 1$ (see Figure 1.4). Let us comment on this fact in more detail. The external functional

$$I : (\alpha, y) \mapsto \int_0^\alpha (y-1)^2\,dx, \quad y \in H^1(]0,\alpha[),\ \alpha \in U^{ad},$$

is continuously differentiable with respect to both variables, while the composite function

$$J : \alpha \mapsto \int_0^\alpha (u_\alpha - 1)^2\,dx = I(\alpha, u_\alpha), \quad \alpha \in U^{ad},$$

is *not*. From this it immediately follows that the inner *control state mapping* $\alpha \mapsto u_\alpha$ *cannot* be continuously differentiable. Indeed, u_α being a solution to the variational inequality (1.12) can be expressed as the projection of an appropriate function from $H^1(]0,\alpha[)$ onto the convex set $K(\alpha)$. A well-known result (see [Céa71]) says that the (nonlinear) projection operator of a Hilbert space on its closed convex subset is Lipschitz continuous. This in short explains the source of the possible nondifferentiability of the control state mappings in the case of state variational inequalities (for more detail on this subject we refer to [SZ92]).

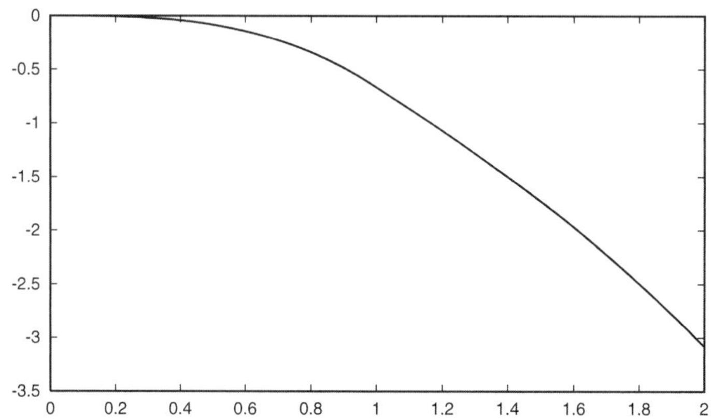

Figure 1.5. *Graph of the cost functional.*

The fact that a minimized function is not differentiable at some points may have practical consequences. A common way of minimizing J is to use classical gradient type methods that require gradient information on the minimized function. The absence of such information at some points may give unsatisfactory or unreliable numerical results. Fortunately there is a class of generalized gradient methods for which the nondifferentiability does not present a serious difficulty and that were developed just to treat this type of problem. This is a nice illustration of how a thorough analysis and understanding of the problem may help with the choice of an appropriate minimization algorithm.

The fact that the control state mapping is not differentiable, however, does not mean that the composite function is automatically nondifferentiable as well. Indeed, let us choose the new cost functional

$$J(\alpha) = \frac{1}{2}\int_0^\alpha (u'_\alpha)^2\,dx - \int_0^\alpha 2u_\alpha\,dx, \quad \alpha \in U^{ad}, \tag{1.15}$$

where $u_\alpha \in K(\alpha)$ solves (1.12). Substituting (1.14) into (1.15) we obtain

$$J(\alpha) = \begin{cases} -\dfrac{2}{3}\alpha^3, & \alpha < 1, \\ -\dfrac{1}{6}\alpha^3 + \dfrac{1}{2\alpha} - \alpha, & \alpha \geq 1 \end{cases} \tag{1.16}$$

(see Figure 1.5). One can easily check that J is a once (*but not twice!*) continuously differentiable function in U^{ad}. Therefore, for a special choice of cost functionals it may happen that the composite function is continuously differentiable despite a possible nondifferentiability of the inner mapping. We shall meet the same situation several times in this textbook (see Chapters 2, 3, 7).

We end this chapter with one shape optimization problem that has *no* solution, showing how this fact manifests itself during computations.

Let $\Omega(\alpha) = \{(x_1, x_2) \in \mathbb{R}^2 \mid 0 < x_1 < \alpha(x_2),\ x_2 \in]0,1[\}$ be a "curved rectangle" (see Figure 2.2) with the curved side $\Gamma(\alpha)$ being the graph of a function $\alpha \in U^{ad}$, where

$$U^{ad} = \left\{ \alpha \in C([0,1]) \mid 0 < \alpha_{\min} \leq \alpha \leq \alpha_{\max} \text{ in } [0,1],\ \int_0^1 \alpha(x_2)\,dx_2 = \gamma \right\}, \tag{1.17}$$

with $0 < \alpha_{\min} < \alpha_{\max}$ and $\gamma > 0$ given. Further denote by $\mathcal{O} = \{\Omega(\alpha) \mid \alpha \in U^{ad}\}$ the set of all admissible domains; i.e., $\mathcal{O}$ is realized by domains whose curved part $\Gamma(\alpha)$ of the boundary lies within the strip bounded by $\alpha_{\min}, \alpha_{\max}$ and have the same area equal to γ. On any $\Omega(\alpha)$, $\alpha \in U^{ad}$, we shall consider the following Dirichlet–Neumann state problem:

$$\begin{cases} -\Delta u(\alpha) = C & \text{in } \Omega(\alpha), \\ \dfrac{\partial u(\alpha)}{\partial \nu} = 0 & \text{on } \Gamma_1 = \partial\Omega(\alpha) \setminus \overline{\Gamma(\alpha)}, \\ u(\alpha) = 0 & \text{on } \Gamma(\alpha), \end{cases} \tag{1.18}$$

where $C > 0$ is a given constant. We define the following optimal shape design problem:

$$\begin{cases} \text{Find } \alpha^* \in U^{ad} \quad \text{such that} \\ J(\alpha^*, u(\alpha^*)) \leq J(\alpha, u(\alpha)) \quad \forall \alpha \in U^{ad}, \end{cases} \tag{$\mathbb{P}_2$}$$

where

$$J(\alpha, y) = \frac{1}{2} \int_{\Omega(\alpha)} y^2\,dx \tag{1.19}$$

($dx = dx_1 dx_2$) and $u(\alpha)$ solves (1.18) in $\Omega(\alpha)$.

Problem ($\mathbb{P}_2$) was solved with the following values of the parameters: $C = 2$, $\alpha_{\min} = 0.1$, $\alpha_{\max} = 0.9$, and $\gamma = 0.5$. Shapes were discretized by piecewise quadratic Bézier functions. The number of segments is $d = 10, 18$. Computed results are shown in Figure 1.6. We see that the designed part $\Gamma(\alpha)$ oscillates and the oscillation becomes faster and faster for increasing values of d. Now a question arises: What does the optimal shape look like? It is difficult to imagine such a domain on the basis of the obtained results since there is no "understandable" domain from $\mathcal{O}$ that can be deduced from them. In fact the oscillatory pattern is a consequence of the *nonexistence* of solutions to ($\mathbb{P}_2$). The rigorous mathematical analysis of this problem is far from easy and goes beyond the scope of this textbook. Let us try, however, to give a naive "explanation." Boundary value problem (1.18) describes the stationary temperature distribution in the body represented by $\Omega(\alpha)$, which is isolated on Γ_1 and cooled along $\Gamma(\alpha)$. The right-hand side C characterizes the heat source. Since the area of all $\Omega(\alpha)$ is the same, the only way to decrease the temperature in $\Omega(\alpha)$ is to increase the length of the cooling part $\Gamma(\alpha)$. This is just what the solution found mimics: $\Gamma(\alpha)$ begins to oscillate. The constant volume constraint is added to avoid the trivial solution $\alpha^* = \alpha_{\min}$.

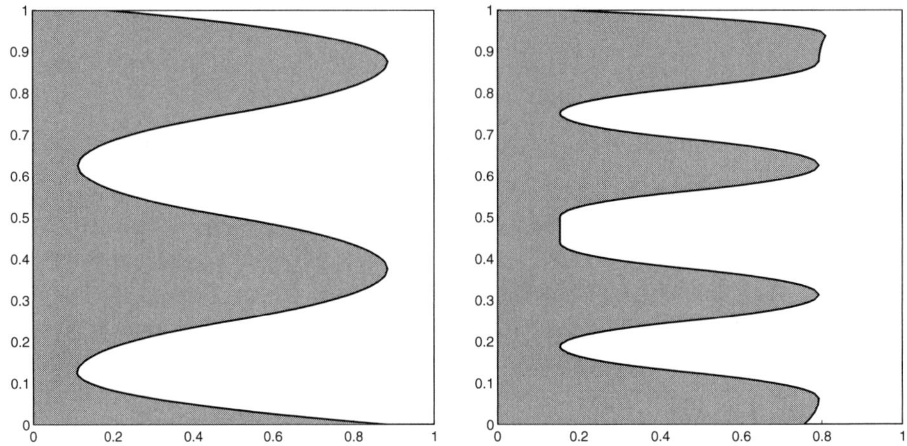

Figure 1.6. *Oscillating boundaries.*

A possible way to ensure the solvability of $(\mathbb{P}_2)$ is to impose additional constraints, keeping the length of $\Gamma(\alpha)$ bounded. Thus instead of (1.17) one can consider another (narrower) system of admissible domains defined by

$$U^{ad} = \Big\{ \alpha \in C^{0,1}([0,1]) \mid 0 < \alpha_{\min} \le \alpha \le \alpha_{\max} \text{ in } [0,1]; \\ |\alpha'| \le L_0 \text{ almost everywhere in }]0,1[;\ \int_0^1 \alpha(x_2)\,dx_2 = \gamma \Big\}, \quad (1.20)$$

where $L_0 > 0$ is given. As we shall see later on, problem $(\mathbb{P}_2)$ with U^{ad} defined by (1.20) already has a solution.

Since shape optimization problems are generally nonconvex they may have more than one solution (see Problem 1.2).

Problems

PROBLEM 1.1. Prove that $u(\alpha)$ solves the variational inequality (1.12) iff it satisfies (1.13).

PROBLEM 1.2. [BG75] Consider the optimal shape design problem $(\mathbb{P}_2)$ with the state problem (1.18) and the cost functional (1.19). Prove that if $\Omega(\alpha^*)$ solving $(\mathbb{P}_2)$ ($\alpha^* \in U^{ad}$ given by (1.17)) is not symmetric with respect to the line $x_2 = 1/2$, then $\Omega(\alpha^{**})$, where $\alpha^{**}(x_2) :=$ (composite function) $\alpha^*(1 - x_2)$, $x_2 \in]0,1[$, is a solution of $(\mathbb{P}_2)$, too.

Chapter 2

A Mathematical Introduction to Sizing and Shape Optimization

The aim of this chapter is to present ideas that are used in existence and convergence analysis in sizing and shape optimization (SSO). As we shall see, the basic ideas are more or less the same: we first prove that solutions of state problems depend *continuously* on design variables (and we shall specify in which sense). Then, imposing appropriate *continuity* (or better *lower semicontinuity*) assumptions on a cost functional, we immediately arrive at an existence result. The same scheme remains more or less true when doing the convergence analysis. Before we give an abstract setting for optimal sizing and shape design problems and prove abstract existence and convergence results, we show how to proceed in particular model examples. The same ideas will be used later on in the abstract form.

2.1 Thickness optimization of an elastic beam: Existence and convergence analysis

Let us consider a clamped elastic beam of variable thickness e subject to a vertical load f. The beam is represented by an interval $I = [0, \ell]$, $\ell > 0$. We want to find the thickness distribution in I minimizing the compliance of the beam, given by the value $J(u(e))$, where

$$J(y) = \int_I f\, y\, dx \tag{2.1}$$

and $u(e)$ is the solution of the following boundary value problem:

$$\begin{cases} (\beta e^3 u'')''(x) = f(x) \quad \forall x \in]0, \ell[, \\ u(0) = u'(0) = u(\ell) = u'(\ell) = 0. \end{cases} \tag{2.2}$$

Here $\beta \in L^\infty(I)$, $\beta \geq \beta_0 = \text{const.} > 0$, is a function depending on material properties and on the shape of the cross-sectional area of the beam. The solution of (2.2) is assumed to be a function of e, playing the role of a *control variable*. To define an optimization problem, one has to specify a class U^{ad} of *admissible thicknesses*. As we already know, the result of the optimization process depends on, among other factors, how large U^{ad} is.

Let U^{ad} be given by

$$U^{ad} = \Big\{ e \in C^{0,1}(I) \mid 0 < e_{\min} \le e \le e_{\max} \text{ in } I, \\ |e(x_1) - e(x_2)| \le L_0|x_1 - x_2| \quad \forall x_1, x_2 \in I, \quad \int_I e(x)\,dx = \gamma \Big\}; \tag{2.3}$$

i.e., U^{ad} consists of functions that are *uniformly bounded* and *uniformly Lipschitz continuous* in I and preserve the beam volume. The positive constants $e_{\min}$, $e_{\max}$, L_0, and γ are chosen in such a way that $U^{ad} \neq \emptyset$.

REMARK 2.1. The uniform Lipschitz constraint

$$|e(x_1) - e(x_2)| \le L_0|x_1 - x_2| \quad \forall x_1, x_2 \in I$$

appearing in the definition of U^{ad} prevents thickness oscillations and plays an important role in the forthcoming mathematical analysis.

We are now ready to formulate the following thickness optimization problem:

$$\begin{cases} \text{Find } e^* \in U^{ad} \text{ such that} \\ J(u(e^*)) = \min_{e \in U^{ad}} J(u(e)), \end{cases} \tag{$\mathbb{P}$}$$

where J is the *cost functional* (2.1) and $u(e)$ is the displacement, solving (2.2). In what follows we shall prove that ($\mathbb{P}$) has at least one solution e^*. The key point of the existence analysis is to show that the solution u of (2.2) depends continuously on the control variable e. But first we have to specify what the word *continuous* means. This notion differs for u and e.

CONVENTION: *For the sake of simplicity of our notation throughout the book we shall use the same symbol for subsequences and the respective original sequences.*

Our analysis starts with a weak formulation of (2.2). Let $H_0^2(I)$ be the Sobolev space of functions v whose derivatives up to the second order are square integrable in I and satisfy the boundary conditions $v(0) = v'(0) = v(\ell) = v'(\ell) = 0$ (see Appendix A). The weak formulation of (2.2) reads as follows:

$$\begin{cases} \text{Find } u := u(e) \in H_0^2(I) \text{ such that} \\ \displaystyle\int_I \beta e^3 u'' v''\,dx = \int_I f v\,dx \quad \forall v \in H_0^2(I), \end{cases} \tag{$\mathcal{P}(e)$}$$

where $e \in U^{ad}$ and $f \in L^2(I)$ are given. We adopt the notation $(\mathcal{P}(e))$, $u(e), \dots$ to stress the dependence on $e \in U^{ad}$. The continuous dependence of u on e mentioned above will be understood in the sense of the following lemma.

LEMMA 2.1. *Let e_n, $e \in U^{ad}$, be such that $e_n \rightrightarrows e$ (uniformly) in I and let $u_n := u(e_n)$ be solutions to $(\mathcal{P}(e_n))$, $n = 1, 2, \dots$. Then*

$$u_n \to u(e) \quad \text{in } H_0^2(I),$$

and $u(e)$ is the solution of $(\mathcal{P}(e))$.

Proof. Let $u_n \in H_0^2(I)$ solve $(\mathcal{P}(e_n))$; i.e.,

$$\int_I \beta e_n^3 u_n'' v''\,dx = \int_I f v\,dx \quad \forall v \in H_0^2(I). \tag{2.4}$$

Inserting $v := u_n$ into (2.4) and using the definition of U^{ad} we obtain

$$\beta_0 e_{\min}^3 |u_n|_{2,I}^2 \le \|f\|_{0,I} \|u_n\|_{2,I} \quad \forall n \in \mathbb{N}.$$

From this and the Friedrichs inequality we see that the sequence $\{u_n\}$ is bounded:

$$\exists\, c > 0 : \quad \|u_n\|_{2,I} \le c \quad \forall n \in \mathbb{N}. \tag{2.5}$$

Thus one can pass to a subsequence such that

$$u_n \rightharpoonup u \quad \text{(weakly) in } H_0^2(I), \tag{2.6}$$

where u is an element of $H_0^2(I)$. It remains to show that u solves $(\mathcal{P}(e))$. But this follows from (2.4) by letting $n \to \infty$. It is readily seen that

$$\int_I \beta e_n^3 u_n'' v''\,dx \to \int_I \beta e^3 u'' v''\,dx, \quad n \to \infty. \tag{2.7}$$

Indeed,

$$\begin{aligned} &\left| \int_I \beta e_n^3 u_n'' v''\,dx - \int_I \beta e^3 u'' v''\,dx \right| \\ &\le \int_I \beta |e_n^3 - e^3|\, |u_n''|\, |v''|\,dx + \left| \int_I \beta e^3 (u_n'' - u'') v''\,dx \right| \to 0 \quad \text{as } n \to \infty, \end{aligned} \tag{2.8}$$

taking into account that $e_n \rightrightarrows e$ in I and (2.6). From (2.7) it follows that u solves $(\mathcal{P}(e))$. Since $(\mathcal{P}(e))$ has a unique solution, not only this subsequence but the whole sequence $\{u_n\}$ tends weakly to u in $H_0^2(I)$. To prove strong convergence, it is sufficient to show that

$$|||u_n||| \to |||u|||, \quad n \to \infty,$$

where $|||\cdot|||$ is a norm with respect to which $H_0^2(I)$ is complete. Clearly the energy norm

$$|||v|||^2 := \int_I \beta e^3 (v'')^2\,dx \tag{2.9}$$

with e being the limit of $\{e_n\}$ possesses this property. Then

$$\begin{aligned} |||u_n|||^2 = \int_I \beta e^3 (u_n'')^2\,dx &= \int_I \beta (e^3 - e_n^3)(u_n'')^2\,dx + \int_I \beta e_n^3 (u_n'')^2\,dx \\ &= \int_I \beta (e^3 - e_n^3)(u_n'')^2\,dx + \int_I f u_n\,dx \to \int_I f u\,dx = |||u|||^2, \end{aligned}$$

as follows from the definition of $(\mathcal{P}(e_n))$, $(\mathcal{P}(e))$, (2.6), and uniform convergence of $\{e_n\}$ to e in I proving strong convergence of $\{u_n\}$ to $u(e) := u$ in the $H_0^2(I)$-norm. $\square$

Having this result at our disposal we easily arrive at the following existence result for $(\mathbb{P})$.

THEOREM 2.1. *Let U^{ad} be as in (2.3). Then $(\mathbb{P})$ has a solution.*

Proof. Denote

$$q := \inf_{e \in U^{ad}} J(u(e)) = \lim_{n \to \infty} J(u(e_n)), \tag{2.10}$$

where $\{e_n\}$, $e_n \in U^{ad}$, is a minimizing sequence. Since U^{ad} is a compact subset of $C(I)$, as follows from the Arzelà–Ascoli theorem, one can pass to a subsequence of $\{e_n\}$ such that

$$e_n \rightrightarrows e^* \in U^{ad} \text{ in } I, \quad n \to \infty,$$

for an element $e^* \in U^{ad}$. At the same time

$$u(e_n) \to u(e^*) \quad \text{in } H_0^2(I), \tag{2.11}$$

as follows from Lemma 2.1. From (2.10), (2.11), and continuity of J in $H_0^2(I)$ we see that

$$q = J(u(e^*));$$

i.e., $e^* \in U^{ad}$ solves $(\mathbb{P})$. □

DEFINITION 2.1. *A pair $(e^*, u(e^*))$, where e^* solves $(\mathbb{P})$ and $u(e^*)$ is a solution of the respective state problem $(\mathcal{P}(e^*))$, is called an optimal pair of $(\mathbb{P})$.*

CONVENTION: *The wording "optimal pair" will be used in the same sense in all other optimization problems.*

COMMENTS 2.1.

(i) There are three substantial properties used in the previous existence proof: the continuous dependence of $u(e)$ on the control variable e; the compactness of U^{ad}, enabling us to pass to a convergent subsequence of any minimizing sequence; and continuity of J.

(ii) Some extensions in the setting of $(\mathbb{P})$ are possible: instead of the load $f \in L^2(I)$ we can take any $f \in H^{-2}(I)$ (the dual of $H_0^2(I)$), enabling us to consider concentrated loads $f = \delta_{x_0}$, e.g., where the Dirac distribution δ_{x_0} at a point $x_0 \in I$ is defined by

$$\delta_{x_0}(v) = v(x_0) \quad \forall v \in H_0^2(I).$$

The functional J given by (2.1) depends on the state variable $y \in V$ ($V = H_0^2(I)$ in our example) *but not explicitly* on the control variable $e \in U^{ad}$. It becomes a function of e only through the composition of J with the *control state* mapping $u : e \mapsto u(e)$. In many situations, however, cost functionals depend on *both* variables of $(e, y) \in U^{ad} \times V$. For this reason the abstract setting of these problems considers cost functionals defined on $U^{ad} \times V$, i.e., *depending explicitly* on both variables. From the proof of Theorem 2.1 we see that, for a general $J : U^{ad} \times H_0^2(I) \to \mathbb{R}$, the existence of solutions to $(\mathbb{P})$ is guaranteed provided that J is *lower semicontinuous* in the following sense:

$$\left.\begin{aligned} e_n &\rightrightarrows e \quad \text{in } I \\ y_n &\to y \quad \text{in } H_0^2(I) \end{aligned}\right\} \Longrightarrow \liminf_{n\to\infty} J(e_n, y_n) \geq J(e, y). \tag{2.12}$$

(iii) The admissible set U^{ad} defined by (2.3) is quite narrow. It would be possible to extend it by omitting the uniform Lipschitz constraint, i.e., to take U^{ad} as

$$U^{ad} = \left\{ e \in C(I) \mid 0 < e_{\min} \leq e \leq e_{\max} \text{ in } I, \int_I e(x)\,dx = \gamma \right\}. \tag{2.13}$$

Unfortunately, U^{ad} is *not* a compact subset of $C(I)$. But one can still extend U^{ad} as follows:

$$\widetilde{U}^{ad} = \left\{ e \in L^\infty(I) \,|\, 0 < e_{\min} \leq e \leq e_{\max} \text{ almost everywhere (a.e.) in } I, \int_I e(x)\,dx = \gamma \right\}. \tag{2.14}$$

Then $\widetilde{U}^{ad}$ is a compact subset of $L^\infty(I)$ with respect to $L^\infty *$ weak convergence. On the other hand the mapping $u : e \mapsto u(e),\ e \in \widetilde{U}^{ad}$, is *not* continuous with respect to this convergence. More precisely, if $\overline{e}_n,\ \overline{e} \in \widetilde{U}^{ad}$ are such that

$$\overline{e}_n \rightharpoonup \overline{e} \quad L^\infty(I) * \text{ weakly}, \quad n \to \infty,$$

then the sequence $\{u(\overline{e}_n)\}$ tends weakly to an element $\overline{u}$ in $H_0^2(I)$, but not necessarily $\overline{u} = u(\overline{e})$; i.e., the limit element $\overline{u}$ is not a solution of $(\mathcal{P}(\overline{e}))$. In order to analyze $(\mathbb{P})$ with $\widetilde{U}^{ad}$ defined by (2.14), deeper mathematical tools, going beyond the scope of this book, have to be used.

(iv) Let us again consider $(\mathbb{P})$ with the cost functional (2.1). Since $(\mathcal{P}(e))$ is the problem with a symmetric bilinear form, its solution can be equivalently characterized as a minimizer of the total potential energy functional E,

$$E(e, u(e)) = \min_{v \in H_0^2(I)} E(e, v),$$

where

$$E(e, v) = \int_I \beta e^3 (v'')^2\,dx - 2\int_I f v\,dx.$$

It is easy to verify that

$$E(e, u(e)) = -\int_I f u(e)\,dx = -J(u(e)) \quad \forall e \in U^{ad}.$$

Consequently

$$\begin{aligned} \min_{e \in U^{ad}} J(u(e)) &= \min_{e \in U^{ad}} \left[-\min_{v \in H_0^2(I)} E(e, v) \right] \\ &= -\max_{e \in U^{ad}} \min_{v \in H_0^2(I)} E(e, v). \end{aligned}$$

Thus the original, minimum compliance problem can be reformulated as a max-min problem for the functional $E : U^{ad} \times H_0^2(I) \to \mathbb{R}$. This form of ($\mathbb{P}$) is closely related to a *saddle-point type problem* for E on the Cartesian product $U^{ad} \times H_0^2(I)$ enabling us to extend U^{ad} even more than has been done so far. Indeed, instead of U^{ad}, $\widetilde{U}^{ad}$, given by (2.3), (2.14), respectively, one can also consider

$$U_{\#}^{ad} = \left\{ e \in L^{\infty}(I) \mid 0 \le e \le e_{\max} \quad \text{a.e. in } I, \int_I e(x)\, dx = \gamma \right\};$$

i.e., we allow zero thickness of the beam. The problem of finding a saddle point of E on $U_{\#}^{ad} \times H_0^2(I)$ still has good mathematical meaning, regardless of the fact that elements of $U_{\#}^{ad}$ may vanish in subsets of I, implying the degeneracy of $(\mathcal{P}(e))$. The explanation is very simple: in the saddle-point formulation $(\mathcal{P}(e))$ is not solved explicitly. This approach is used in the so-called free topology optimization of structures, whose goal is, roughly speaking, to find an optimal distribution of the material permitting the presence of voids, i.e., the absence of any material in some parts of structures.

We now pass to a discretization of ($\mathbb{P}$), i.e., to its transformation into a new problem, defined by a finite number of parameters.

We start with a discretization of U^{ad}. Let $d \in \mathbb{N}$ be given and $\Delta_h : 0 = a_0 < a_1 < \cdots < a_d = \ell$ be an equidistant partition of I with the step $h = \ell/d$, $a_i = ih$, $i = 0, \ldots, d$. Instead of general functions from U^{ad} we shall consider only those that are *continuous* and *piecewise linear* on Δ_h; i.e., we define

$$U_h^{ad} = \left\{ e_h \in C(I) \mid e_{h|[a_{i-1}, a_i]} \in P_1([a_{i-1}, a_i]),\ i = 1, \ldots, d(h) \right\} \cap U^{ad}. \tag{2.15}$$

For a discretization of $(\mathcal{P}(e))$ we use a finite element approach. Let V_h be the finite dimensional subspace of $H_0^2(I)$ defined as follows:

$$\begin{aligned} V_h = \{ v_h \in C^1(I) \mid v_{h|[a_{i-1}, a_i]} \in P_3([a_{i-1}, a_i]),\ i = 1, \ldots, d(h), \\ v_h(0) = v_h'(0) = v_h(\ell) = v_h'(\ell) = 0 \}; \end{aligned} \tag{2.16}$$

i.e., V_h contains all piecewise cubic polynomials that are continuous together with their first derivatives in I and that satisfy the same boundary conditions as functions from $H_0^2(I)$.

Let $e_h \in U_h^{ad}$ be given. Then the *discretized* state problem $(\mathcal{P}_h(e_h))$ reads as follows:

$$\left\{ \begin{array}{l} \text{Find } u_h := u_h(e_h) \in V_h \text{ such that} \\ \displaystyle\int_I \beta e_h^3 u_h'' v_h''\, dx = \int_I f v_h\, dx \quad \forall v_h \in V_h. \end{array} \right. \tag{$\mathcal{P}_h(e_h)$}$$

We now define the discretization of ($\mathbb{P}$) as follows:

$$\left\{ \begin{array}{l} \text{Find } e_h^* \in U_h^{ad} \text{ such that} \\ J(u_h(e_h^*)) = \displaystyle\min_{e_h \in U_h^{ad}} J(u_h(e_h)), \end{array} \right. \tag{$\mathbb{P}_h$}$$

where $u_h(e_h) \in V_h$ solves $(\mathcal{P}_h(e_h))$.

It is left as an easy exercise to prove the following theorem.

Theorem 2.2. *Problem* $(\mathbb{P}_h)$ *has a solution for any* $h > 0$.

In what follows we shall show how problem $(\mathbb{P}_h)$, characterized by a finite number of degrees of freedom, can be realized numerically. To this end we derive its algebraic form.

Let $h > 0$ be fixed. Any function $e_h \in U_h^{ad}$ can be identified with a vector $\boldsymbol{e} = (e_0, \ldots, e_d) \in \mathcal{U}$ whose components are the nodal values of e_h; i.e., $e_i = e_h(a_i)$, $i = 0, \ldots, d$. It is easy to see that

$$\mathcal{U} = \Bigg\{ \boldsymbol{e} \in \mathbb{R}^{d+1} \mid e_{\min} \le e_i \le e_{\max},\ i = 0, \ldots, d; \\ |e_{i+1} - e_i| \le L_0 h,\ i = 0, \ldots, d-1; \quad h \sum_{i=0}^{d-1} (e_i + e_{i+1}) = 2\gamma \Bigg\}. \tag{2.17}$$

The discrete state problem $(\mathcal{P}_h(e_h))$ transforms into a system of linear algebraic equations,

$$\boldsymbol{K}(\boldsymbol{e})\,\boldsymbol{q}(\boldsymbol{e}) = \boldsymbol{f}, \tag{2.18}$$

where $\boldsymbol{K}(\boldsymbol{e}) = \left(k_{ij}(\boldsymbol{e})\right)_{i,j=1}^n$ is the (symmetric) stiffness matrix, $\boldsymbol{f} = (f_i)_{i=1}^n$ is the force vector, $\boldsymbol{q}(\boldsymbol{e}) \in \mathbb{R}^n$ is the nodal vector representing the finite element solution $u_h(e_h) \in V_h$ of $(\mathcal{P}_h(e_h))$, and $n := n(h) = \dim V_h$. Any function $v_h \in V_h$ is uniquely defined by the vector $\boldsymbol{v}$ of nodal values $v_h(a_i)$, $v_h'(a_i)$, $i = 1, \ldots, d-1$, so that $n = 2(d-1)$. Thus the components of $\boldsymbol{q}(\boldsymbol{e})$ are the values of u_h and u_h' at the nodal points $a_i \in \Delta_h$, $i = 1, \ldots, d-1$, which will be arranged as follows:

$$\boldsymbol{q}(\boldsymbol{e}) = (u_h(a_1), u_h'(a_1), u_h(a_2), u_h'(a_2), \ldots, u_h(a_{d-1}), u_h'(a_{d-1})). \tag{2.19}$$

The elements of $\boldsymbol{K}(\boldsymbol{e})$ and $\boldsymbol{f}$ are computed in the standard way:

$$k_{ij}(\boldsymbol{e}) = \int_I \beta e_h^3 \varphi_i'' \varphi_j''\, dx, \qquad f_i = \int_I f \varphi_i\, dx, \quad i, j = 1, \ldots, n, \tag{2.20}$$

where $\{\varphi_j\}_{j=1}^n$ are the basis functions of V_h. The one-to-one correspondences $e_h \leftrightarrow \boldsymbol{e}$ and $v_h \leftrightarrow \boldsymbol{v}$ between elements of U_h^{ad} and $\mathcal{U}$ and V_h and $\mathbb{R}^n$, respectively, defines the following isomorphisms $\mathcal{T}_D$ and $\mathcal{T}_S$:

$$\mathcal{T}_D : U_h^{ad} \to \mathcal{U}; \quad \mathcal{T}_D(e_h) = \boldsymbol{e}, \quad e_h \in U_h^{ad}; \tag{2.21}$$

$$\mathcal{T}_S : V_h \to \mathbb{R}^n; \quad \mathcal{T}_S(v_h) = \boldsymbol{v}, \quad v_h \in V_h, \tag{2.22}$$

enabling us to identify U_h^{ad} with $\mathcal{U}$ and V_h with $\mathbb{R}^n$. In a similar way one can identify the cost functional J restricted to V_h with a function $\mathcal{J} : \mathbb{R}^n \to \mathbb{R}$:

$$\mathcal{J}(\boldsymbol{v}) := J(\mathcal{T}_S^{-1}(\boldsymbol{v})),$$

where $\mathcal{T}_S^{-1}$ denotes the inverse of $\mathcal{T}_S$ (if $J : U^{ad} \times H_0^2(I) \to \mathbb{R}$ depends explicitly on both variables, then $\mathcal{J} : \mathcal{U} \times \mathbb{R}^n \to \mathbb{R}$, where $\mathcal{J}(\boldsymbol{e}, \boldsymbol{v}) := J(\mathcal{T}_D^{-1}(\boldsymbol{e}), \mathcal{T}_S^{-1}(\boldsymbol{v}))$). Therefore the discrete thickness optimization problem $(\mathbb{P}_h)$ leads to the following *nonlinear mathematical programming problem*:

$$\begin{cases} \text{Find } \boldsymbol{e}^* \in \mathcal{U} \text{ such that} \\ \mathcal{J}(\boldsymbol{q}(\boldsymbol{e}^*)) \leq \mathcal{J}(\boldsymbol{q}(\boldsymbol{e})) \quad \forall \boldsymbol{e} \in \mathcal{U}, \end{cases} \tag{$\mathbb{P}_d$}$$

where $\boldsymbol{q}(\boldsymbol{e}) \in \mathbb{R}^n$ solves (2.18). Problem $(\mathbb{P}_d)$ can be solved using nonlinear mathematical programming methods. Some of them will be presented in brief in Chapter 4.

Until now, the mesh parameter $h > 0$ has been fixed. A natural question arises: What happens if $h \to 0+$, meaning that finer and finer meshes are used for the discretization of U^{ad} and $(\mathcal{P}(e))$? Is there any relation between solutions to $(\mathbb{P})$ and $(\mathbb{P}_h)$ as $h \to 0+$? This is what we shall study now.

Let $\{\Delta_h\}$, $h \to 0+$, be a *family* of equidistant partitions of I whose norms tend to zero and consider the respective families $\{U_h^{ad}\}$, $\{V_h\}$, $h \to 0+$. The next lemma, which is a counterpart of Lemma 2.1 in the continuous setting, plays a major role in the convergence analysis.

Lemma 2.2. *Let $\{e_h\}$, $e_h \in U_h^{ad}$, be such that $e_h \rightrightarrows e$ in I and let $\{u_h(e_h)\}$ be the sequence of solutions to $(\mathcal{P}_h(e_h), h \to 0+)$. Then*

$$u_h(e_h) \to u(e) \quad \textit{in } H_0^2(I), \quad h \to 0+,$$

and $u(e)$ solves $(\mathcal{P}(e))$.

Proof. As in Lemma 2.1 one can prove that $\{u_h(e_h)\}$ is bounded:

$$\exists\, c > 0 : \quad \|u_h(e_h)\|_{2,I} \leq c \quad \forall h > 0.$$

Passing to a subsequence, there exists $u \in H_0^2(I)$ such that

$$u_h(e_h) \rightharpoonup u \quad \text{in } H_0^2(I), \quad h \to 0+. \tag{2.23}$$

Next, we show that u solves $(\mathcal{P}(e))$. Let $\overline{v} \in H_0^2(I)$ be arbitrary but fixed. Then there is a sequence $\{\overline{v}_h\}$, $\overline{v}_h \in V_h$, such that

$$\overline{v}_h \to \overline{v} \quad \text{in } H_0^2(I), \quad h \to 0+. \tag{2.24}$$

From the definition of $(\mathcal{P}_h(e_h))$ it follows that

$$\int_I \beta e_h^3 u_h'' \overline{v}_h''\, dx = \int_I f \overline{v}_h\, dx. \tag{2.25}$$

Passing to the limit with $h \to 0+$ in (2.25) we obtain

$$\int_I \beta e_h^3 u_h'' \overline{v}_h''\, dx \to \int_I \beta e^3 u'' \overline{v}''\, dx, \tag{2.26}$$

$$\int_I f \overline{v}_h\, dx \to \int_I f \overline{v}\, dx. \tag{2.27}$$

Indeed,

$$\left|\int_I (\beta e_h^3 u_h'' \overline{v}_h'' - \beta e^3 u'' \overline{v}'')\, dx\right| \le c \int_I |e_h^3 - e^3|\, |u_h'' \overline{v}_h''|\, dx + \left|\int_I \beta e^3 (u_h'' \overline{v}_h'' - u'' \overline{v}'')\, dx\right|$$
$$\le c \|e_h^3 - e^3\|_{C(I)} \|u_h\|_{2,I} \|\overline{v}_h\|_{2,I} + \left|\int_I \beta e^3 (u_h'' \overline{v}_h'' - u'' \overline{v}'')\, dx\right| \to 0,$$

making use of (2.23), (2.24), and uniform convergence of e_h to e in I. The limit passage (2.27) is trivial. From (2.25), (2.26), and (2.27) it follows that

$$\int_I \beta e^3 u'' \overline{v}''\, dx = \int_I f \overline{v}\, dx$$

is satisfied for any $\overline{v} \in H_0^2(I)$; i.e., u solves $(\mathcal{P}(e))$. Due to its uniqueness, the whole sequence $\{u_h(e_h)\}$ tends weakly to $u(e) := u$ in $H_0^2(I)$. Strong convergence of $\{u_h(e_h)\}$ to $u(e)$ can be established in the same way as was done in Lemma 2.1. $\square$

A direct consequence of the previous lemma is the following convergence result.

Theorem 2.3. *Let $\{(e_h^*, u_h(e_h^*))\}$ be a sequence of optimal pairs of $(\mathbb{P}_h)$, $h \to 0+$. Then one can find a subsequence of $\{(e_h^*, u_h(e_h^*))\}$ such that*

$$\begin{cases} e_h^* \rightrightarrows e^* & \text{in } I, \\ u_h(e_h^*) \to u(e^*) & \text{in } H_0^2(I),\ h \to 0+, \end{cases} \tag{2.28}$$

where $(e^*, u(e^*))$ is an optimal pair of $(\mathbb{P})$. In addition, any accumulation point of $\{(e_h^*, u_h(e_h^*))\}$ in the sense of (2.28) *possesses this property.*

Proof. Let $\{e_h^*\}$ be a sequence of solutions to $(\mathbb{P}_h)$, $h \to 0+$ (if $(\mathbb{P}_h)$ had more than one solution, we would choose one of them). Since $U_h^{ad} \subset U^{ad}$ $\forall h > 0$ and U^{ad} is compact in $C(I)$, one can pass to a subsequence of $\{e_h^*\}$ such that

$$e_h^* \rightrightarrows e^* \in U^{ad} \text{ in } I, \quad h \to 0+.$$

At the same time

$$u_h(e_h^*) \to u(e^*) \quad \text{in } H_0^2(I), \tag{2.29}$$

as follows from Lemma 2.2.

Let $\overline{e} \in U^{ad}$ be given. Then one can find a sequence $\{\overline{e}_h\}$, $\overline{e}_h \in U_h^{ad}$, such that (see Problem 2.2)

$$\overline{e}_h \rightrightarrows \overline{e} \quad \text{in } I, \quad h \to 0+, \tag{2.30}$$

and

$$u_h(\overline{e}_h) \to u(\overline{e}) \quad \text{in } H_0^2(I), \quad h \to 0+, \tag{2.31}$$

using Lemma 2.2 once again. From the definition of $(\mathbb{P}_h)$ it follows that

$$J(u_h(e_h^*)) \leq J(u_h(\overline{e}_h)).$$

Letting $h \to 0+$ in this inequality, using continuity of J, (2.29), and (2.31), we arrive at

$$J(u(e^*)) \leq J(u(\overline{e}));$$

i.e., $(e^*, u(e^*))$ is an optimal pair of $(\mathbb{P})$. From the proof we also see that any other accumulation point of $\{(e_h^*, u_h(e_h^*))\}$ is also an optimal pair of $(\mathbb{P})$. □

COMMENTS 2.2.

(i) Theorem 2.3 says that $(\mathbb{P}_h)$, $h \to 0+$, and $(\mathbb{P})$ are close in the sense of subsequences only. This is due to the fact that $(\mathbb{P})$ may have more than one solution. This theorem, however, ensures the existence of at least one subsequence of $\{(e_h^*, u_h(e_h^*))\}$ tending to an optimal pair of $(\mathbb{P})$. In addition, any convergent subsequence of $\{(e_h^*, u_h(e_h^*))\}$ in the sense of (2.28) tends only to an optimal pair of $(\mathbb{P})$.

(ii) Besides the continuous dependence (established in Lemma 2.2) and the compactness of $\{U_h^{ad}\}$ in U^{ad}, yet another property plays a role in the convergence analysis: one needs the density of $\{U_h^{ad}\}$ and $\{V_h\}$ in U^{ad}, $H_0^2(I)$, respectively.

(iii) Since functions from U_h^{ad} are continuous, their $C(I)$- and $L^\infty(I)$-norms coincide. Nevertheless the assertion of Lemma 2.2 remains true even if $U_h^{ad} \subset L^\infty(I)$, but uniform convergence of $\{e_h\}$ in I has to be replaced by convergence in the $L^\infty(I)$-norm. This elementary result will be used below.

Besides the mathematical aspects, practical aspects should also be taken into account. The set U_h^{ad} defined by (2.15) is very simple from a mathematical point of view but usually not acceptable for practical purposes. Rather than a piecewise linear thickness distribution, engineers prefer stepped beams, which are characterized by a *piecewise constant* thickness distribution. In what follows we shall consider this case by changing the definition of U_h^{ad}.

Let Δ_h be the partition of I as before. The new set $\widetilde{U}_h^{ad}$ of discrete thickness distributions is defined as follows:

$$\widetilde{U}_h^{ad} = \Big\{ \widetilde{e}_h \in L^\infty(I) \mid \widetilde{e}_h^i \in P_0([a_{i-1}, a_i]),\ i = 1, \dots, d,$$
$$e_{\min} \leq \widetilde{e}_h \leq e_{\max} \text{ a.e. in } I,\ \int_I \widetilde{e}_h \, dx = \gamma,$$
$$|\widetilde{e}_h^{i+1} - \widetilde{e}_h^i| \leq L_0 h,\ i = 1, \dots, d-1 \Big\}, \tag{2.32}$$

where $\widetilde{e}_h^i := \widetilde{e}_{h|[a_{i-1}, a_i]}$ (see Figure 2.1).

This means that $\widetilde{U}_h^{ad}$ consists of all piecewise constant functions on Δ_h satisfying the same uniform boundedness and volume constraints as functions from the original set U^{ad}. The uniform Lipschitz constraint is satisfied by the discrete values $\widetilde{e}_h^i$, $i = 1, \dots, d(h)$.

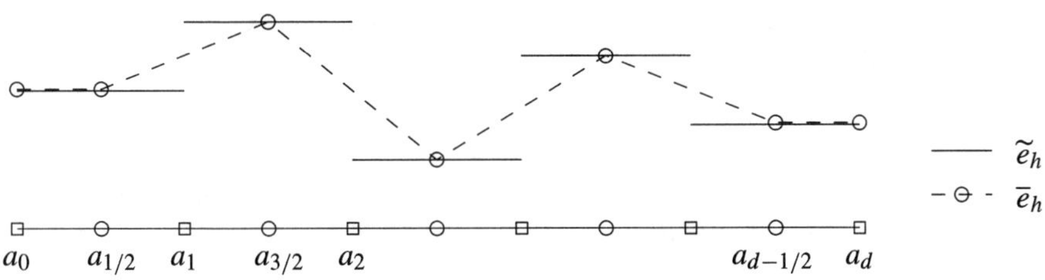

Figure 2.1. *Piecewise constant approximation of* U^{ad}.

This time, however, $\widetilde{U}_h^{ad}$ is *not* a subset of U^{ad} since it contains *discontinuous* functions. Next, we shall study the following discrete thickness optimization problem:

$$\begin{cases} \text{Find } \widetilde{e}_h^* \in \widetilde{U}_h^{ad} \quad \text{such that} \\ J(u_h(\widetilde{e}_h^*)) \le J(u_h(\widetilde{e}_h)) \quad \forall \widetilde{e}_h \in \widetilde{U}_h^{ad}, \end{cases} \qquad (\widetilde{\mathbb{P}}_h)$$

with $u_h(\widetilde{e}_h) \in V_h$ being the solution to $(\mathcal{P}_h(\widetilde{e}_h))$, $\widetilde{e}_h \in \widetilde{U}_h^{ad}$, and J given by (2.1).

It is not surprising that a convergence result similar to Theorem 2.3 also remains valid for $(\widetilde{\mathbb{P}}_h)$ with a minor change concerning convergence of $\{\widetilde{e}_h^*\}$ to e^*.

Theorem 2.4. *Let* $\{(\widetilde{e}_h^*, u_h(\widetilde{e}_h^*))\}$ *be a sequence of optimal pairs of* $(\widetilde{\mathbb{P}}_h)$, $h \to 0+$. *Then from any sequence* $\{(\widetilde{e}_h^*, u_h(\widetilde{e}_h^*))\}$ *one can pass to its subsequence such that*

$$\begin{cases} \widetilde{e}_h^* \to e^* \quad \text{in } L^\infty(I), \\ u_h(\widetilde{e}_h^*) \to u(e^*) \quad \text{in } H_0^2(I), \quad h \to 0+, \end{cases} \tag{2.33}$$

where $(e^*, u(e^*))$ *is an optimal pair of* $(\mathbb{P})$. *In addition, any accumulation point of* $\{(\widetilde{e}_h^*, u_h(\widetilde{e}_h^*))\}$ *in the sense of* (2.33) *possesses this property.*

Proof. In view of Comments 2.2(iii) we know that Lemma 2.2 remains true if we change "$e_h \rightrightarrows e$ in I" to "$\widetilde{e}_h \to e$ in $L^\infty(I)$". We only have to verify the compactness of any sequence $\{\widetilde{e}_h\}$, $\widetilde{e}_h \in \widetilde{U}_h^{ad}$, and the density type result for $\{\widetilde{U}_h^{ad}\}$ (see Comments 2.2(ii)).

Let $\{\widetilde{e}_h\}$, $h \to 0+$, $\widetilde{e}_h \in \widetilde{U}_h^{ad}$, be an arbitrary sequence. With any $\widetilde{e}_h$, the following continuous, piecewise linear function $\overline{e}_h$ defined on the partition $\overline{\Delta}_h : 0 = a_0 < a_{1/2} < a_{3/2} < \cdots < a_{d-1/2} < a_d = \ell$ will be associated (see Figure 2.1):

$$\overline{e}_h(a_{i+1/2}) = \widetilde{e}_h^{i+1}, \ i = 0, \ldots, d-1,$$
$$\overline{e}_h(a_0) = \widetilde{e}_h^1, \quad \overline{e}_h(a_d) = \widetilde{e}_h^d,$$

where $a_{i-1/2}$ denotes the midpoint of the interval $[a_{i-1}, a_i]$, $i = 1, \ldots, d$. From the construction of $\overline{e}_h$ we see that

$$e_{\min} \le \overline{e}_h \le e_{\max} \quad \text{in } I$$

and

$$|\overline{e}_h'| \le L_0 \quad \text{a.e. in } I.$$

Thus $\{\overline{e}_h\}$ is compact in $C(I)$ so that there exists a subsequence of $\{\overline{e}_h\}$ and an element $\overline{e} \in C(I)$ such that

$$\overline{e}_h \rightrightarrows \overline{e} \quad \text{in } I, \quad h \to 0+, \tag{2.34}$$

satisfying $e_{\min} \le \overline{e} \le e_{\max}$ in I, $|\overline{e}'| \le L_0$ a.e. in I. The function $\widetilde{e}_h$ can be viewed as the piecewise constant interpolant of $\overline{e}_h \in W^{1,\infty}(I)$ implying that

$$\|\widetilde{e}_h - \overline{e}_h\|_{L^\infty(I)} \le ch|\overline{e}_h|_{W^{1,\infty}(I)} \le ch,$$

as follows from the well-known approximation properties and the fact that the sequence $\left\{\|\overline{e}_h\|_{W^{1,\infty}(I)}\right\}$ is bounded. This, (2.34), and the triangle inequality yield

$$\|\widetilde{e}_h - \overline{e}\|_{L^\infty(I)} \le \|\widetilde{e}_h - \overline{e}_h\|_{L^\infty(I)} + \|\overline{e}_h - \overline{e}\|_{L^\infty(I)} \to 0$$

as $h \to 0+$. At the same time $\int_I \overline{e}\, dx = \gamma$ so that $\overline{e} \in U^{ad}$.

The density of $\{\widetilde{U}_h^{ad}\}$ in U^{ad} in the $L^\infty(I)$-norm is easy to prove. Indeed, let $e \in U^{ad}$ be given and define $\widetilde{e}_h$ as follows:

$$\widetilde{e}_h = \sum_i \left(\frac{1}{h}\int_{a_{i-1}}^{a_i} e\, dx\right)\chi_i, \tag{2.35}$$

where χ_i is the characteristic function of $[a_{i-1}, a_i]$, $i = 1, \dots, d$. Clearly $\widetilde{e}_h \in \widetilde{U}_h^{ad}$ and $\widetilde{e}_h \to e$ in $L^\infty(I)$, $h \to 0+$. $\square$

2.2 A model optimal shape design problem

In this section we present the main ideas that will be used in the existence analysis of optimal shape design problems. One of the main difficulties we face in any shape optimization problem is that functions are defined in variable domains whose shapes are the object of optimization. One of the possible ways to handle this difficulty is to *extend* functions from their domain of definition to a *larger (fixed) set* containing all admissible domains. This extension is straightforward for homogeneous Dirichlet boundary value problems formulated in $H_0^1(\Omega)$, since any function from $H_0^1(\Omega)$ can be extended by zero outside of Ω, preserving its norm. For this reason our analysis starts with just this type of state problem. For the sake of simplicity we restrict ourselves to the case when only a *part* of the boundary is subject to optimization and, in addition, this part is represented by the *graph of a function*.

Let

$$\mathcal{O} = \{\Omega(\alpha) \mid \alpha \in U^{ad}\}$$

be a family of *admissible domains*, where

$$U^{ad} = \Big\{\alpha \in C^{0,1}([0,1]) \mid 0 < \alpha_{\min} \le \alpha \le \alpha_{\max} \text{ in } [0,1],$$
$$|\alpha'| \le L_0 \text{ a.e. in }]0,1[, \quad \int_0^1 \alpha(x_2)\, dx_2 = \gamma\Big\}$$

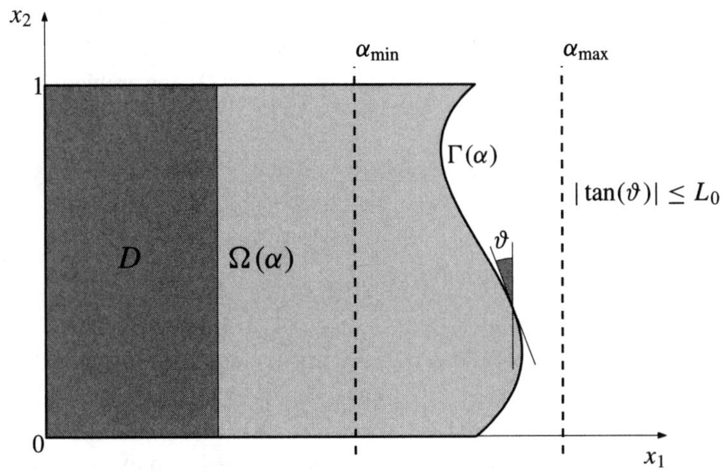

Figure 2.2. *Parameters characterizing admissible domains.*

and

$$\Omega(\alpha) = \left\{(x_1, x_2) \in \mathbb{R}^2 \mid 0 < x_1 < \alpha(x_2),\ x_2 \in]0, 1[\right\}, \quad \alpha \in U^{ad}$$

(see Figure 2.2). Here $C^{0,1}([0, 1])$ is the set of all Lipschitz continuous functions in $[0, 1]$ and the parameters $\alpha_{\min}$, $\alpha_{\max}$, L_0, and γ are such that $U^{ad} \neq \emptyset$.

The family $\mathcal{O}$ consists of all "curved rectangles" whose curved sides $\Gamma(\alpha)$, represented by the graph of $\alpha \in U^{ad}$, will be the object of optimization. Since the shape of any $\Omega(\alpha) \in \mathcal{O}$ is characterized solely by $\alpha \in U^{ad}$, there exists a one-to-one correspondence between $\mathcal{O}$ and U^{ad}: $\Omega(\alpha) \in \mathcal{O} \leftrightarrow \alpha \in U^{ad}$. Elements of U^{ad} will be called *design variables*, determining the shape of $\Omega \in \mathcal{O}$ in a unique way.

Remark 2.2. From the definition of U^{ad} it follows that all $\Omega(\alpha) \in \mathcal{O}$ are domains with Lipschitz boundaries (see Appendix A), with the same area equal to γ, and such that all $\Gamma(\alpha)$ remain in the strip bounded by $\alpha_{\min}$, $\alpha_{\max}$ (see Figure 2.2). In addition, the uniform Lipschitz constraint prevents oscillations of $\Gamma(\alpha)$ (see also Remark 2.1).

Remark 2.3. The admissible domains depicted in Figure 2.2 are parametrized in a very simple way. In real-life problems, however, the parametrization of the geometry may represent a difficult task.

On any $\Omega(\alpha)$, $\alpha \in U^{ad}$, we shall consider the following state problem:

$$\begin{cases} -\Delta u(\alpha) = f & \text{in } \Omega(\alpha), \\ u(\alpha) = 0 & \text{on } \partial\Omega(\alpha), \end{cases} \qquad (\mathcal{P}(\alpha))$$

where $f \in L^2_{\text{loc}}(\mathbb{R}^2)$. To emphasize that u depends on $\Omega(\alpha)$ (and hence on α), we shall write α in the argument of u.

Let D be a "target set" that is a part of any $\Omega(\alpha) \in \mathcal{O}$, say $D =]0, \alpha_{\min}/2[\times]0, 1[$. We want to find $\Omega(\alpha^*) \in \mathcal{O}$ such that the respective solution $u(\alpha^*)$ of $(\mathcal{P}(\alpha^*))$ is "as close as possible" to a given function z_d in D. Assuming $z_d \in L^2(D)$, the problem can be formulated as follows:

$$\begin{cases} \text{Find } \alpha^* \in U^{ad} \text{ such that} \\ J(u(\alpha^*)) \le J(u(\alpha)) \quad \forall \alpha \in U^{ad}, \end{cases} \tag{$\mathbb{P}$}$$

where $u(\alpha) \in H_0^1(\Omega(\alpha))$ solves $(\mathcal{P}(\alpha))$ and

$$J(y) = \frac{1}{2}\|y - z_d\|_{0,D}^2, \quad y \in L^2(D).$$

Remark 2.4. Since D is the same for all $\alpha \in U^{ad}$, the cost functional J depends explicitly only on the state variable y and becomes a function of $\alpha \in U^{ad}$ only through its composition with the inner control state mapping $u : \alpha \mapsto u(\alpha)$. Cost functionals, however, may depend *explicitly* on both control and state variables. For this reason, the abstract formulation of optimal shape design problems presented in Section 2.4 uses functionals depending on both variables.

We start our analysis by introducing convergence in $\mathcal{O}$.

Definition 2.2. *Let $\{\Omega(\alpha_n)\}$, $\Omega(\alpha_n) \in \mathcal{O}$, be a sequence of domains. We say that $\{\Omega(\alpha_n)\}$ tends to $\Omega(\alpha) \in \mathcal{O}$ (and write $\Omega(\alpha_n) \to \Omega(\alpha)$) iff*

$$\alpha_n \rightrightarrows \alpha \quad \text{in } [0, 1], \quad n \to \infty.$$

In other words, taking an arbitrary δ-neighborhood $B_\delta(\Gamma(\alpha))$ of $\Gamma(\alpha)$, there exists $n_0 := n_0(\delta) \in \mathbb{N}$ such that $\Gamma(\alpha_n) \subset B_\delta(\Gamma(\alpha))$ $\forall n \ge n_0$.

From the definition of U^{ad} we have the following result.

Lemma 2.3. *The family $\mathcal{O}$ is compact for the convergence of sets introduced in Definition* 2.2.

Proof. The proof follows from the Ascoli–Arzelà theorem. □

Lemma 2.4. *Let $\Omega(\alpha_n) \to \Omega(\alpha)$, $n \to \infty$, $\Omega(\alpha_n)$, $\Omega(\alpha) \in \mathcal{O}$, and χ_n, χ be the characteristic functions of $\Omega(\alpha_n)$, $\Omega(\alpha)$, respectively. Then*

$$\chi_n \to \chi \quad \text{in } L^p(\widehat{\Omega}) \quad \forall\, p \in [1, \infty[,$$

where $\widehat{\Omega}$ is such that $\widehat{\Omega} \supseteq \Omega(\alpha)$ $\forall \alpha \in U^{ad}$. In what follows we shall take $\widehat{\Omega} =]0, 2\alpha_{\max}[\times]0, 1[$.

Another point has yet to be clarified, namely, how to define convergence of functions belonging to $H_0^1(\Omega(\alpha))$ for different $\alpha \in U^{ad}$. Let $\widehat{\Omega}$ be the same as in Lemma 2.4.

If $v \in H_0^1(\Omega(\alpha))$ for some $\alpha \in U^{ad}$, then the function $\widetilde{v}$ defined by

$$\widetilde{v} = \begin{cases} v & \text{in } \Omega(\alpha), \\ 0 & \text{in } \widehat{\Omega} \setminus \overline{\Omega(\alpha)} \end{cases} \tag{2.36}$$

belongs to $H_0^1(\widehat{\Omega})$ and $\|\widetilde{v}\|_{1,\widehat{\Omega}} = \|v\|_{1,\Omega(\alpha)}$.

Convention: *In the rest of this section the symbol* $\tilde{}$ *above a function stands for its extension by zero to* $\widehat{\Omega}$.

Definition 2.3. *Let* $v_n \in H_0^1(\Omega(\alpha_n))$, $v \in H_0^1(\Omega(\alpha))$, $\alpha_n, \alpha \in U^{ad}$, $n \to \infty$. *We say that*

$$v_n \to v \quad \textit{iff} \quad \widetilde{v}_n \to \widetilde{v} \textit{ in } H_0^1(\widehat{\Omega}), \tag{2.37}$$

$$v_n \rightharpoonup v \quad \textit{iff} \quad \widetilde{v}_n \rightharpoonup \widetilde{v} \textit{ in } H_0^1(\widehat{\Omega}), \quad n \to \infty. \tag{2.38}$$

The symbols $\to$ *and* $\rightharpoonup$ *on the right of* (2.37) *and* (2.38) *denote classical strong and weak convergence, respectively, in* $H_0^1(\widehat{\Omega})$.

As in the previous section, the following continuous dependence type result plays a key role in the forthcoming analysis.

Lemma 2.5. *Let* $\Omega(\alpha_n), \Omega(\alpha) \in \mathcal{O}$ *be such that* $\Omega(\alpha_n) \to \Omega(\alpha)$ *and let* $u_n := u(\alpha_n)$ *be the respective solution to* $(\mathcal{P}(\alpha_n))$, $n \to \infty$. *Then*

$$\widetilde{u}_n \to u \text{ in } H_0^1(\widehat{\Omega}), \quad n \to \infty,$$

and $u(\alpha) := u_{|\Omega(\alpha)}$ *solves* $(\mathcal{P}(\alpha))$.

Proof. The function $u_n \in H_0^1(\Omega(\alpha_n))$ being the solution of $(\mathcal{P}(\alpha_n))$ satisfies

$$\int_{\Omega_n} \nabla u_n \cdot \nabla \varphi \, dx = \int_{\Omega_n} f \varphi \, dx \quad \forall \varphi \in H_0^1(\Omega(\alpha_n)), \tag{2.39}$$

where $\Omega_n := \Omega(\alpha_n)$ and $dx := dx_1 dx_2$. Inserting $\varphi := u_n$ into (2.39) and using (2.36) we see that $\{\widetilde{u}_n\}$ is bounded in $H_0^1(\widehat{\Omega})$. Indeed,

$$|\widetilde{u}_n|_{1,\widehat{\Omega}}^2 = |u_n|_{1,\Omega_n}^2 \le \|f\|_{0,\widehat{\Omega}} \|\widetilde{u}_n\|_{1,\widehat{\Omega}}.$$

From this and the Friedrichs inequality in $H_0^1(\widehat{\Omega})$ the existence of a constant $c > 0$ such that

$$\|\widetilde{u}_n\|_{1,\widehat{\Omega}} \le c \quad \forall n \in \mathbb{N} \tag{2.40}$$

follows. One can pass to a subsequence of $\{\widetilde{u}_n\}$ such that

$$\widetilde{u}_n \rightharpoonup u \text{ in } H_0^1(\widehat{\Omega}). \tag{2.41}$$

We first prove that $u_{|\widehat{\Omega} \setminus \overline{\Omega(\alpha)}} = 0$, implying that $u_{|\Omega(\alpha)} \in H_0^1(\Omega(\alpha))$.

Let $\bar{x} \in \widehat{\Omega} \setminus \overline{\Omega(\alpha)}$ be arbitrary. Since $\widehat{\Omega} \setminus \overline{\Omega(\alpha)}$ is open, there exists a δ-neighborhood $B_\delta(\bar{x})$ of $\bar{x}$ such that $B_\delta(\bar{x}) \subset B_{2\delta}(\bar{x}) \subset \widehat{\Omega} \setminus \overline{\Omega(\alpha)}$ and at the same time $\Gamma(\alpha_n) \subset B_\delta(\Gamma(\alpha))$ for $n \in \mathbb{N}$ large enough, as follows from Definition 2.2. Therefore $\widetilde{u}_n = 0$ in $B_\delta(\bar{x})$ so that $u = 0$ in $B_\delta(\bar{x})$, as follows from (2.41). Since $\bar{x} \in \widehat{\Omega} \setminus \overline{\Omega(\alpha)}$ is arbitrary we have $u = 0$ in $\widehat{\Omega} \setminus \overline{\Omega(\alpha)}$. To prove that $u_{|\Omega(\alpha)}$ solves $(\mathcal{P}(\alpha))$ we rewrite (2.39) as follows:

$$\int_{\widehat{\Omega}} \chi_n \nabla \widetilde{u}_n \cdot \nabla \widetilde{\varphi}\, dx = \int_{\widehat{\Omega}} \chi_n f \widetilde{\varphi}\, dx \quad \forall \varphi \in H_0^1(\Omega(\alpha_n)), \tag{2.42}$$

where χ_n is the characteristic function of $\Omega(\alpha_n)$. Let $\xi \in C_0^\infty(\Omega(\alpha))$ be fixed. Then ξ vanishes in a neighborhood of $\partial\Omega(\alpha)$ so that $\widetilde{\xi}_{|\Omega(\alpha_n)} \in C_0^\infty(\Omega(\alpha_n))$ for $n \in \mathbb{N}$ large enough and $\widetilde{\xi}_{|\Omega(\alpha_n)}$ can be used as a test function in (2.42):

$$\int_{\widehat{\Omega}} \chi_n \nabla \widetilde{u}_n \cdot \nabla \widetilde{\xi}\, dx = \int_{\widehat{\Omega}} \chi_n f \widetilde{\xi}\, dx. \tag{2.43}$$

Passing to the limit with $n \to \infty$, using (2.41) and Lemma 2.4, we easily obtain

$$\int_{\widehat{\Omega}} \chi \nabla u \cdot \nabla \widetilde{\xi}\, dx = \int_{\widehat{\Omega}} \chi f \widetilde{\xi}\, dx,$$

where χ is the characteristic function of $\Omega(\alpha)$. This is equivalent to

$$\int_{\Omega(\alpha)} \nabla u \cdot \nabla \xi\, dx = \int_{\Omega(\alpha)} f \xi\, dx. \tag{2.44}$$

From the density of $C_0^\infty(\Omega(\alpha))$ in $H_0^1(\Omega(\alpha))$, it follows that (2.44) holds for any $\xi \in H_0^1(\Omega(\alpha))$; i.e., $u_{|\Omega(\alpha)}$ solves $(\mathcal{P}(\alpha))$. Thus we proved that any accumulation point u of the weakly convergent sequence $\{\widetilde{u}_n\}$ is such that $u_{|\widehat{\Omega}\setminus\overline{\Omega(\alpha)}} = 0$ and $u_{|\Omega(\alpha)}$ solves $(\mathcal{P}(\alpha))$. Such a function is unique so that not only a subsequence but the whole sequence tends weakly to u in $H_0^1(\widehat{\Omega})$. Let us prove strong convergence. From the definition of $(\mathcal{P}(\alpha_n))$ and (2.41) it follows that

$$|\widetilde{u}_n|^2_{1,\widehat{\Omega}} = \int_{\widehat{\Omega}} \chi_n |\nabla \widetilde{u}_n|^2\, dx = \int_{\widehat{\Omega}} \chi_n f \widetilde{u}_n\, dx \to \int_{\widehat{\Omega}} \chi f u\, dx = \int_{\widehat{\Omega}} \chi |\nabla u|^2\, dx = |u|^2_{1,\widehat{\Omega}}.$$

Indeed,

$$\int_{\widehat{\Omega}} \chi_n f \widetilde{u}_n\, dx - \int_{\widehat{\Omega}} \chi f u\, dx = \int_{\widehat{\Omega}} \chi_n f(\widetilde{u}_n - u)\, dx + \int_{\widehat{\Omega}} (\chi_n - \chi) f u\, dx \to 0, \quad n \to \infty. \tag{2.45}$$

The first and second integrals on the right of (2.45) tend to zero because of (2.41) implying $\widetilde{u}_n \to u$ in $L^2(\widehat{\Omega})$ and the Lebesgue-dominated convergence theorem, respectively. $\square$

Remark 2.5. Since $u = 0$ in $\widehat{\Omega} \setminus \overline{\Omega(\alpha)}$ and $u(\alpha) := u_{|\Omega(\alpha)}$ solves $(\mathcal{P}(\alpha))$, we have $u = \widetilde{u}(\alpha)$ in $\widehat{\Omega}$. Thus the statement of Lemma 2.5 can be written in the form

$$\widetilde{u}_n \to \widetilde{u}(\alpha) \text{ in } H_0^1(\widehat{\Omega}).$$

From Lemma 2.5 the existence of a solution to problem ($\mathbb{P}$) easily follows.

THEOREM 2.5. *Problem* ($\mathbb{P}$) *has a solution.*

Proof. Denote

$$q := \inf_{\alpha \in U^{ad}} J(u(\alpha)) = \lim_{n \to \infty} J(u(\alpha_n)),$$

where $\{\alpha_n\}$, $\alpha_n \in U^{ad}$, is a minimizing sequence. Since U^{ad} is compact in $C([0, 1])$, one can pass to a subsequence such that

$$\alpha_n \rightrightarrows \alpha^* \text{ in } [0, 1]$$

and $\alpha^* \in U^{ad}$. At the same time

$$\widetilde{u}(\alpha_n) \to \widetilde{u}(\alpha^*) \text{ in } H_0^1(\widehat{\Omega}),$$

as follows from Lemma 2.5. This and continuity of J yield

$$q = \lim_{n \to \infty} J(u(\alpha_n)) = J(\widetilde{u}(\alpha^*)) = J(u(\alpha^*));$$

i.e., α^* is a solution of ($\mathbb{P}$). $\square$

COMMENTS 2.3.

(i) The family $\mathcal{O} = \left\{\Omega(\alpha) \mid \alpha \in U^{ad}\right\}$ is a special case of domains satisfying the so-called uniform cone property. This class of domains possesses a very important *uniform extension property* (see Appendix A). Since most of the domains that we meet in real-life problems belong to this class, we restrict ourselves to them throughout the book.

(ii) The continuity assumption on J used in the proof of Theorem 2.5 is too strong and can be weakened. Consider a cost functional J depending on both variables α and y. The existence of solutions to ($\mathbb{P}$) will be guaranteed provided that J is lower semicontinuous as follows:

$$\left.\begin{aligned} \alpha_n \rightrightarrows \alpha \quad & \text{in } [0, 1],\ \alpha_n,\ \alpha \in U^{ad} \\ y_n \to y \quad & \text{in } H_0^1(\widehat{\Omega}),\ y_n, y \in H_0^1(\widehat{\Omega}) \end{aligned}\right\} \\ \Longrightarrow \liminf_{n \to \infty} J(\alpha_n, y_{n|\Omega(\alpha_n)}) \geq J(\alpha, y_{|\Omega(\alpha)}). \quad (2.46)$$

We now pass to an approximation of ($\mathbb{P}$); i.e., we shall discretize both the admissible family $\mathcal{O}$ and state problem ($\mathcal{P}(\alpha)$), $\alpha \in U^{ad}$. We start with the former. The simplest way would be to take piecewise linear approximations of searched boundaries. This type of approximation has, however, serious drawbacks, as mentioned in Appendix B. Engineers prefer to discretize admissible domains by domains that are smooth enough and at the same

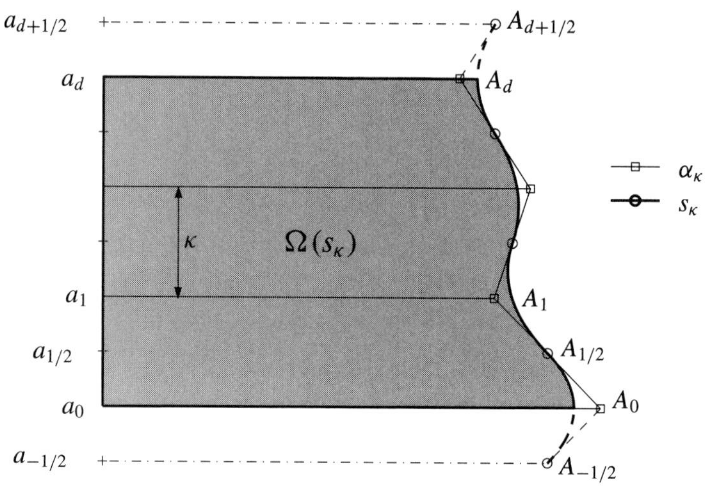

Figure 2.3. *Approximation of the boundary.*

time are defined by a finite number of parameters. For this reason piecewise spline approximations of $\Gamma(\alpha)$ locally realized by quadratic Bézier functions will be used throughout this book. For more detail we refer to Appendix B.

Let $d \in \mathbb{N}$ be given; $\Delta_\varkappa : 0 = a_0 < a_1 < \cdots < a_d = 1$ be an equidistant partition of $[0, 1]$, $a_i = i\varkappa$, $\varkappa = 1/d$, $i = 0, \ldots, d$; and $a_{i+1/2}$ be the midpoint of $[a_i, a_{i+1}]$. Further let $A_i = (\alpha_i, i\varkappa)$, $\alpha_i \in \mathbb{R}$, $i = 0, \ldots, d$, be *design nodes* and $A_{i+1/2} = \frac{1}{2}(A_i + A_{i+1})$ be the midpoint of the segment $\overline{A_i A_{i+1}}$, $i = 0, \ldots, d-1$. In addition let $a_{-1/2} = -\frac{\varkappa}{2}$, $a_{d+1/2} = 1 + \frac{\varkappa}{2}$, $A_{-1/2} = (\frac{1}{2}\alpha_0 + \frac{1}{2}\alpha_1, a_{-1/2})$, $A_{d(\varkappa)+1/2} = (\frac{1}{2}\alpha_{d-1} + \frac{1}{2}\alpha_d, a_{d+1/2})$ (see Figure 2.3). Introduce the sets

$$\widetilde{U}_\varkappa = \Big\{\widetilde{s}_\varkappa \in C^1([-\tfrac{\varkappa}{2}, 1 + \tfrac{\varkappa}{2}]) \mid \widetilde{s}_{\varkappa|[a_{i-1/2}, a_{i+1/2}]} \text{ is a quadratic Bézier function}$$
$$\text{determined by } \{A_{i-1/2},\ A_i,\ A_{i+1/2}\},\ i = 0, \ldots, d\Big\},$$

$$U_\varkappa = \widetilde{U}_{\varkappa|[0,1]} = \left\{s_\varkappa \in C^1([0, 1]) \mid \exists \widetilde{s}_\varkappa \in \widetilde{U}_\varkappa :\ s_\varkappa = \widetilde{s}_{\varkappa|[0,1]}\right\}.$$

Remark 2.6. The triple $\{A_{i-1/2},\ A_i,\ A_{i+1/2}\}$ is termed the control points of the Bézier function (see Appendix B).

In order to define a family of admissible shapes locally realized by Bézier functions, it is necessary to specify $\alpha_i \in \mathbb{R}$ defining the position of A_i, $i = 0, \ldots, d$. With the partition $\Delta_\varkappa$ we associate the set $\mathcal{Q}^{ad}_\varkappa \subset U^{ad}$ of continuous, piecewise linear functions over $\Delta_\varkappa$:

$$\mathcal{Q}^{ad}_\varkappa = \left\{\alpha_\varkappa \in C([0, 1]) \mid \alpha_{\varkappa|[a_{i-1}, a_i]} \in P_1([a_{i-1}, a_i])\ \forall i = 1, \ldots, d\right\} \cap U^{ad}. \tag{2.47}$$

From the previous section we already know that $\mathcal{Q}_{\varkappa}^{ad}$ can be identified with the convex, compact subset $\mathcal{U} \subset \mathbb{R}^{d+1}$:

$$\mathcal{U} = \Bigg\{ \boldsymbol{\alpha} = (\alpha_0, \dots, \alpha_d) \in \mathbb{R}^{d+1} \mid \alpha_{\min} \le \alpha_i \le \alpha_{\max},\ i = 0, \dots, d;$$

$$\frac{|\alpha_{i+1} - \alpha_i|}{\varkappa} \le L_0,\ i = 0, \dots, d-1; \quad \varkappa \sum_{i=0}^{d-1} (\alpha_i + \alpha_{i+1}) = 2\gamma \Bigg\}, \tag{2.48}$$

where $\alpha_i = \alpha_{\varkappa}(a_i),\ i = 0, \dots, d;\ \alpha_{\varkappa} \in \mathcal{Q}_{\varkappa}^{ad}$.

The family of *admissible discretized design domains* is now represented by

$$\mathcal{O}_{\varkappa} = \left\{ \Omega(s_{\varkappa}) \mid s_{\varkappa} \in U_{\varkappa}^{ad} \right\}, \tag{2.49}$$

where

$$U_{\varkappa}^{ad} = \big\{ s_{\varkappa} \in U_{\varkappa} \mid \text{ the design nodes } A_i = (\alpha_i, i\varkappa),\ i = 0, \dots, d,$$
$$\text{are such that } \boldsymbol{\alpha} = (\alpha_0, \dots, \alpha_d) \in \mathcal{U} \big\}. \tag{2.50}$$

We now turn to an approximation of $(\mathcal{P}(\alpha))$. We use the finite element method with continuous, piecewise linear polynomials over a triangulation of the so-called computational domain, an appropriate approximation of $\Omega(s_{\varkappa}) \in \mathcal{O}_{\varkappa}$. We now describe its construction. To this end we introduce another family of regular partitions $\{\Delta_h\}$, $h \to 0+$, of $[0, 1]$, $\Delta_h : 0 = b_0 < b_1 < \cdots < b_{\overline{d}(h)} = 1$ (not necessarily equidistant), whose norm will be denoted by h. *Next we shall suppose that* $h \to 0+$ *iff* $\varkappa \to 0+$. Let $r_h s_{\varkappa}$ be the piecewise linear Lagrange interpolant of $s_{\varkappa}$ on Δ_h:

$$r_h s_{\varkappa}(b_i) = s_{\varkappa}(b_i) \quad \forall i = 0, \dots, \overline{d}(h);$$
$$r_h s_{\varkappa |[b_{i-1}, b_i]} \in P_1([b_{i-1}, b_i]) \quad \forall i = 1, \dots, \overline{d}(h).$$

The computational domain related to $\Omega(s_{\varkappa})$ will be represented by $\Omega(r_h s_{\varkappa})$; i.e., the curved side $\Gamma(s_{\varkappa})$, the graph of $s_{\varkappa} \in U_{\varkappa}^{ad}$, is replaced by its piecewise linear Lagrange approximation $r_h s_{\varkappa}$ on Δ_h. The system of all $\Omega(r_h s_{\varkappa})$, $s_{\varkappa} \in U_{\varkappa}^{ad}$, will be denoted by $\mathcal{O}_{\varkappa h}$ in what follows:

$$\mathcal{O}_{\varkappa h} = \left\{ \Omega(r_h s_{\varkappa}) \mid s_{\varkappa} \in U_{\varkappa}^{ad} \right\}. \tag{2.51}$$

Since $\Omega(r_h s_{\varkappa})$ is already *polygonal*, one can construct its triangulation $\mathcal{T}(h, s_{\varkappa})$ with the norm $h > 0$ (the same as above) and depending on $s_{\varkappa} \in U_{\varkappa}^{ad}$.

Convention: *The domain $\Omega(r_h s_{\varkappa})$ with a given triangulation $\mathcal{T}(h, s_{\varkappa})$ will be denoted by $\Omega_h(s_{\varkappa})$ in what follows.*

With any $\Omega_h(s_{\varkappa})$ the space of continuous, piecewise linear functions over $\mathcal{T}(h, s_{\varkappa})$ will be associated:

$$V_h(s_{\varkappa}) = \left\{ v_h \in C(\overline{\Omega}_h(s_{\varkappa})) \mid v_{h|T} \in P_1(T),\ T \in \mathcal{T}(h, s_{\varkappa});\ v_h = 0 \text{ on } \partial\Omega_h(s_{\varkappa}) \right\}.$$

The state problem is discretized in a standard way. For any $s_\varkappa \in U_\varkappa^{ad}$ fixed, we define it as follows:

$$\begin{cases} \text{Find } u_h := u_h(s_\varkappa) \in V_h(s_\varkappa) \text{ such that} \\ \displaystyle\int_{\Omega_h(s_\varkappa)} \nabla u_h \cdot \nabla \varphi_h \, dx = \int_{\Omega_h(s_\varkappa)} f \varphi_h \, dx \quad \forall \varphi_h \in V_h(s_\varkappa). \end{cases} \tag{$\mathcal{P}_h(s_\varkappa)$}$$

Finally, the discretization of ($\mathbb{P}$) reads as follows:

$$\begin{cases} \text{Find } s_\varkappa^* \in U_\varkappa^{ad} \text{ such that} \\ J(u_h(s_\varkappa^*)) \le J(u_h(s_\varkappa)) \quad \forall s_\varkappa \in U_\varkappa^{ad}, \end{cases} \tag{$\mathbb{P}_{h\varkappa}$}$$

where $u_h(s_\varkappa)$ is the solution of $(\mathcal{P}_h(s_\varkappa))$. The approximate optimal shape is given by $\Omega(s_\varkappa^*)$.

Next, we shall analyze

- *the existence of solutions to* $(\mathbb{P}_{h\varkappa})$,
- *the relation between solutions of* ($\mathbb{P}$) *and* $(\mathbb{P}_{h\varkappa})$ *as* $h, \varkappa \to 0+$.

In order to establish these results, we have to impose additional assumptions on the family of triangulations $\{\mathcal{T}(h, s_\varkappa)\}$, $h, \varkappa \to 0+$, which are listed below.

We shall suppose that, for any $h, \varkappa > 0$ fixed, the system $\{\mathcal{T}(h, s_\varkappa)\}$, $s_\varkappa \in U_\varkappa^{ad}$, consists of *topologically equivalent* triangulations, meaning that

($\mathcal{T}1$) the triangulation $\mathcal{T}(h, s_\varkappa)$ has the *same number* of nodes and the nodes still have the *same* neighbors for any $s_\varkappa \in U_\varkappa^{ad}$,

($\mathcal{T}2$) the positions of the nodes of $\mathcal{T}(h, s_\varkappa)$ depend *solely* and *continuously* on variations of the design nodes $\{A_i\}_{i=0}^d$.

For $h, \varkappa \to 0+$ we suppose that

($\mathcal{T}3$) the family $\{\mathcal{T}(h, s_\varkappa)\}$ is *uniformly regular* with respect to h, $\varkappa$, and $s_\varkappa \in U_\varkappa^{ad}$: there is $\vartheta_0 > 0$ such that $\vartheta(h, s_\varkappa) \ge \vartheta_0$ $\forall h, \varkappa > 0$, $\forall s_\varkappa \in U_\varkappa^{ad}$, where $\vartheta(h, s_\varkappa)$ is the minimal interior angle of all triangles from $\mathcal{T}(h, s_\varkappa)$.

In general boundary value problems, boundaries of designed domains usually consist of several nonoverlapping parts where different types of boundary conditions are prescribed. In this case we suppose that

($\mathcal{T}4$) the family $\{\mathcal{T}(h, s_\varkappa)\}$ is *consistent* with the respective decomposition of boundaries (see Appendix A) for any $s_\varkappa \in U_\varkappa^{ad}$ and for any $\varkappa, h > 0$.

CONVENTION: *Throughout the book we shall consider only such families* $\{\mathcal{T}(h, s_\varkappa)\}$ *that satisfy* ($\mathcal{T}1$)–($\mathcal{T}4$).

As in the previous section one can show that $(\mathbb{P}_{h\varkappa})$ leads to the following *nonlinear mathematical programming problem*:

$$\begin{cases} \text{Find } \boldsymbol{\alpha}^* \in \mathcal{U} \text{ such that} \\ \mathcal{J}(\boldsymbol{q}(\boldsymbol{\alpha}^*)) \le \mathcal{J}(\boldsymbol{q}(\boldsymbol{\alpha})) \quad \forall \boldsymbol{\alpha} \in \mathcal{U}, \end{cases} \tag{$\mathbb{P}_d$}$$

where $\boldsymbol{q}(\boldsymbol{\alpha}) \in \mathbb{R}^n$ is the unique solution of a linear algebraic system

$$\boldsymbol{K}(\boldsymbol{\alpha})\,\boldsymbol{q}(\boldsymbol{\alpha}) = \boldsymbol{f}(\boldsymbol{\alpha}), \qquad (\mathcal{P}(\boldsymbol{\alpha}))$$

$\mathcal{U}$ is defined by (2.48), and $\mathcal{J}$ is the algebraic representation of J.

Remark 2.7. From $(\mathcal{T}1)$ it follows that $n = \dim V_h(s_\varkappa)$ does not depend on $s_\varkappa \in U_\varkappa^{ad}$ or equivalently on $\boldsymbol{\alpha} \in \mathcal{U}$. The elements $k_{ij}(\boldsymbol{\alpha})$ and $f_i(\boldsymbol{\alpha})$ of the stiffness matrix $\boldsymbol{K}(\boldsymbol{\alpha})$ and the force vector $\boldsymbol{f}(\boldsymbol{\alpha})$, respectively, are given by

$$k_{ij}(\boldsymbol{\alpha}) = \int_{\Omega_h(s_\varkappa)} \nabla\varphi_i \cdot \nabla\varphi_j \, dx, \quad f_i(\boldsymbol{\alpha}) = \int_{\Omega_h(s_\varkappa)} f\varphi_i \, dx, \quad i, j = 1, \dots, n,$$

where $\{\varphi_i\}_{i=1}^n$ is the Courant basis of $V_h(s_\varkappa)$.

A simple but important consequence of assumptions $(\mathcal{T}1)$ and $(\mathcal{T}2)$ is the following.

Lemma 2.6. *Let $(\mathcal{T}1)$ and $(\mathcal{T}2)$ be satisfied. Then the mapping*

$$\boldsymbol{K} : \boldsymbol{\alpha} \mapsto \boldsymbol{K}(\boldsymbol{\alpha}), \quad \boldsymbol{\alpha} \in \mathcal{U},$$

is continuous.

Proof. Let $\{N_j\}_{j=1}^n$ be the set of all the interior nodes of $\mathcal{T}(h, s_\varkappa)$, $N_j = (x_1(j), x_2(j))$. Then from $(\mathcal{T}1)$ and $(\mathcal{T}2)$ it follows that

$$N_j = \Phi_j(\boldsymbol{\alpha}), \quad j = 1, \dots, n, \ \forall \boldsymbol{\alpha} \in \mathcal{U}, \tag{2.52}$$

where $\Phi_j : \mathcal{U} \to \mathbb{R}^2$ are continuous functions. Let $\varphi_i \in V_h(s_\varkappa)$ be the Courant basis function associated with the node N_i, i.e., $\varphi_i(N_j) = \delta_{ij}$, where δ_{ij} denotes the Kronecker symbol. Finally, let $T \in \mathcal{T}(h, s_\varkappa)$ be a triangle sharing N_i as one of its vertices and let N_j, N_k be the remaining vertices. It is known that

$$\varphi_{i|T} = \frac{\begin{vmatrix} x_1(j) & x_1(k) \\ x_2(j) & x_2(k) \end{vmatrix} - \begin{vmatrix} 1 & x_2(j) \\ 1 & x_2(k) \end{vmatrix} x_1 + \begin{vmatrix} 1 & x_1(j) \\ 1 & x_1(k) \end{vmatrix} x_2}{2 \operatorname{meas} T}$$

and

$$\operatorname{meas} T = \frac{1}{2} \begin{vmatrix} 1 & x_1(i) & x_2(i) \\ 1 & x_1(j) & x_2(j) \\ 1 & x_1(k) & x_2(k) \end{vmatrix}.$$

From this and (2.52) we see that

$$\varphi_{i|T} = a_i(\boldsymbol{\alpha})x_1 + b_i(\boldsymbol{\alpha})x_2 + c_i(\boldsymbol{\alpha}), \tag{2.53}$$

where the coefficients $a_i(\boldsymbol{\alpha})$, $b_i(\boldsymbol{\alpha})$, and $c_i(\boldsymbol{\alpha})$ are continuous functions of $\boldsymbol{\alpha}$ in $\mathcal{U}$. Thus $\varphi_{i|T}$ and $\nabla_x \varphi_{i|T}$ are continuous functions of $\boldsymbol{\alpha} \in \mathcal{U}$. Since meas T depends continuously on $\boldsymbol{\alpha}$, we easily deduce that

$$k_{ij} : \boldsymbol{\alpha} \mapsto k_{ij}(\boldsymbol{\alpha}), \quad \boldsymbol{\alpha} \in \mathcal{U},$$

is a continuous mapping for any $i, j = 1, \dots, n$. $\square$

Suppose now that the right-hand side f is continuous in $\overline{\widehat{\Omega}}$. The elements of the force vector $\boldsymbol{f}$ will be computed using the following simple quadrature formula:

$$\int_T f\varphi_i\,dx \approx \operatorname{meas} T f(Q_T)\varphi_i(Q_T), \tag{2.54}$$

where Q_T is the center of gravity of T. Then one has the following result.

Lemma 2.7. *Let* ($\mathcal{T}1$) *and* ($\mathcal{T}2$) *be satisfied,* $f \in C(\overline{\widehat{\Omega}})$, *and the force vector* $\boldsymbol{f}(\boldsymbol{\alpha})$ *be computed by means of* (2.54). *Then the mapping* $\boldsymbol{f} : \boldsymbol{\alpha} \mapsto \boldsymbol{f}(\boldsymbol{\alpha})$, $\boldsymbol{\alpha} \in \mathcal{U}$, *is continuous.*

Remark 2.8. If $\boldsymbol{f}(\boldsymbol{\alpha})$ were computed exactly, i.e., no numerical integration were used, then the assertion of Lemma 2.7 would be true even if $f \notin C(\overline{\widehat{\Omega}})$. From the analysis presented above we also see that if the mappings Φ_i, $i = 1, \dots, n$, from (2.52) belonged to the class C^ℓ for an integer ℓ, then the respective coefficients $a_i(\boldsymbol{\alpha})$, $b_i(\boldsymbol{\alpha})$, and $c_i(\boldsymbol{\alpha})$ in (2.53) would share the same smoothness property and the mapping $\boldsymbol{K} : \boldsymbol{\alpha} \mapsto \boldsymbol{K}(\boldsymbol{\alpha})$ would preserve the C^ℓ-regularity. The same holds for the mapping $\boldsymbol{f} : \boldsymbol{\alpha} \mapsto \boldsymbol{f}(\boldsymbol{\alpha})$ provided that f is smooth enough in $\overline{\widehat{\Omega}}$.

From Lemmas 2.6 and 2.7 we easily get the following continuity type result.

Lemma 2.8. *Let all the assumptions of Lemmas* 2.6 *and* 2.7 *be satisfied. Then there exists a constant* $c > 0$ *such that*

$$\|\boldsymbol{q}(\boldsymbol{\alpha}) - \boldsymbol{q}(\boldsymbol{\beta})\| \le c\big\{\|\boldsymbol{K}(\boldsymbol{\alpha}) - \boldsymbol{K}(\boldsymbol{\beta})\| + \|\boldsymbol{f}(\boldsymbol{\alpha}) - \boldsymbol{f}(\boldsymbol{\beta})\|\big\} \tag{2.55}$$

holds for any $\boldsymbol{\alpha}, \boldsymbol{\beta} \in \mathcal{U}$, *where* $\boldsymbol{q}(\boldsymbol{\alpha})$, $\boldsymbol{q}(\boldsymbol{\beta})$ *are solutions of* ($\mathcal{P}(\boldsymbol{\alpha})$), ($\mathcal{P}(\boldsymbol{\beta})$), *respectively.*

An easy consequence of Lemma 2.8 is the following.

Theorem 2.6. *Problem* ($\mathbb{P}_d$) *has at least one solution.*

Proof. Since the composite mapping $\boldsymbol{\alpha} \mapsto \boldsymbol{q}(\boldsymbol{\alpha}) \mapsto \mathcal{J}(\boldsymbol{q}(\boldsymbol{\alpha}))$ is continuous in $\mathcal{U}$, as follows from (2.55) and continuity of $\mathcal{J}$, and $\mathcal{U}$ is the compact subset of $\mathbb{R}^{d+1}$, the existence of a minimizer follows from the classical result of the calculus of variations. □

Next we shall pay attention to the convergence analysis. We start with two auxiliary lemmas, which will be used in subsequent parts.

Lemma 2.9. *Let* $s_\varkappa \in U_\varkappa^{ad}$. *Then* $s_\varkappa \in W^{2,\infty}\big([0,1]\big)$ *and*

(i) $\alpha_{\min} \le s_\varkappa(x_2) \le \alpha_{\max} \quad \forall x_2 \in [0,1]$;
(ii) $|s'_\varkappa(x_2)| \le L_0 \quad \forall x_2 \in [0,1]$;
(iii) *the segment* $\overline{A_i A_{i+1}}$ *is tangent to the graph of* $s_\varkappa$ *at* $A_{i+1/2}$, $i = 0, \dots, d-1$;

(iv) *the graph of* $s^i_\varkappa := s_{\varkappa|[a_{i-1/2},a_{i+1/2}]} \in \operatorname{conv}\{A_{i-1/2}, A_i, A_{i+1/2}\}$:= *the convex hull of* $\{A_{i-1/2}, A_i, A_{i+1/2}\}$ *for* $i = 1, \ldots, d-1$ *with the following modifications for* $i = 0, d$:

$$s^0_\varkappa := s_{\varkappa|[a_0,a_{1/2}]} \in \operatorname{conv}\{A_{-1/2}, A_0, A_{1/2}\},$$
$$s^d_\varkappa := s_{\varkappa|[a_{d-1/2},a_d]} \in \operatorname{conv}\{A_{d-1/2}, A_d, A_{d+1/2}\};$$

(v) *there exists a constant* $c > 0$ *that does not depend on* $\varkappa$, h, *and* $s_\varkappa \in U^{ad}_\varkappa$ *such that*

$$\|r_h s_\varkappa - s_\varkappa\|_{C([0,1])} \le ch, \ h \to 0+;$$

(vi) *there exists a constant* $c > 0$ *that does not depend on* $\varkappa$ *and* $s_\varkappa \in U^{ad}_\varkappa$ *such that*

$$\left| \int_0^1 s_\varkappa(x_2)\, dx_2 - \gamma \right| \le c\varkappa, \ \varkappa \to 0+,$$

where $\alpha_{\min}$, $\alpha_{\max}$, L_0, *and* γ *are the same as in the definition of* U^{ad}.

Proof. The fact that $s_\varkappa \in W^{2,\infty}([0,1])$ is straightforward, as are (i)–(iv), which follow from the basic properties of Bézier functions (see Appendix B). The error estimate (v) follows from (ii) and

$$\|r_h s_\varkappa - s_\varkappa\|_{C([0,1])} \le ch\|s_\varkappa\|_{C^1([0,1])} \le ch.$$

It remains to prove (vi). Let $\alpha_\varkappa \in \mathcal{Q}^{ad}_\varkappa$ be the piecewise linear function determined by the design nodes A_i, $i = 0, \ldots, d$. Then

$$\|s_\varkappa - \alpha_\varkappa\|_{C([a_i,a_{i+1}])} \le c\varkappa\|s_\varkappa\|_{C^1([a_i,a_{i+1}])} \le c\varkappa, \quad i = 0, \ldots, d-1, \tag{2.56}$$

as follows from (ii) and the fact that $\alpha_{\varkappa|[a_i,a_{i+1}]}$ is tangent to $s_\varkappa$ at $A_{i+1/2}$. From this, (vi) easily follows. Indeed,

$$\left| \int_0^1 s_\varkappa(x_2)\, dx_2 - \gamma \right| \le \int_0^1 |s_\varkappa(x_2) - \alpha_\varkappa(x_2)|\, dx_2$$
$$= \sum_{i=0}^{d(\varkappa)-1} \int_{a_i}^{a_{i+1}} |s_\varkappa(x_2) - \alpha_\varkappa(x_2)|\, dx_2 \le c\varkappa, \ \varkappa \to 0+.$$

□

The next lemma collects the basic properties of the system $\{\mathcal{O}_\varkappa\}$, $\varkappa \to 0+$. Recall that $\mathcal{O}_\varkappa = \{\Omega(s_\varkappa) \mid s_\varkappa \in U^{ad}_\varkappa\}$.

Lemma 2.10. *It holds that*

(i) *there exists a system* $\widetilde{\mathcal{O}}$ *of domains, compact with respect to convergence, introduced in Definition* 2.2 *such that*

$$\mathcal{O}_\varkappa \subseteq \widetilde{\mathcal{O}} \quad \forall \varkappa > 0;$$

(ii) *for any sequence* $\{\Omega(s_\varkappa)\}$, $\Omega(s_\varkappa) \in \mathcal{O}_\varkappa$, $\varkappa \to 0+$, *there exist its subsequence and an element* $\Omega(\alpha) \in \mathcal{O}$ *such that*

$$\Omega(s_\varkappa) \to \Omega(\alpha), \quad \varkappa \to 0+;$$

(iii) *for any* $\Omega(\alpha) \in \mathcal{O}$ *there exists a sequence* $\{\Omega(s_\varkappa)\}$, $\Omega(s_\varkappa) \in \mathcal{O}_\varkappa$, *such that*

$$\Omega(s_\varkappa) \to \Omega(\alpha), \quad \varkappa \to 0+;$$

(iv) *if* $\Omega(s_\varkappa) \to \Omega(\alpha)$, $\varkappa \to 0+$, *then*

$$\Omega(r_h s_\varkappa) \to \Omega(\alpha), \quad h,\, \varkappa \to 0+.$$

Proof. Since the constant area constraint is slightly violated by functions $s_\varkappa \in U_\varkappa^{ad}$, as follows from Lemma 2.9(vi), we see that $\mathcal{O}_\varkappa \not\subset \mathcal{O}$. On the other hand, omitting this constraint in the definition of $\mathcal{O}$ and keeping all the remaining parameters unchanged, we obtain a new family $\widetilde{\mathcal{O}}$ possessing the compactness property as stated in (i). From Lemma 2.9(i)–(ii) it follows that $\mathcal{O}_\varkappa \subseteq \widetilde{\mathcal{O}}\ \forall \varkappa > 0$.

Let $\{\Omega(s_\varkappa)\}$, $\Omega(s_\varkappa) \in \mathcal{O}_\varkappa$, be given. Then from (i) it follows that there exists its subsequence and $\Omega(\alpha) \in \widetilde{\mathcal{O}}$ such that

$$\Omega(s_\varkappa) \to \Omega(\alpha), \quad \varkappa \to 0+.$$

From Lemma 2.9(vi) we see that $\int_0^1 \alpha(x_2)\,dx_2 = \gamma$, meaning that $\Omega(\alpha) \in \mathcal{O}$; i.e., (ii) holds.

Let $\overline{\alpha} \in U^{ad}$ be given. Then there exists a sequence $\{\overline{\alpha}_\varkappa\}$, $\overline{\alpha}_\varkappa \in \mathcal{Q}_\varkappa^{ad}$ (see Problem 2.2), such that

$$\overline{\alpha}_\varkappa \rightrightarrows \overline{\alpha} \quad \text{in } [0,1], \quad \varkappa \to 0+. \tag{2.57}$$

With any $\overline{\alpha}_\varkappa$ we associate a unique $\overline{s}_\varkappa \in U_\varkappa^{ad}$ defined by the design points $\overline{A}_i = (\overline{\alpha}_\varkappa(a_i), i\varkappa)$, $i = 0, \ldots, d$. Then

$$\|\overline{s}_\varkappa - \overline{\alpha}\|_{C([0,1])} \le \|\overline{s}_\varkappa - \overline{\alpha}_\varkappa\|_{C([0,1])} + \|\overline{\alpha}_\varkappa - \overline{\alpha}\|_{C([0,1])} \to 0,$$

as follows from (2.56) and (2.57), proving (iii).

Finally, (iv) is a direct consequence of Lemma 2.9(v). $\square$

The following continuity type result for approximate solutions is important in the forthcoming convergence analysis.

LEMMA 2.11. *Let* $s_\varkappa \rightrightarrows \alpha$ *in* $[0,1]$, $\varkappa \to 0+$, $s_\varkappa \in U_\varkappa^{ad}$, $\alpha \in U^{ad}$, *and* $u_h := u_h(s_\varkappa)$ *be a solution to* $(\mathcal{P}_h(s_\varkappa))$, $h \to 0+$. *Then*

$$\widetilde{u}_h(s_\varkappa) \to u \text{ in } H_0^1(\widehat{\Omega}), \quad h, \varkappa \to 0+,$$

and $u(\alpha) := u_{|\Omega(\alpha)}$ *solves* $(\mathcal{P}(\alpha))$.

Proof. Using the same approach as in Lemma 2.5 one can show that $\{\|\widetilde{u}_h\|_{1,\widehat{\Omega}}\}$ is bounded so that

$$\widetilde{u}_h \rightharpoonup u \text{ in } H_0^1(\widehat{\Omega}) \tag{2.58}$$

holds for an appropriate subsequence and $u \in H_0^1(\widehat{\Omega})$. Since $r_h s_\varkappa \rightrightarrows \alpha$ in $[0,1]$ as $h,\ \varkappa \to 0+$ we easily get that $u = 0$ in $\widehat{\Omega} \setminus \overline{\Omega(\alpha)}$ so that $u(\alpha) := u_{|\Omega(\alpha)} \in H_0^1(\Omega(\alpha))$. This can be proved just as in Lemma 2.5. We now show that $u(\alpha)$ solves $(\mathcal{P}(\alpha))$. Let $\xi \in C_0^\infty(\Omega(\alpha))$ be given and ξ_h be the piecewise linear Lagrange interpolant of $\widetilde{\xi}_{|\Omega_h(s_\varkappa)}$ on $\mathcal{T}(h, s_\varkappa)$. For $h, \varkappa > 0$ small enough, the graph of $r_h s_\varkappa$ has an empty intersection with $\operatorname{supp}\widetilde{\xi}$. This means that $\xi_h \in V_h(s_\varkappa)$ and it can be used as a test function in $(\mathcal{P}_h(s_\varkappa))$. In addition,

$$\|\widetilde{\xi}_h - \widetilde{\xi}\|_{W^{1,\infty}(\widehat{\Omega})} = \|\xi_h - \widetilde{\xi}\|_{W^{1,\infty}(\Omega_h(s_\varkappa))} \le ch\|\widetilde{\xi}\|_{C^2(\widehat{\Omega})}, \tag{2.59}$$

where $c > 0$ is a constant that does not depend on h, $\varkappa$, and $s_\varkappa$, as follows from the well-known approximation results and the uniform regularity assumption $(\mathcal{T}3)$ on $\{\mathcal{T}(h, s_\varkappa)\}$. Finally,

$$\chi_{h\varkappa} \to \chi \text{ in } L^p(\widehat{\Omega}) \quad \forall p \in [1, \infty[,\ h, \varkappa \to 0+, \tag{2.60}$$

where $\chi_{h\varkappa}$ and χ are the characteristic functions of $\Omega(s_\varkappa)$ and $\Omega(\alpha)$, respectively. The definition of $(\mathcal{P}_h(s_\varkappa))$ yields

$$\int_{\widehat{\Omega}} \chi_{h\varkappa} \nabla \widetilde{u}_h \cdot \nabla \widetilde{\xi}_h \, dx = \int_{\widehat{\Omega}} \chi_{h\varkappa} f \widetilde{\xi}_h \, dx.$$

Passing here to the limit with $\varkappa, h \to 0+$ and using (2.58), (2.59), and (2.60) we obtain

$$\int_{\widehat{\Omega}} \chi \nabla u \cdot \nabla \widetilde{\xi} \, dx = \int_{\widehat{\Omega}} \chi f \widetilde{\xi} \, dx;$$

i.e., $u_{|\Omega(\alpha)}$ is the solution of $(\mathcal{P}(\alpha))$. Since $(\mathcal{P}(\alpha))$ has a unique solution, the whole sequence $\{\widetilde{u}_h(s_\varkappa)\}$ tends weakly to u in $H_0^1(\widehat{\Omega})$. Arguing as in Lemma 2.5, one can prove strong convergence. □

On the basis of this lemma we have the following result.

Theorem 2.7. *Let $\{(s_\varkappa^*, u_h(s_\varkappa^*))\}$ be a sequence of optimal pairs of $(\mathbb{P}_{h\varkappa})$, $h \to 0+$. Then one can find a subsequence of $\{(s_\varkappa^*, u_h(s_\varkappa^*))\}$ such that*

$$s_\varkappa^* \rightrightarrows \alpha^* \quad \text{in } [0,1], \tag{2.61}$$

$$\widetilde{u}_h(s_\varkappa^*) \to u^* \quad \text{in } H_0^1(\widehat{\Omega}),\ h, \varkappa \to 0+, \tag{2.62}$$

where $(\alpha^*, u^*{}_{|\Omega(\alpha^*)})$ is an optimal pair of $(\mathbb{P})$. In addition, any accumulation point of $\{(s_\varkappa^*, u_h(s_\varkappa^*))\}$ in the sense of (2.61), (2.62) possesses this property.

Proof. Let $\overline{\alpha} \in U^{ad}$ be arbitrary. Then there exists a sequence $\{\overline{s}_\varkappa\}$, $\overline{s}_\varkappa \in U_\varkappa^{ad}$, such that (see Lemma 2.10(iii))

$$\overline{s}_\varkappa \rightrightarrows \overline{\alpha} \quad \text{in } [0,1], \quad \varkappa \to 0+,$$

and

$$\widetilde{u}_h(\overline{s}_\varkappa) \to \overline{u} \quad \text{in } H_0^1(\widehat{\Omega}), \quad h, \varkappa \to 0+, \tag{2.63}$$

where $u(\overline{\alpha}) := \overline{u}_{|\Omega(\overline{\alpha})}$ solves $(\mathcal{P}(\overline{\alpha}))$, as follows from Lemma 2.11. From Lemma 2.10(ii) it follows that one can pass to a subsequence of $\{s^*_\varkappa\}$ such that

$$s^*_\varkappa \rightrightarrows \alpha^* \quad \text{in } [0,1], \quad \varkappa \to 0+,$$

and

$$\widetilde{u}_h(s^*_\varkappa) \to u^* \quad \text{in } H^1_0(\widehat{\Omega}), \quad h, \varkappa \to 0+, \tag{2.64}$$

where $u(\alpha^*) := u^*_{|\Omega(\alpha^*)}$ solves $(\mathcal{P}(\alpha^*))$, applying Lemma 2.11 once again. Using the definition of $(\mathbb{P}_{h\varkappa})$ we have

$$J(u_h(s^*_\varkappa)) \le J(u_h(\overline{s}_\varkappa)).$$

Letting $h, \varkappa \to 0+$, using (2.63), (2.64), and continuity of J, we arrive at

$$J(u(\alpha^*)) \le J(u(\overline{\alpha})) \quad \forall \overline{\alpha} \in U^{ad}. \qquad \square$$

2.3 Abstract setting of sizing optimization problems: Existence and convergence results

The aim of this section is to give an abstract formulation of a large class of sizing optimization problems and their approximations. We present sufficient conditions under which they have at least one solution and discretized problems are close in the sense of subsequences to the respective continuous setting. We shall follow ideas used in the particular sizing optimization problem studied in Section 2.1. However, in view of other possible applications, we shall consider a more general class of state problems given by *variational inequalities*.

Let V be a real Hilbert space, $\|\cdot\|$ its norm, V' the dual space of V with the norm $\|\cdot\|_*$, $\langle\cdot,\cdot\rangle$ the duality between V' and V, and $K \subset V$ a *nonempty, closed*, and *convex* set. Let U be another Banach space and $U^{ad} \subset U$ be its *compact* subset, representing the set of all *admissible controls*. With any $e \in U^{ad}$ we associate a bilinear form $a_e : V \times V \to \mathbb{R}$. Next we shall suppose that the system $\{a_e\}$, $e \in U^{ad}$, satisfies the following assumptions:

$(\mathcal{A}1)$ *uniform boundedness with respect to* U^{ad}:

$$\exists M > 0 : \quad |a_e(y,v)| \le M\|y\|\|v\| \quad \forall y, v \in V, \ \forall e \in U^{ad};$$

$(\mathcal{A}2)$ *uniform V-ellipticity with respect to* U^{ad}:

$$\exists \alpha > 0 : \quad a_e(v,v) \ge \alpha\|v\|^2 \quad \forall v \in V, \ \forall e \in U^{ad};$$

$(\mathcal{A}3)$ *symmetry condition*:

$$a_e(y,v) = a_e(v,y) \quad \forall y, v \in V, \ \forall e \in U^{ad}.$$

For any $e \in U^{ad}$ we shall consider the following *state problem*:

$$\begin{cases} \text{Find } u := u(e) \in K \text{ such that} \\ a_e(u, v-u) \ge \langle f, v-u \rangle \quad \forall v \in K, \end{cases} \tag{$\mathcal{P}(e)$}$$

where $f \in V'$ is given. From $(\mathcal{A}1)$ and $(\mathcal{A}2)$ it follows that $(\mathcal{P}(e))$ has a unique solution $u(e)$ for any $e \in U^{ad}$ (see Appendix A). Finally, let $J : U^{ad} \times V \to \mathbb{R}$ be a *cost functional.*

An *abstract sizing optimization problem* reads as follows:

$$\begin{cases} \text{Find } e^* \in U^{ad} \text{ such that} \\ J(e^*, u(e^*)) \leq J(e, u(e)) \quad \forall e \in U^{ad}, \end{cases} \tag{$\mathbb{P}$}$$

where $u(e)$ solves $(\mathcal{P}(e))$.

In order to guarantee the existence of solutions to $(\mathbb{P})$, we need additional continuity assumptions on the mappings $a : e \mapsto a_e$, $J : (e, y) \mapsto J(e, y)$, $e \in U^{ad}$, $y \in V$. Next we suppose that

$$(\mathcal{A}4)\quad e_n \to e \text{ in } U,\ e_n, e \in U^{ad}$$
$$\implies \sup_{\substack{\|y\| \leq 1 \\ \|v\| \leq 1}} |a_{e_n}(y, v) - a_e(y, v)| \to 0,\ n \to \infty;$$

$$(\mathcal{A}5)\quad y_n \to y \text{ in } V, \quad e_n \to e \text{ in } U,\ e_n, e \in U^{ad}$$
$$\implies \liminf_{n\to\infty} J(e_n, y_n) \geq J(e, y).$$

Remark 2.9. Assumption $(\mathcal{A}3)$ is not necessary but it simplifies our presentation. Assumption $(\mathcal{A}4)$ can be equivalently expressed in a more explicit form. Let $A(e) \in \mathcal{L}(V, V')$ be the mapping defined by

$$\langle A(e)y, v\rangle = a_e(y, v) \quad \forall y, v \in V,\ \forall e \in U^{ad}.$$

Then $(\mathcal{A}4)$ is *equivalent* to saying that

$$(\mathcal{A}6)\quad e_n \to e \text{ in } U,\ e_n, e \in U^{ad} \implies A(e_n) \to A(e) \text{ in } \mathcal{L}(V, V').$$

This form of the continuous dependence of a_e on e will be used in what follows.

We start our analysis by proving that the mapping $u : e \mapsto u(e)$, where $e \in U^{ad}$ and $u(e)$ solves $(\mathcal{P}(e))$, is *continuous*. As we already know, this property plays a major role in the existence analysis.

Lemma 2.12. *Suppose that* $(\mathcal{A}1)$–$(\mathcal{A}4)$ *are satisfied. Let* $e_n \to e$ *in* U, $e_n, e \in U^{ad}$, *and* $u_n := u(e_n) \in K$ *be a solution of* $(\mathcal{P}(e_n))$, $n \to \infty$. *Then*

$$u_n \to u(e) \quad \text{in } V, \quad n \to \infty,$$

and $u(e)$ *solves* $(\mathcal{P}(e))$.

Proof. Let us fix $\overline{v} \in K$. Then, from the definition of $(\mathcal{P}(e_n))$, $(\mathcal{A}1)$, and $(\mathcal{A}2)$, it follows that

$$\begin{aligned} \alpha\|u_n\|^2 &\leq a_{e_n}(u_n, u_n) \leq a_{e_n}(u_n, \overline{v}) + \langle f, u_n - \overline{v}\rangle \\ &\leq M\|u_n\|\|\overline{v}\| + \|f\|_* \left(\|u_n\| + \|\overline{v}\|\right), \end{aligned}$$

implying the boundedness of $\{u_n\}$:

$$\exists c = \text{const.} > 0 : \quad \|u_n\| \le c \quad \forall n \in \mathbb{N}. \tag{2.65}$$

One can pass to a subsequence of $\{u_n\}$ such that

$$u_n \rightharpoonup u \quad \text{in } V, \quad n \to \infty. \tag{2.66}$$

Let us show that u solves $(\mathcal{P}(e))$. The convex set K, being closed, is weakly closed, so that $u \in K$, as follows from (2.66). From the definition of $(\mathcal{P}(e_n))$ we have that

$$\lim_{n\to\infty} a_{e_n}(u_n, v - u_n) \ge \lim_{n\to\infty} \langle f, v - u_n\rangle \quad \forall v \in K \tag{2.67}$$

provided that both limits exist. From (2.66) we see that

$$\lim_{n\to\infty} \langle f, v - u_n\rangle = \langle f, v - u\rangle. \tag{2.68}$$

The evaluation of the limit on the left of (2.67) is not so straightforward. We first prove that

$$\lim_{n\to\infty} a_{e_n}(u_n, v) = a_e(u, v) \tag{2.69}$$

holds for any $v \in K$. Indeed,

$$\begin{aligned}
|a_{e_n}(u_n, v) - a_e(u, v)| &= |\langle A(e_n)u_n, v\rangle - \langle A(e)u, v\rangle| \\
&\le |\langle A(e_n)u_n, v\rangle - \langle A(e)u_n, v\rangle| + |\langle A(e)u_n, v\rangle - \langle A(e)u, v\rangle| \\
&\le \|A(e_n) - A(e)\|_{\mathcal{L}(V,V')}\|u_n\|\|v\| + |\langle u_n - u, A(e)v\rangle| \to 0,\ n \to \infty,
\end{aligned}$$

making use of $(\mathcal{A}3)$, $(\mathcal{A}6)$, (2.65), and (2.66).

Next we show that

$$\lim_{n\to\infty} \langle A(e_n)u_n, u_n - u\rangle = 0. \tag{2.70}$$

Inserting $v := u$ into $(\mathcal{P}(e_n))$ we obtain

$$\limsup_{n\to\infty} \langle A(e_n)u_n, u_n - u\rangle \le \limsup_{n\to\infty} \langle f, u_n - u\rangle = 0. \tag{2.71}$$

On the other hand,

$$\begin{aligned}
\liminf_{n\to\infty} \langle A(e_n)u_n, u_n - u\rangle &\ge \liminf_{n\to\infty} \langle A(e_n)u, u_n - u\rangle \\
&= \liminf_{n\to\infty} \{\langle u, A(e_n)u_n\rangle - \langle u, A(e_n)u\rangle\} \\
&= \langle u, A(e)u\rangle - \langle u, A(e)u\rangle = 0,
\end{aligned}$$

taking into account $(\mathcal{A}3)$, $(\mathcal{A}6)$, and (2.69). From this and (2.71) we arrive at (2.70). Using (2.69) and (2.70) we see that

$$\lim_{n\to\infty} \langle A(e_n)u_n, u_n\rangle = \langle A(e)u, u\rangle.$$

From this, (2.67), (2.68), and (2.69) it follows that the limit element $u \in K$ satisfies

$$a_e(u, v - u) \geq \langle f, v - u \rangle \quad \forall v \in K;$$

i.e., u solves $(\mathcal{P}(e))$. Since u is unique, the whole sequence $\{u_n\}$ tends weakly to u in V. Strong convergence of $\{u_n\}$ to $u(e) := u$ now easily follows:

$$\begin{aligned}\alpha \|u_n - u\|^2 &\leq \langle A(e_n)(u_n - u), u_n - u \rangle \\ &= \langle A(e_n)u_n, u_n - u \rangle - \langle A(e_n)u, u_n - u \rangle \to 0, \ n \to \infty,\end{aligned}$$

taking into account $(\mathcal{A}6)$, (2.66), and (2.70). $\square$

We are now ready to prove the following existence result.

Theorem 2.8. *Let U^{ad} be a compact subset of U and let $(\mathcal{A}1)$–$(\mathcal{A}5)$ be satisfied. Then $(\mathbb{P})$ has a solution.*

Proof. Let $\{e_n\}$, $e_n \in U^{ad}$, be a minimizing sequence of $(\mathbb{P})$:

$$q := \inf_{e \in U^{ad}} J(e, u(e)) = \lim_{n \to \infty} J(e_n, u(e_n)). \tag{2.72}$$

Since U^{ad} is compact, one can find a subsequence of $\{e_n\}$ and an element $e^* \in U^{ad}$ such that

$$e_n \to e^* \text{ in } U, \quad n \to \infty.$$

From the previous lemma we know that

$$u(e_n) \to u(e^*) \quad \text{in } V, \quad n \to \infty,$$

where $u(e_n)$, $u(e^*)$ are solutions of $(\mathcal{P}(e_n))$, $(\mathcal{P}(e^*))$, respectively. This, (2.72), and lower semicontinuity of J yield

$$q = \liminf_{n \to \infty} J(e_n, u(e_n)) \geq J(e^*, u(e^*)) \geq q;$$

i.e., e^* is a solution of $(\mathbb{P})$. $\square$

We now pass to an approximation of $(\mathbb{P})$. Let $h > 0$ be a discretization parameter tending to zero (for example, a mesh norm in finite element methods). With any $h > 0$ finite dimensional spaces $V_h \subset V$, $U_h \subset \widetilde{U}$ will be associated. The symbol $\widetilde{U}$ stands for another Banach space such that $U \subseteq \widetilde{U}$. The reason for introducing $\widetilde{U}$ is simple: sometimes it is more convenient to work with approximations of U^{ad} that do not belong to the original space U (cf. $\widetilde{U}_h^{ad}$ given by (2.32), which is a subset of $\widetilde{U} = L^\infty(I)$, but not of $U = C(I)$). The nonempty, closed, convex set K will be approximated by *nonempty, closed, convex subsets* $K_h \subset V_h$, not necessarily being a part of K, i.e., $K_h \not\subset K$ in general. Similarly, the set of admissible controls U^{ad} will be replaced by *compact* subsets $U_h^{ad} \subset U_h$. Again, we do not require that $U_h^{ad} \subset U^{ad}$! *Next we shall suppose that the bilinear forms a_e are also defined for any $e \in \bigcup_{h>0} U_h^{ad}$.* The state problem will be approximated by means of a Galerkin type method on K_h, using elements of U_h^{ad} as controls. Thus, for any $e_h \in U_h^{ad}$ we define the following problem:

$$\begin{cases} \text{Find } u_h := u_h(e_h) \in K_h \text{ such that} \\ a_{e_h}(u_h, v_h - u_h) \geq \langle f, v_h - u_h \rangle \quad \forall v_h \in K_h. \end{cases} \tag{$\mathcal{P}_h(e_h)$}$$

The *approximation* of ($\mathbb{P}$) is now stated as follows:

$$\begin{cases} \text{Find } e_h^* \in U_h^{ad} \text{ such that} \\ J(e_h^*, u_h(e_h^*)) \leq J(e_h, u_h(e_h)) \quad \forall e_h \in U_h^{ad}, \end{cases} \tag{$\mathbb{P}_h$}$$

where $u_h(e_h)$ solves ($\mathcal{P}_h(e_h)$). In what follows we present an abstract convergence theory; i.e., we shall study the mutual relation between solutions of ($\mathbb{P}$) and ($\mathbb{P}_h$) as $h \to 0+$.

To this end we shall need the following additional assumptions:

$(\mathcal{A}1)_h$ $\exists \widetilde{M} > 0: \quad |a_{e_h}(y, v)| \leq \widetilde{M}\|y\|\|v\| \quad \forall y, v \in V, \quad \forall e_h \in \bigcup_{h>0} U_h^{ad};$

$(\mathcal{A}2)_h$ $\exists \widetilde{\alpha} > 0: \quad a_{e_h}(v, v) \geq \widetilde{\alpha}\|v\|^2 \quad \forall v \in V, \quad \forall e_h \in \bigcup_{h>0} U_h^{ad};$

$(\mathcal{A}3)_h$ $a_{e_h}(y, v) = a_{e_h}(v, y) \quad \forall y, v \in V, \quad \forall e_h \in \bigcup_{h>0} U_h^{ad};$

$(\mathcal{A}4)_h$ $e_h \to e$ in $\widetilde{U}$, $e_h \in U_h^{ad}$, $e \in U^{ad}$

$$\implies A(e_h) \to A(e) \text{ in } \mathcal{L}(V, V');$$

$(\mathcal{A}5)_h$ $\forall v \in K \quad \exists \{v_h\},\ v_h \in K_h: \quad v_h \to v$ in V;

$(\mathcal{A}6)_h$ $v_h \rightharpoonup v$ in V, $v_h \in K_h \implies v \in K$;

$(\mathcal{A}7)_h$ $\forall e \in U^{ad} \quad \exists \{e_h\},\ e_h \in U_h^{ad}: \quad e_h \to e$ in $\widetilde{U}$;

$(\mathcal{A}8)_h$ for any sequence $\{e_h\}$, $e_h \in U_h^{ad}$, there exists its subsequence and an element $e \in U^{ad}$ such that $e_h \to e$ in $\widetilde{U}$;

$(\mathcal{A}9)_h$ $e_h \to e$ in $\widetilde{U}$, $y_h \to y$ in V, where $e_h \in U_h^{ad}$, $e \in U^{ad}$, $y_h \in V_h$, $y \in V$

$$\implies \lim_{h \to 0+} J(e_h, y_h) = J(e, y).$$

COMMENTS 2.4.

(i) If $\widetilde{U} = U$ and $U_h^{ad} \subset U^{ad}$ $\forall h > 0$, then assumptions $(\mathcal{A}1)_h$–$(\mathcal{A}4)_h$ follow from $(\mathcal{A}1)$–$(\mathcal{A}4)$ and $\widetilde{M} = M$, $\widetilde{\alpha} = \alpha$;

(ii) $(\mathcal{A}5)_h$ and $(\mathcal{A}7)_h$ are standard density assumptions;

(iii) If $K_h \subset K$ $\forall h \to 0+$, i.e., K_h is the so-called inner approximation of K, then $(\mathcal{A}6)_h$ is automatically satisfied;

(iv) $(\mathcal{A}8)_h$ is a compactness type assumption.

The convergence analysis starts with the following auxiliary result.

Lemma 2.13. *Let* $(\mathcal{A}1)_h$–$(\mathcal{A}6)_h$ *be satisfied and* $\{e_h\}$, $e_h \in U_h^{ad}$, *be a sequence such that* $e_h \to e \in U^{ad}$, $h \to 0+$, *in* $\widetilde{U}$. *Then*

$$u_h(e_h) \to u(e) \quad \textit{in } V, \quad h \to 0+,$$

where $u_h(e_h)$, $u(e)$ *are the solutions of* $(\mathcal{P}_h(e_h))$, $(\mathcal{P}(e))$, *respectively.*

Proof. We first prove that the sequence $\{u_h\}$, $u_h := u_h(e_h) \in K_h$, is bounded in V. Let $\overline{v} \in K$ be a fixed element. From $(\mathcal{A}5)_h$ the existence of $\{\overline{v}_h\}$, $\overline{v}_h \in K_h$, such that

$$\overline{v}_h \to \overline{v} \quad \text{in } V, \quad h \to 0+, \tag{2.73}$$

follows. From the definition of $(\mathcal{P}_h(e_h))$, $(\mathcal{A}1)_h$, and $(\mathcal{A}2)_h$ we have

$$\begin{aligned}\widetilde{\alpha}\|u_h\|^2 &\le a_{e_h}(u_h, u_h) \le a_{e_h}(u_h, \overline{v}_h) + \langle f, u_h - \overline{v}_h\rangle \\ &\le \widetilde{M}\|u_h\|\|\overline{v}_h\| + \|f\|_*\big(\|u_h\| + \|\overline{v}_h\|\big),\end{aligned}$$

implying the boundedness of $\{u_h\}$. Therefore one can pass to a subsequence of $\{u_h\}$ such that

$$u_h \rightharpoonup u \quad \text{in } V, \quad h \to 0+. \tag{2.74}$$

From $(\mathcal{A}6)_h$ it follows that $u \in K$. Next we show that u solves the limit problem $(\mathcal{P}(e))$.

Let $\overline{v} \in K$ be an arbitrary element and $\{\overline{v}_h\}$, $\overline{v}_h \in K_h$, be a sequence satisfying (2.73). Then from the definition of $(\mathcal{P}_h(e_h))$ it follows that

$$a_{e_h}(u_h, \overline{v}_h - u_h) \ge \langle f, \overline{v}_h - u_h\rangle. \tag{2.75}$$

Passing to the limit with $h \to 0+$ and using (2.73) and (2.74) we obtain

$$\langle f, \overline{v}_h - u_h\rangle \to \langle f, \overline{v} - u\rangle, \quad h \to 0+. \tag{2.76}$$

We now pass to the limit on the left of (2.75) by modifying the approach used in the proof of Lemma 2.12. It is easy to show that

$$\lim_{h\to 0+} a_{e_h}(u_h, \overline{v}_h) = a_e(u, \overline{v}). \tag{2.77}$$

Indeed,

$$\begin{aligned}|a_{e_h}(u_h, \overline{v}_h) - a_e(u, \overline{v})| \le\, &|\langle A(e_h)u_h, \overline{v}_h\rangle - \langle A(e)u_h, \overline{v}_h\rangle| \\ &+ |\langle A(e)u_h, \overline{v}_h\rangle - \langle A(e)u, \overline{v}\rangle| \to 0, \ h \to 0+,\end{aligned}$$

as follows from $(\mathcal{A}4)_h$, (2.73), and (2.74). Also the analogy with (2.70) remains true:

$$\lim_{h\to 0+} \langle A(e_h)u_h, \overline{u}_h - u_h\rangle = 0 \tag{2.78}$$

holds for any sequence $\{\overline{u}_h\}$, $\overline{u}_h \in K_h$, such that $\overline{u}_h \to u$, $h \to 0+$, whose existence follows from $(\mathcal{A}5)_h$, implying that

$$\lim_{h\to 0+} a_{e_h}(u_h, u_h) = a_e(u, u)$$

and consequently

$$\lim_{h\to 0+} a_{e_h}(u_h, \overline{v}_h - u_h) = a_e(u, \overline{v} - u).$$

From this and (2.76) we conclude that

$$a_e(u, \overline{v} - u) \geq \langle f, \overline{v} - u\rangle \quad \forall \overline{v} \in K;$$

i.e., $u(e) := u$ solves $(\mathcal{P}(e))$. Since $u(e)$ is unique, the whole sequence $\{u_h\}$ tends weakly to $u(e)$ in V. It remains to show strong convergence. Let $\{\overline{u}_h\}$, $\overline{u}_h \in K_h$, be a sequence such that

$$\overline{u}_h \to u(e) \quad \text{in } V, \quad h \to 0+. \tag{2.79}$$

Then

$$\begin{aligned}\widetilde{\alpha}\|u_h - \overline{u}_h\|^2 &\leq a_{e_h}(u_h - \overline{u}_h, u_h - \overline{u}_h)\\ &\leq \langle f, u_h - \overline{u}_h\rangle - a_{e_h}(\overline{u}_h, u_h - \overline{u}_h) \to 0,\ h \to 0+,\end{aligned}$$

as follows from the definition of $(\mathcal{P}_h(e_h))$, (2.74), (2.77), and (2.79). From this and the triangle inequality we arrive at the assertion. □

On the basis of the previous lemma, we prove the following convergence result.

THEOREM 2.9. *Let $(\mathcal{A}1)_h$–$(\mathcal{A}9)_h$ be satisfied. Then for any sequence $\{(e_h^*, u_h(e_h^*))\}$ of optimal pairs of $(\mathbb{P}_h)$, $h \to 0+$, there exists its subsequence such that*

$$\left\{\begin{aligned} e_{h_j}^* &\to e^* \quad \text{in } \widetilde{U},\\ u_{h_j}(e_{h_j}^*) &\to u(e^*) \quad \text{in } V, \quad j \to \infty.\end{aligned}\right. \tag{2.80}$$

In addition, $(e^*, u(e^*))$ is an optimal pair of $(\mathbb{P})$. Furthermore, any accumulation point of $\{(e_h^*, u_h(e_h^*))\}$ in the sense of (2.80) possesses this property.

Proof. Let $\overline{e} \in U^{ad}$ be an arbitrary element. From $(\mathcal{A}7)_h$ the existence of a sequence $\{\overline{e}_h\}$, $\overline{e}_h \in U_h^{ad}$, such that

$$\overline{e}_h \to \overline{e} \quad \text{in } \widetilde{U}, \quad h \to 0+,$$

follows. Using $(\mathcal{A}8)_h$ one can find a subsequence $\{e_{h_j}^*\}$ of $\{e_h^*\}$ such that

$$e_{h_j}^* \to e^* \in U^{ad} \quad \text{in } \widetilde{U}, \quad j \to \infty.$$

At the same time

$$u_{h_j}(e_{h_j}^*) \to u(e^*), \quad u_{h_j}(\overline{e}_{h_j}) \to u(\overline{e}) \quad \text{in } V, \quad j \to \infty,$$

where $u(e^*)$, $u(\overline{e})$ are solutions of $(\mathcal{P}(e^*))$, $(\mathcal{P}(\overline{e}))$, respectively, as follows from Lemma 2.13. The definition of $(\mathbb{P}_{h_j})$ yields

$$J(e^*_{h_j}, u_{h_j}(e^*_{h_j})) \leq J(\overline{e}_{h_j}, u_{h_j}(\overline{e}_{h_j})) \quad \forall j \in \mathbb{N}. \tag{2.81}$$

Letting $h_j \to 0+$ in (2.81), using previous convergences of $\{(e^*_{h_j}, u_{h_j}(e^*_{h_j}))\}$, $\{(\overline{e}_{h_j}, u_{h_j}(\overline{e}_{h_j}))\}$, and $(\mathcal{A}9)_h$, we may conclude that $J(e^*, u(e^*)) \leq J(\overline{e}, u(\overline{e})) \quad \forall \overline{e} \in U^{ad}$. $\square$

Remark 2.10. It would be possible to consider more general approximations of $(\mathbb{P})$ in which the bilinear form a, the linear term f, and the cost functional J are replaced by suitable approximations a_h, f_h, and J_h, respectively. Such approximations in practical applications occur when a numerical integration is used for the evaluation of integrals defining a, f, and J. In this case one can also formulate sufficient conditions on the families $\{a_h\}$, $\{f_h\}$, and $\{J_h\}$ under which the continuous and discrete problems are close in the sense of subsequences.

2.4 Abstract setting of optimal shape design problems and their approximations

In contrast to sizing optimization, when control variables appear in coefficients of differential operators, the situation in shape optimization is different and more involved: this time domains themselves, in which state problems are solved, are the object of optimization. Two basic questions immediately appear, namely how to define convergence of sets and convergence of functions with variable domains of their definition. We have already met and solved these problems for the particular example in Section 2.2. Convergences of sets and of functions defined in variable domains, however, can be introduced in many different ways, depending on the type of state problems we consider. For this reason we shall not specify any particular choice of these convergences in the abstract setting.

Any formulation of the problem starts with introducing a *family $\mathcal{O}$ of admissible domains*; i.e., $\mathcal{O}$ contains all possible candidates among which an optimal one is sought. The choice of $\mathcal{O}$ depends on the particular problems that we solve. It should reflect all technological constraints characterizing the problem. Further, let $\widetilde{\mathcal{O}}$ be a larger system containing $\mathcal{O}$. The reason for introducing $\widetilde{\mathcal{O}}$ will be explained later on in this section.

Let $\{\Omega_n\}$, $\Omega_n \in \widetilde{\mathcal{O}}$, be a sequence and $\Omega \in \widetilde{\mathcal{O}}$. We first define a rule that enables us to say that $\{\Omega_n\}$ tends to Ω. This fact will be denoted by

$$\Omega_n \overset{\widetilde{\mathcal{O}}}{\to} \Omega, \quad n \to \infty. \tag{2.82}$$

Next we shall suppose that the convergence $\overset{\widetilde{\mathcal{O}}}{\to}$ satisfies the following (very natural) assumption:

For any subsequence $\{\Omega_{n_k}\}$ of $\{\Omega_n\}$ satisfying (2.82) *it holds that*

$$\Omega_{n_k} \overset{\widetilde{\mathcal{O}}}{\to} \Omega, \quad k \to \infty; \tag{2.83}$$

i.e., any subsequence of the convergent sequence tends to the same element as the original one.

With any $\Omega \in \widetilde{\mathcal{O}}$ we associate a function space $V(\Omega)$ of real functions defined in Ω. Next we introduce convergence of functions belonging to $V(\Omega)$ for different $\Omega \in \mathcal{O}$. In particular, if $\{y_n\}$, $y_n \in V(\Omega_n)$, $\Omega_n \in \widetilde{\mathcal{O}}$, and $y \in V(\Omega)$, $\Omega \in \widetilde{\mathcal{O}}$, we have to specify convergence of $\{y_n\}$ to y:

$$y_n \rightsquigarrow y. \tag{2.84}$$

We use the notation $\rightsquigarrow$ in order to distinguish this convergence from standard ones in the space $V(\Omega)$ with $\Omega \in \widetilde{\mathcal{O}}$ being fixed. Again we suppose that abstract convergence (2.84) satisfies an assumption similar to (2.83):

For any subsequence $\{y_{n_k}\}$ of $\{y_n\}$ satisfying (2.84) *it holds that*

$$y_{n_k} \rightsquigarrow y, \quad k \to \infty. \tag{2.85}$$

In any $\Omega \in \widetilde{\mathcal{O}}$ we solve a *state problem*. In this way we define a mapping u that with any $\Omega \in \widetilde{\mathcal{O}}$ associates an element $u(\Omega) \in V(\Omega)$ that is the solution of a partial differential equation (PDE), inequality, etc., describing the behavior of a physical system represented by Ω:

$$u : \Omega \mapsto u(\Omega) \in V(\Omega), \quad \Omega \in \widetilde{\mathcal{O}}. \tag{$\mathcal{P}(\Omega)$}$$

CONVENTION: *Throughout this book we suppose that $(\mathcal{P}(\Omega))$ has a unique solution for any $\Omega \in \widetilde{\mathcal{O}}$.*

Let $\mathcal{G}$ be the graph of the mapping $(u(\cdot))$ restricted to $\mathcal{O}$; i.e.,

$$\mathcal{G} = \{(\Omega, u(\Omega)) \mid \Omega \in \mathcal{O}\}.$$

Finally, let $J : (\Omega, y) \mapsto J(\Omega, y) \in \mathbb{R}$, $\Omega \in \widetilde{\mathcal{O}}$, $y \in V(\Omega)$, be a *cost functional.*

An *abstract optimal shape design problem* reads as follows:

$$\begin{cases} \text{Find } \Omega^* \in \mathcal{O} \text{ such that} \\ J(\Omega^*, u(\Omega^*)) \le J(\Omega, u(\Omega)) \quad \forall \Omega \in \mathcal{O}, \end{cases} \tag{$\mathbb{P}$}$$

where $u(\Omega) \in V(\Omega)$ solves $(\mathcal{P}(\Omega))$.

The existence of solutions to $(\mathbb{P})$ will be ensured by an appropriate *compactness* property of $\mathcal{G}$ and *lower semicontinuity* of J. Next we shall suppose that

$(\mathcal{B}1)$ (*compactness of* $\mathcal{G}$)
for any sequence $\{(\Omega_n, u(\Omega_n))\}$, $(\Omega_n, u(\Omega_n)) \in \mathcal{G}$, there exists its subsequence $\{(\Omega_{n_k}, u(\Omega_{n_k}))\}$ and an element $(\Omega, u(\Omega)) \in \mathcal{G}$ such that

$$\begin{aligned} &\Omega_{n_k} \xrightarrow{\widetilde{\mathcal{O}}} \Omega, \\ &u(\Omega_{n_k}) \rightsquigarrow u(\Omega), \quad k \to \infty; \end{aligned}$$

$(\mathcal{B}2)$ (*lower semicontinuity of* J)

$$\left.\begin{aligned} &\Omega_n \xrightarrow{\widetilde{\mathcal{O}}} \Omega, \ \Omega_n, \ \Omega \in \widetilde{\mathcal{O}} \\ &y_n \rightsquigarrow y, \ y_n \in V(\Omega_n), \ y \in V(\Omega) \end{aligned}\right\} \implies \liminf_{n\to\infty} J(\Omega_n, y_n) \ge J(\Omega, y).$$

REMARK 2.11. Convergences of sets and functions in $(\mathcal{B}1)$, $(\mathcal{B}2)$ are understood in the sense introduced above. The usual way of verifying $(\mathcal{B}1)$ is as follows: we first prove that $\mathcal{O}$ is compact in $\widetilde{\mathcal{O}}$; i.e., for any sequence $\{\Omega_n\}$, $\Omega_n \in \mathcal{O}$, there is a subsequence $\{\Omega_{n_k}\}$ and an element $\Omega \in \mathcal{O}$ such that

$$\Omega_{n_k} \stackrel{\widetilde{\mathcal{O}}}{\to} \Omega, \quad k \to \infty. \tag{2.86}$$

Next we show that solutions $u(\Omega)$ of $(\mathcal{P}(\Omega))$ depend continuously on variations of $\Omega \in \mathcal{O}$:

$$\Omega_n \stackrel{\widetilde{\mathcal{O}}}{\to} \Omega,\ \Omega_n, \Omega \in \mathcal{O} \implies u(\Omega_n) \rightsquigarrow u(\Omega).$$

This together with (2.86) implies $(\mathcal{B}1)$.

We are now ready to prove the following result.

THEOREM 2.10. *Let $(\mathcal{B}1)$, $(\mathcal{B}2)$ be satisfied. Then $(\mathbb{P})$ has at least one solution.*

Proof. Denote

$$q := \inf_{\Omega \in \mathcal{O}} J(\Omega, u(\Omega)) = \lim_{n\to\infty} J(\Omega_n, u(\Omega_n)),$$

where $\{\Omega_n\}$, $\Omega_n \in \mathcal{O}$, is a minimizing sequence of $(\mathbb{P})$. From $(\mathcal{B}1)$ the existence of $\{(\Omega_{n_k}, u(\Omega_{n_k}))\} \subset \{(\Omega_n, u(\Omega_n))\}$ and $(\Omega^*, u(\Omega^*)) \in \mathcal{G}$ such that

$$\begin{aligned} &\Omega_{n_k} \stackrel{\widetilde{\mathcal{O}}}{\to} \Omega^*, \\ &u(\Omega_{n_k}) \rightsquigarrow u(\Omega^*),\ k \to \infty, \end{aligned}$$

follows. From this and $(\mathcal{B}2)$ we arrive at

$$q = \lim_{n\to\infty} J(\Omega_n, u(\Omega_n)) = \liminf_{k\to\infty} J(\Omega_{n_k}, u(\Omega_{n_k})) \geq J(\Omega^*, u(\Omega^*));$$

i.e., $(\Omega^*, u(\Omega^*))$ is an optimal pair of $(\mathbb{P})$. □

We now turn to an abstract formulation of approximations of $(\mathbb{P})$. We shall follow the same ideas used in Section 2.2 when approximating the particular problem. Two types of discretized domains will be introduced: *discrete design* and *computational domains.* The boundaries of discrete design domains are usually realized by smooth piecewise spline functions (for example, Bézier curves). The optimal discrete design domain is the main output of the computational process on the basis of which a designer makes decisions. On the other hand, computational domains represent an auxiliary tool that simplifies the numerical realization of state problems. If, for example, standard straight finite elements for the approximation of $(\mathcal{P}(\Omega))$ are used, computational domains are represented by polygonal approximations of discrete design domains where all computations will be performed.

Let $\varkappa > 0$ be a *discretization* parameter and $n(\varkappa)$ be the number of parameters defining the shape of discrete design domains $\Omega_\varkappa$. For $\varkappa > 0$ fixed, the set of all *admissible discrete design domains* will be denoted by $\mathcal{O}_\varkappa$. *We shall suppose that the number $n(\varkappa)$ is the same for all $\Omega_\varkappa \in \mathcal{O}_\varkappa$, $\varkappa > 0$ fixed, and $\mathcal{O}_\varkappa \subset \widetilde{\mathcal{O}}$ for any $\varkappa > 0$ but not necessarily $\mathcal{O}_\varkappa \subset \mathcal{O}$.*

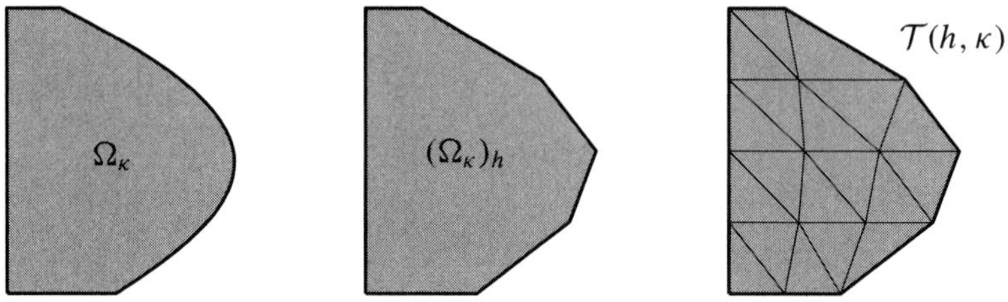

Figure 2.4. *Design and computational domains with triangulation.*

Remark 2.12. Due to the presence of complicated constraints appearing in the definition of $\mathcal{O}$, it is sometimes difficult to construct $\mathcal{O}_\varkappa$ as a subset of $\mathcal{O}$. Some constraints for $\varkappa > 0$ given might be violated. For this reason we introduced a larger system $\widetilde{\mathcal{O}}$. Any $\mathcal{O}_\varkappa$, $\varkappa > 0$, can be viewed as an *external approximation* of $\mathcal{O}$. We shall formulate assumptions under which the gap between $\mathcal{O}$ and $\mathcal{O}_\varkappa$ disappears when $\varkappa \to 0+$.

With any $\Omega_\varkappa \in \mathcal{O}_\varkappa$ we associate in a *unique way* a computational domain $(\Omega_\varkappa)_h$, where $h := h(\varkappa) > 0$ is another discretization parameter such that there exists a one-to-one relation between h and $\varkappa$ such that

$$h \to 0+ \iff \varkappa \to 0+.$$

In other words, from the knowledge of $\Omega_\varkappa$ one can construct $(\Omega_\varkappa)_h$ and vice versa. To clarify—if computational domains $(\Omega_\varkappa)_h$ are realized by polygonal approximations of $\Omega_\varkappa$, then h is related to the number of sides of the respective polygonal domain approximating $\Omega_\varkappa$ and also to the norm of a triangulation $\mathcal{T}(h,\varkappa)$ of $(\Omega_\varkappa)_h$ used for the construction of finite element spaces (see Figure 2.4). Instead of $(\Omega_\varkappa)_h$ we shall write $\Omega_{\varkappa h}$ in brief. The set of all computational domains corresponding to $\mathcal{O}_\varkappa$ will be denoted by $\mathcal{O}_{\varkappa h}$. *In what follows we shall also suppose that* $\mathcal{O}_{\varkappa h} \subset \widetilde{\mathcal{O}}$ for any $\varkappa > 0$, $h(\varkappa) > 0$.

With any $\Omega_{\varkappa h} \in \mathcal{O}_{\varkappa h}$ a finite dimensional space $V_h(\Omega_{\varkappa h}) \subset V(\Omega_{\varkappa h})$ will be associated. *We shall suppose that* $\dim V_h(\Omega_{\varkappa h})$ *is the same for all* $\Omega_{\varkappa h} \in \mathcal{O}_{\varkappa h}$, $h, \varkappa > 0$ fixed. Since both $\mathcal{O}_\varkappa$ and $\mathcal{O}_{\varkappa h}$ belong to $\widetilde{\mathcal{O}}$, the same convergences of sets and functions as before can be used. Finally we define a mapping

$$u_h : \Omega_{\varkappa h} \mapsto u_h(\Omega_{\varkappa h}) \in V_h(\Omega_{\varkappa h}) \qquad (\mathcal{P}_h(\Omega_{\varkappa h}))$$

associating with any $\Omega_{\varkappa h} \in \mathcal{O}_{\varkappa h}$ a unique element $u_h(\Omega_{\varkappa h})$ from $V_h(\Omega_{\varkappa h})$ and defining an *approximation* of the state problem.

Thus, starting from $\Omega_\varkappa \in \mathcal{O}_\varkappa$, we have the following chain of mappings:

$$\Omega_\varkappa \mapsto \Omega_{\varkappa h} \mapsto u_h(\Omega_{\varkappa h}) \mapsto J(\Omega_{\varkappa h}, u_h(\Omega_{\varkappa h})).$$

The *approximation* of $(\mathbb{P})$ reads as follows:

$$\begin{cases} \text{Find } \Omega^*_{\varkappa h} \in \mathcal{O}_{\varkappa h} \text{ such that} \\ J(\Omega^*_{\varkappa h}, u_h(\Omega^*_{\varkappa h})) \le J(\Omega_{\varkappa h}, u_h(\Omega_{\varkappa h})) \quad \forall \Omega_{\varkappa h} \in \mathcal{O}_{\varkappa h}. \end{cases} \qquad (\mathbb{P}_{h\varkappa})$$

The set $\Omega^*_{\varkappa h}$ will be called the *optimal computational domain*, while $\Omega^*_{\varkappa}$, which can be reconstructed from $\Omega^*_{\varkappa h}$, is termed the *optimal discrete design domain.*

Solutions to discrete state problems are available only on computational domains so that only these can be used when evaluating J. From what has been said before, however, it follows that not the optimal computational but the respective optimal discrete design domain is of primary interest.

CONVENTION: *By a solution of* $(\mathbb{P}_{h\varkappa})$ *we understand* $\Omega^*_{\varkappa}$ *corresponding to* $\Omega^*_{\varkappa h}$. *An optimal pair of* $(\mathbb{P}_{h\varkappa})$ *will be denoted by* $(\Omega^*_{\varkappa}, u_h(\Omega^*_{\varkappa h}))$ *in what follows.*

Next we shall analyze convergence properties of $(\mathbb{P}_{h\varkappa})$ as $\varkappa, h \to 0+$. To this end we introduce the following assumptions on approximated data:

$(\mathcal{B}1)_{\varkappa}$ $\forall \Omega \in \mathcal{O}$ $\exists \{\Omega_{\varkappa}\}$, $\Omega_{\varkappa} \in \mathcal{O}_{\varkappa}$, *such that*

$$\Omega_{\varkappa} \overset{\tilde{\mathcal{O}}}{\to} \Omega \quad \text{and} \quad \Omega_{\varkappa h} \overset{\tilde{\mathcal{O}}}{\to} \Omega, \qquad h, \varkappa \to 0+;$$

$(\mathcal{B}2)_{\varkappa}$ *for any sequence* $\{\Omega_{\varkappa}\}$, $\Omega_{\varkappa} \in \mathcal{O}_{\varkappa}$, *there exists a subsequence* $\{\Omega_{\varkappa_j}\}$ *and an element* $\Omega \in \mathcal{O}$ *such that*

$$\begin{gathered} \Omega_{\varkappa_j} \overset{\tilde{\mathcal{O}}}{\to} \Omega \quad \text{and} \quad \Omega_{\varkappa_j h_j} \overset{\tilde{\mathcal{O}}}{\to} \Omega, \\ u_{h_j}(\Omega_{\varkappa_j h_j}) \rightsquigarrow u(\Omega), \ j \to \infty; \end{gathered}$$

$$(\mathcal{B}3)_{\varkappa} \quad \left.\begin{array}{l} \Omega_{\varkappa h} \overset{\tilde{\mathcal{O}}}{\to} \Omega, \quad \Omega_{\varkappa h} \in \mathcal{O}_{\varkappa h}, \ \Omega \in \mathcal{O} \\ u_h(\Omega_{\varkappa h}) \rightsquigarrow u(\Omega) \end{array}\right\}$$

$$\Longrightarrow J(\Omega_{\varkappa h}, u_h(\Omega_{\varkappa h})) \to J(\Omega, u(\Omega)), \qquad \varkappa, h \to 0+.$$

COMMENTS 2.5.

(i) $(\mathcal{B}1)_{\varkappa}$ is the density type assumption for the systems $\{\mathcal{O}_{\varkappa}\}$, $\{\mathcal{O}_{\varkappa h}\}$ in $\mathcal{O}$. In addition it says that if $\{\Omega_{\varkappa}\}$ tends to $\Omega \in \mathcal{O}$, then $\{\Omega_{\varkappa h}\}$ tends to the *same* element.

(ii) $(\mathcal{B}2)_{\varkappa}$ is the compactness type assumption saying that the limit pair $(\Omega, u(\Omega)) \in \mathcal{G}$.

(iii) $(\mathcal{B}3)_{\varkappa}$ is the continuity property of J.

We conclude this section with the following result.

THEOREM 2.11. *Let* $(\mathcal{B}1)_{\varkappa}$–$(\mathcal{B}3)_{\varkappa}$ *be satisfied. Then for any sequence* $\{(\Omega^*_{\varkappa}, u_h(\Omega^*_{\varkappa h}))\}$ *of optimal pairs of* $(\mathbb{P}_{h\varkappa})$, $h \to 0+$, *there exists its subsequence such that*

$$\left\{\begin{array}{l} \Omega^*_{\varkappa_j} \overset{\tilde{\mathcal{O}}}{\to} \Omega^*, \\ u_{h_j}(\Omega^*_{\varkappa_j h_j}) \rightsquigarrow u(\Omega^*), \qquad j \to \infty. \end{array}\right. \tag{2.87}$$

In addition, $(\Omega^*, u(\Omega^*))$ *is an optimal pair of* $(\mathbb{P})$. *Any accumulation point of* $\{(\Omega^*_{\varkappa}, u_h(\Omega^*_{\varkappa h}))\}$ *in the sense of* (2.87) *possesses this property.*

Proof. From $(\mathcal{B}2)_\varkappa$ it follows that there exists a subsequence $\{\Omega^*_{\varkappa_j}\}$ of $\{\Omega^*_\varkappa\}$ and an element $\Omega^* \in \mathcal{O}$ such that

$$\begin{cases} \Omega^*_{\varkappa_j} \overset{\widetilde{\mathcal{O}}}{\to} \Omega^*, \quad \Omega^*_{\varkappa_j h_j} \overset{\widetilde{\mathcal{O}}}{\to} \Omega^*, \\ u_{h_j}(\Omega^*_{\varkappa_j h_j}) \rightsquigarrow u(\Omega^*), \quad j \to \infty. \end{cases} \tag{2.88}$$

Let $\widetilde{\Omega} \in \mathcal{O}$ be arbitrary and $\{\widetilde{\Omega}_\varkappa\}$, $\widetilde{\Omega}_\varkappa \in \mathcal{O}_\varkappa$, be a sequence such that

$$\widetilde{\Omega}_\varkappa \overset{\widetilde{\mathcal{O}}}{\to} \widetilde{\Omega}.$$

The existence of such a sequence follows from $(\mathcal{B}1)_\varkappa$. Applying $(\mathcal{B}2)_\varkappa$ once again to the sequence $\{(\widetilde{\Omega}_{\varkappa h}, u_h(\widetilde{\Omega}_{\varkappa h}))\}$ we have a similar result to (2.88) with $\widetilde{\Omega}_{\varkappa_j}$, $\widetilde{\Omega}_{\varkappa_j h_j}$ instead of $\Omega^*_{\varkappa_j}$, $\Omega^*_{\varkappa_j h_j}$, respectively. (Observe that due to (2.83) and (2.85) we can take the common filter of indices $\{\varkappa_j, h_j\}$ for which all these convergences hold true.) The rest of the proof now easily follows from the definition of $(\mathbb{P}_{h_j \varkappa_j})$:

$$J(\Omega^*_{\varkappa_j h_j}, u_{h_j}(\Omega^*_{\varkappa_j h_j})) \leq J(\widetilde{\Omega}_{\varkappa_j h_j}, u_{h_j}(\widetilde{\Omega}_{\varkappa_j h_j})).$$

Passing to the limit with $h_j, \varkappa_j \to 0+$ and using $(\mathcal{B}3)_\varkappa$ we arrive at

$$J(\Omega^*, u(\Omega^*)) \leq J(\widetilde{\Omega}, u(\widetilde{\Omega})).$$

Since $\widetilde{\Omega} \in \mathcal{O}$ is an arbitrary element we see that $(\Omega^*, u(\Omega^*))$ is an optimal pair of $(\mathbb{P})$. From the proof it is also readily seen that any accumulation point possesses such a property. □

Remark 2.13. It would be possible to consider the case in which both the computational and discrete design domains coincide.

2.5 Applications of the abstract results

2.5.1 Thickness optimization of an elastic unilaterally supported beam

This section presents an example of sizing optimization with a *state variational inequality*. We shall prove the existence of a solution to this problem and analyze its approximation by using the abstract results of Section 2.3.

Let us consider the same elastic beam of variable thickness e, subject to a vertical load f, as in Section 2.1, with the following minor change: the deflection u of the beam is now restricted from below by a rigid obstacle described by a function $\varphi \in C(I)$ (we use the notation of Section 2.1). We want again to find the thickness distribution $e \in U^{ad}$, with U^{ad} defined by (2.3), minimizing the compliance (2.1). The state problem is now represented by the following elliptic inequality:

$$\begin{cases} \text{Find } u := u(e) \in K \text{ such that} \\ \displaystyle\int_I \beta e^3 u''(v'' - u'')\,dx \geq \int_I f(v-u)\,dx \quad \forall v \in K, \end{cases} \tag{$\mathcal{P}(e)$}$$

where

$$K = \{v \in H_0^2(I) \mid v \geq \varphi \text{ in } I\}.$$

Next we shall suppose that $\varphi(0) \leq 0$, $\varphi(\ell) \leq 0$, implying that K is nonempty. It is readily seen that K is a closed, convex subset of $H_0^2(I)$. Furthermore U^{ad} is a compact subset of $U = C(I)$, as follows from the Ascoli–Arzelà theorem.

The thickness optimization problem reads as follows:

$$\begin{cases} \text{Find } e^* \in U^{ad} \text{ such that} \\ J(u(e^*)) = \min\limits_{e \in U^{ad}} J(u(e)), \end{cases} \tag{$\mathbb{P}$}$$

where $u(e) \in K$ solves $(\mathcal{P}(e))$.

We prove the following result.

Theorem 2.12. *Problem* $(\mathbb{P})$ *has a solution.*

Proof. We use Theorem 2.8. Clearly $(\mathcal{A}1)$, $(\mathcal{A}2)$, $(\mathcal{A}3)$, and $(\mathcal{A}5)$ are satisfied. Let us check $(\mathcal{A}4)$:

$$\sup_{\substack{\|y\|_{2,I} \leq 1 \\ \|v\|_{2,I} \leq 1}} \left| \int_I \beta(e_n^3 - e^3) y'' v'' \, dx \right| \leq \|\beta\|_{L^\infty(I)} \|e_n^3 - e^3\|_{C(I)} \to 0$$

as $n \to \infty$, provided $e_n \rightrightarrows e$ in I, $e_n, e \in U^{ad}$. The existence of a solution to $(\mathbb{P})$ now follows from Theorem 2.8. □

Let us pass to the approximation of $(\mathbb{P})$. The convex set K will be replaced by

$$K_h = \{v_h \in V_h \mid v_h(a_i) \geq \varphi(a_i),\ i = 1, \dots, d-1\},$$

with V_h as in (2.16); i.e., K_h contains all functions from V_h satisfying the unilateral constraints at all the interior nodes of Δ_h. Observe that K_h is *not* a subset of K, in general. As far as the approximation of U^{ad} is concerned, we shall distinguish two cases:

(i) U_h^{ad} given by (2.15). Then $U = \widetilde{U} = C(I)$.

(ii) $U_h^{ad} = \widetilde{U}_h^{ad}\ \forall h > 0$ defined by (2.32). In this case $U = C(I)$, $\widetilde{U} = L^\infty(I)$.

For $e_h \in U_h^{ad}$ fixed, we define the following discrete state problem:

$$\begin{cases} \text{Find } u_h := u_h(e_h) \in K_h \text{ such that} \\ \displaystyle\int_I \beta e_h^3 u_h''(v_h'' - u_h'') \, dx \geq \int_I f(v_h - u_h) \, dx \quad \forall v_h \in K_h. \end{cases} \tag{$\mathcal{P}_h(e_h)$}$$

The approximation of $(\mathbb{P})$ is given as follows:

$$\begin{cases} \text{Find } e_h^* \in U_h^{ad} \text{ such that} \\ J(u_h(e_h^*)) = \min\limits_{e_h \in U_h^{ad}} J(u_h(e_h)), \end{cases} \tag{$\mathbb{P}_h$}$$

with $u_h(e_h) \in K_h$ being the solution of $(\mathcal{P}_h(e_h))$.

Next we shall prove that $(\mathbb{P})$ and $(\mathbb{P}_h)$, $h \to 0+$, are close in the sense of subsequences as follows from Theorem 2.9. To this end we shall verify assumptions $(\mathcal{A}1)_h$–$(\mathcal{A}9)_h$ of Section 2.3. It is readily seen that $(\mathcal{A}1)_h$–$(\mathcal{A}4)_h$, $(\mathcal{A}9)_h$ are satisfied as well as $(\mathcal{A}7)_h$ and $(\mathcal{A}8)_h$ for U_h^{ad} given by (2.15) (see Problem 2.2); for $U_h^{ad} = \widetilde{U}_h^{ad}$ defined by (2.32) we refer to the proof of Theorem 2.4 and Problem 2.5. It remains to verify $(\mathcal{A}5)_h$ and $(\mathcal{A}6)_h$. These assumptions follow from a lemma.

Lemma 2.14. *Let $v \in K$ be given. Then there exists a sequence $\{v_h\}$, $v_h \in K_h$, such that*

$$v_h \to v \quad \text{in } H_0^2(I), \quad h \to 0+.$$

If $\{v_h\}$, $v_h \in K_h$, is such that

$$v_h \rightharpoonup v \quad \text{in } H_0^2(I), \quad h \to 0+,$$

then $v \in K$.

Proof. We shall show that the set $K \cap C_0^\infty(I)$ is dense in K in the $H_0^2(I)$-norm. To simplify our presentation we suppose that the function φ describing the obstacle is such that

$$\varphi(0) < 0 \quad \text{and} \quad \varphi(\ell) < 0. \tag{2.89}$$

The case for which the value of φ at one of the endpoints of I is equal to zero is left as an exercise (see Problem 2.10).

Choose a function $\Phi \in H_0^2(I)$ such that $\Phi(x) > 0\ \forall x \in]0, \ell[$. Let $v \in K$. For any $\varepsilon > 0$ we define

$$v_\varepsilon := v + \varepsilon \Phi.$$

It is easy to see that $v_\varepsilon \in H_0^2(I)$,

$$v_\varepsilon \to v \quad \text{in } H_0^2(I), \quad \varepsilon \to 0+, \tag{2.90}$$

and

$$v_\varepsilon > \varphi \quad \text{in } I \text{ for any } \varepsilon > 0. \tag{2.91}$$

The last property is a consequence of (2.89). Fix $\varepsilon > 0$. Since $C_0^\infty(I)$ is dense in $H_0^2(I)$ there exists a sequence $\{v_{\varepsilon n}\}$, $v_{\varepsilon n} \in C_0^\infty(I)$, such that

$$v_{\varepsilon n} \to v_\varepsilon \quad \text{in } H_0^2(I) \text{ as } n \to \infty \tag{2.92}$$

and consequently

$$v_{\varepsilon n} \rightrightarrows v_\varepsilon \quad \text{in } I \text{ as } n \to \infty,$$

making use of the embedding of $H_0^2(I)$ in $C(I)$. From this and (2.91) we also have that $v_{\varepsilon n} \geq \varphi$ in I for n sufficiently large; i.e., $v_{\varepsilon n} \in K \cap C_0^\infty(I)$. Finally, the triangle inequality

$$\|v - v_{\varepsilon n}\|_{2,I} \leq \|v - v_\varepsilon\|_{2,I} + \|v_\varepsilon - v_{\varepsilon n}\|_{2,I}$$

together with (2.90) and (2.92) proves the existence of a sequence $\{v_m\}$, $v_m \in K \cap C_0^\infty(I)$, such that

$$v_m \to v \quad \text{in } H_0^2(I),\ m \to \infty. \tag{2.93}$$

Let $\eta > 0$ be an arbitrary number. From (2.93) it follows that there exists $m_0 \in \mathbb{N}$ such that

$$\|v_{m_0} - v\|_{2,I} \le \frac{\eta}{2}. \tag{2.94}$$

Let $r_h : C_0^\infty(I) \to V_h$ be the piecewise Hermite interpolation operator of functions by means of piecewise cubic polynomials on Δ_h. Then

$$\|v_{m_0} - r_h v_{m_0}\|_{2,I} \le \frac{\eta}{2} \tag{2.95}$$

for $h > 0$ sufficiently small, as follows from the classical approximation theory. From the definition of r_h we see that $r_h v_{m_0} \in K_h$. From (2.94), (2.95), and the triangle inequality we conclude that

$$\|v - r_h v_{m_0}\|_{2,I} \le \|v - v_{m_0}\|_{2,I} + \|v_{m_0} - r_h v_{m_0}\|_{2,I} \le \eta,$$

meaning that any function $v \in K$ can be approximated by functions from $\{K_h\}$, $h \to 0+$, as indicated in the assertion of this lemma.

Let $\{v_h\}$, $v_h \in K_h$, be such that $v_h \rightharpoonup v$ in $H_0^2(I)$. Then

$$v_h \rightrightarrows v \quad \text{in } I$$

by virtue of the compact embedding of $H_0^2(I)$ into $C(I)$. It is easy to show that $v \ge \varphi$ in I; i.e., $v \in K$ (see Problem 2.11). □

Since all the assumptions of Theorem 2.9 are satisfied we arrive at the following result.

Theorem 2.13. *Problems* ($\mathbb{P}$) *and* ($\mathbb{P}_h$), $h \to 0+$, *are close in the sense of Theorem* 2.9.

Remark 2.14. The quality of convergence of $\{e_h^*\}$ to e^* depends on the choice of U_h^{ad}:

$$e_h^* \rightrightarrows e^* \text{ in } I,\ h \to 0+, \quad \text{for } U_h^{ad} \text{ given by (2.15)};$$
$$e_h^* \to e^* \text{ in } L^\infty(I),\ h \to 0+, \quad \text{for } U_h^{ad} = \widetilde{U}_h^{ad} \text{ defined by (2.32)}.$$

2.5.2 Shape optimization with Neumann boundary value state problems

The next few sections will be devoted to shape optimization with state problems involving other types of boundary conditions than the one studied in Section 2.2. The existence and convergence analysis will be based on the abstract theory of Section 2.4.

To simplify our presentation we restrict ourselves in all examples to a class of domains whose shapes are as in Figure 2.2; i.e., the designed part $\Gamma(\alpha)$ of the boundary will be parametrized by functions $\alpha : [0, 1] \to \mathbb{R}$. More precisely, we define

$$\widetilde{U}^{ad} = \big\{\alpha \in C^{0,1}([0, 1]) \mid 0 < \alpha_{\min} \le \alpha \le \alpha_{\max} \text{ in } [0, 1], \\ |\alpha'| \le L_0 \text{ a.e. in }]0, 1[\,\big\}, \tag{2.96}$$

where $\alpha_{\min}, \alpha_{\max}, L_0$ are given and

$$U^{ad} = \{\alpha \in \widetilde{U}^{ad} \mid g_i(\alpha) \le 0\ \forall i \in \mathcal{I}_1, \quad g_i(\alpha) = 0\ \forall i \in \mathcal{I}_2\}. \tag{2.97}$$

$\mathcal{I}_1, \mathcal{I}_2$ are index sets (possibly empty) and g_i, $i \in \mathcal{I}_1 \cup \mathcal{I}_2$, are given *continuous* functions of α. With $\widetilde{U}^{ad}$ and U^{ad} two families of domains will be associated:

$$\widetilde{\mathcal{O}} = \{\Omega(\alpha) \mid \alpha \in \widetilde{U}^{ad}\}, \quad \mathcal{O} = \{\Omega(\alpha) \mid \alpha \in U^{ad}\},$$

where

$$\Omega(\alpha) = \{(x_1, x_2) \in \mathbb{R}^2 \mid 0 < x_1 < \alpha(x_2),\ x_2 \in]0, 1[\}.$$

Remark 2.15. The functions g_i, $i \in \mathcal{I}_1 \cup \mathcal{I}_2$, define additional technological constraints in $\widetilde{\mathcal{O}}$, such as meas $\Omega(\alpha) = \gamma$ or $\gamma \le$ meas $\Omega(\alpha) \le \delta$, where γ, δ are given positive numbers. The reason for introducing a larger family $\widetilde{\mathcal{O}}$ has already been explained in Section 2.4.

Next we shall suppose that the data characterizing $\widetilde{U}^{ad}$ and U^{ad} are such that $\widetilde{U}^{ad} \ne \emptyset$, $U^{ad} \ne \emptyset$. Convergence of domains from $\widetilde{\mathcal{O}}$ will be defined as in Section 2.2:

$$\Omega(\alpha_n) \overset{\widetilde{\mathcal{O}}}{\to} \Omega(\alpha),\ \Omega(\alpha_n),\ \Omega(\alpha) \in \widetilde{\mathcal{O}} \iff \alpha_n \rightrightarrows \alpha \text{ in } [0, 1],\ \alpha_n, \alpha \in \widetilde{U}^{ad}.$$

The definition of $\widetilde{U}^{ad}$ contains the minimum of assumptions under which $\widetilde{\mathcal{O}}$ is compact with respect to the above-mentioned convergence of domains. Owing to the continuity of the functions g_i, $i \in \mathcal{I}_1 \cup \mathcal{I}_2$, the family $\mathcal{O}$ is closed in $\widetilde{\mathcal{O}}$ and hence compact as well. *Basic assumptions, such as the continuous dependence of solutions on domain variations, will be verified on* $\widetilde{\mathcal{O}}$.

The convergence of a sequence $\{y_n\}$, where $y_n \in V(\Omega_n) \subseteq H^1(\Omega_n)$ $(\Omega_n := \Omega(\alpha_n))$, will be defined as strong or weak convergence in $V(\widehat{\Omega}) \subseteq H^1(\widehat{\Omega})$ of appropriate extensions of y_n from Ω_n to $\widehat{\Omega}$, where $\widehat{\Omega} \supset \Omega(\alpha)\ \forall \alpha \in \widetilde{U}^{ad}$. In our particular case we may take, for example, $\widehat{\Omega} =]0, 2\alpha_{\max}[\times]0, 1[$. Such an extension is straightforward for homogeneous Dirichlet boundary value problems formulated in $H_0^1(\Omega(\alpha))$: indeed, any function from $H_0^1(\Omega(\alpha))$ can be extended by zero outside of $\Omega(\alpha)$ to a function from $H_0^1(\widehat{\Omega})$, preserving its norm (see (2.36)). For other boundary conditions such a trivial approach is not possible and one has to use more sophisticated extensions $p_{\Omega(\alpha)} \in \mathcal{L}(V(\Omega(\alpha)), V(\widehat{\Omega}))$ whose operator norm can be bounded independently of $\alpha \in \widetilde{U}^{ad}$. Since $\widetilde{\mathcal{O}}$ consists of domains possessing the uniform ε-cone property for some $\varepsilon := \varepsilon(L_0) > 0$, such extensions exist (see Appendix A):

$$\exists c = \text{const.} > 0 : \quad \|p_{\Omega(\alpha)}\|_{\mathcal{L}(V(\Omega(\alpha)), V(\widehat{\Omega}))} \le c \quad \forall \alpha \in \widetilde{U}^{ad}. \tag{2.98}$$

Thus we define

$$y_n \rightsquigarrow y, \quad y_n \in V(\Omega(\alpha_n)),\ y \in V(\Omega(\alpha)) \iff \begin{cases} p_{\Omega(\alpha_n)} y_n \rightharpoonup p_{\Omega(\alpha)} y & \text{in } V(\widehat{\Omega}) \\ \quad \text{or} \\ p_{\Omega(\alpha_n)} y_n \to p_{\Omega(\alpha)} y & \text{in } V(\widehat{\Omega}). \end{cases}$$

Convention: *If $y \in V(\alpha)$, then its extension $p_{\Omega(\alpha)} y$ from $\Omega(\alpha)$ to $\widehat{\Omega}$ with $p_{\Omega(\alpha)}$ satisfying* (2.98) *will be denoted by $\widetilde{y}$ in what follows. Exceptions to this rule will be quoted explicitly.*

This section deals with shape optimization involving Neumann state problems. On any $\Omega(\alpha)$, $\alpha \in \widetilde{U}^{ad}$, we consider the *Neumann boundary value problem*

$$\begin{cases} -\Delta u(\alpha) + u(\alpha) = f & \text{in } \Omega(\alpha), \\ \dfrac{\partial u(\alpha)}{\partial \nu} = 0 & \text{on } \partial\Omega(\alpha), \end{cases} \tag{$\mathcal{P}(\alpha)'$}$$

or, in the weak form:

$$\begin{cases} \text{Find } u := u(\alpha) \in H^1(\Omega(\alpha)) \text{ such that} \\ \displaystyle\int_{\Omega(\alpha)} (\nabla u \cdot \nabla v + uv)\, dx = \int_{\Omega(\alpha)} fv\, dx \quad v \in H^1(\Omega(\alpha)), \end{cases} \tag{$\mathcal{P}(\alpha)$}$$

where $f \in L^2(\widehat{\Omega})$.

Let $D \subset \Omega(\alpha)\ \forall \alpha \in \widetilde{U}^{ad}$ be a target set (for example, $D =]0, \alpha_{\min}/2[\times]0, 1[)$ and $z_d \in L^2(D)$ be a given function. We shall study the following optimal shape design problem:

$$\begin{cases} \text{Find } \alpha^* \in U^{ad} \text{ such that} \\ J(u(\alpha^*)) \le J(u(\alpha)) \quad \forall \alpha \in U^{ad}, \end{cases} \tag{$\mathbb{P}$}$$

where

$$J(y) = \frac{1}{2}\|y - z_d\|_{0,D}^2, \tag{2.99}$$

$u(\alpha) \in H^1(\Omega(\alpha))$ solves $(\mathcal{P}(\alpha))$, and U^{ad} is given by (2.97).

To prove the existence of solutions to $(\mathbb{P})$ we shall verify assumptions $(\mathcal{B}1)$ and $(\mathcal{B}2)$ of Section 2.4.

Lemma 2.15. (*Verification of* $(\mathcal{B}1)$.) *For any sequence $\{(\alpha_n, u_n)\}$, where $\alpha_n \in \widetilde{U}^{ad}$ and $u_n := u(\alpha_n) \in H^1(\Omega(\alpha_n))$ is a solution of $(\mathcal{P}(\alpha_n))$, $n \to \infty$, there exist its subsequence and elements $\alpha \in \widetilde{U}^{ad}$, $u \in H^1(\widehat{\Omega})$ such that*

$$\alpha_n \rightrightarrows \alpha \text{ in } [0, 1], \tag{2.100}$$

$$\widetilde{u}_n \rightharpoonup u \text{ in } H^1(\widehat{\Omega}), \qquad n \to \infty. \tag{2.101}$$

In addition, $u(\alpha) := u_{|\Omega(\alpha)}$ solves $(\mathcal{P}(\alpha))$.

Proof. One can pass to a subsequence of $\{\alpha_n\}$ such that (2.100) holds for some $\alpha \in \widetilde{U}^{ad}$. Inserting $v := u_n$ into the definition of $(\mathcal{P}(\alpha_n))$ we obtain

$$\|u_n\|^2_{1,\Omega_n} = (f, u_n)_{0,\Omega_n} \leq \|f\|_{0,\widehat{\Omega}} \|u_n\|_{0,\Omega_n} \quad (\Omega_n := \Omega(\alpha_n)),$$

implying the boundedness of $\{u_n\}$:

$$\exists c = \text{const.} > 0 : \quad \|u_n\|_{1,\Omega_n} \leq c \quad \forall n \in \mathbb{N}$$

and also

$$\|\widetilde{u}_n\|_{1,\widehat{\Omega}} \leq c \quad \forall n \in \mathbb{N}$$

for another positive constant c independent of n by virtue of (2.98).

Thus one can pass to a subsequence of $\{\widetilde{u}_n\}$ such that (2.101) holds for an element $u \in H^1(\widehat{\Omega})$. It remains to show that $u(\alpha) := u_{|\Omega(\alpha)}$ solves $(\mathcal{P}(\alpha))$.

Let $v \in H^1(\Omega(\alpha))$ be given and $\widetilde{v} \in H^1(\widehat{\Omega})$ be its extension to $\widehat{\Omega}$. Since $\widetilde{v}_{|\Omega_n} \in H^1(\Omega_n)\ \forall n \in \mathbb{N}$, it can be used as a test function in $(\mathcal{P}(\alpha_n))$ written now in the following form:

$$\int_{\widehat{\Omega}} \chi_n (\nabla \widetilde{u}_n \cdot \nabla \widetilde{v} + \widetilde{u}_n \widetilde{v})\, dx = \int_{\widehat{\Omega}} \chi_n f \widetilde{v}\, dx, \tag{2.102}$$

where χ_n is the characteristic function of Ω_n. Letting $n \to \infty$ in (2.102) we arrive at

$$\int_{\widehat{\Omega}} \chi (\nabla u \cdot \nabla \widetilde{v} + u \widetilde{v})\, dx = \int_{\widehat{\Omega}} \chi f \widetilde{v}\, dx, \tag{2.103}$$

where χ is the characteristic function of the limit domain $\Omega(\alpha)$. To see that, let us compute the limit of the first term in (2.102) (and similarly for the remaining ones). The triangle inequality yields

$$\left| \int_{\widehat{\Omega}} \chi_n \nabla \widetilde{u}_n \cdot \nabla \widetilde{v}\, dx - \int_{\widehat{\Omega}} \chi \nabla u \cdot \nabla \widetilde{v}\, dx \right|$$
$$\leq \left| \int_{\widehat{\Omega}} \chi_n (\nabla \widetilde{u}_n - \nabla u) \cdot \nabla \widetilde{v}\, dx \right| + \left| \int_{\widehat{\Omega}} (\chi_n - \chi) \nabla u \cdot \nabla \widetilde{v}\, dx \right| \to 0 \quad \text{as } n \to \infty$$

since $\nabla \widetilde{u}_n - \nabla u \rightharpoonup 0$ and $\chi_n \nabla \widetilde{v} \to \chi \nabla \widetilde{v}$ in $(L^2(\widehat{\Omega}))^2$. Because $v \in H^1(\Omega(\alpha))$ is an arbitrary function, it follows from (2.103) that $u(\alpha)$ solves $(\mathcal{P}(\alpha))$. $\square$

We leave as an easy exercise the proof of the following lemma.

Lemma 2.16. (*Verification of* $(\mathcal{B}2)$.) *The cost functional J given by* (2.99) *is continuous in the following sense:*

$$\left. \begin{array}{l} \alpha_n \rightrightarrows \alpha \text{ in } [0,1],\ \alpha_n, \alpha \in \widetilde{U}^{ad} \\ y_n \rightharpoonup y \text{ in } H^1(\widehat{\Omega}),\ y_n, y \in H^1(\widehat{\Omega}) \end{array} \right\} \Longrightarrow \lim_{n\to\infty} J(y_n) = J(y).$$

From Theorem 2.10 and Lemmas 2.15 and 2.16 the existence of solutions to $(\mathbb{P})$ follows.

The assertion of Lemma 2.15 can be written in the following form: for an appropriate subsequence of $\{\widetilde{u}_n\}$ it holds that

$$\begin{cases} \nabla \widetilde{u}_n \rightharpoonup \nabla u & \text{in } (L^2(\widehat{\Omega}))^2, \\ \widetilde{u}_n \rightharpoonup u & \text{in } L^2(\widehat{\Omega}) \quad \text{as } n \to \infty. \end{cases} \tag{2.104}$$

At the same time $\chi_n \to \chi$ in $L^2(\widehat{\Omega})$ (see Lemma 2.4), implying that

$$\begin{cases} \chi_n \nabla \widetilde{u}_n \rightharpoonup \chi \nabla u & \text{in } (L^2(\widehat{\Omega}))^2, \\ \chi_n \widetilde{u}_n \rightharpoonup \chi u & \text{in } L^2(\widehat{\Omega}). \end{cases} \tag{2.105}$$

Substitution of $v := \widetilde{u}_n$ into $(\mathcal{P}(\alpha_n))$ yields

$$\int_{\widehat{\Omega}} \chi_n \left(|\nabla \widetilde{u}_n|^2 + |\widetilde{u}_n|^2 \right) dx = \int_{\widehat{\Omega}} \chi_n f \widetilde{u}_n \, dx \underset{n\to\infty}{\longrightarrow} \int_{\widehat{\Omega}} \chi f u \, dx = \int_{\widehat{\Omega}} \chi \left(|\nabla u|^2 + |u|^2 \right) dx.$$

Therefore weak convergence in (2.105) can be replaced by strong. Thus one can expect that the assertion of Lemma 2.15 can be improved. Indeed, we have the following lemma.

Lemma 2.17. *Let $\{(\alpha_n, u_n)\}$, $\alpha \in \widetilde{U}^{ad}$, and $u \in H^1(\widehat{\Omega})$ be the same as in Lemma 2.15. Then*

$$u_{n|Q} \to u_{|Q} \text{ (strongly) in } H^1(Q), \qquad n \to \infty, \tag{2.106}$$

in any domain $Q \subset\subset \Omega(\alpha)$ (i.e., $\overline{Q} \subset \Omega(\alpha)$).

Proof. Let $Q \subset\subset \Omega(\alpha)$ be given. For any $m \in \mathbb{N}$ we define

$$G_m(\alpha) = \left\{ x \in \Omega(\alpha) \mid \operatorname{dist}(x, \Gamma(\alpha)) > \frac{1}{m} \right\}. \tag{2.107}$$

Let $m \in \mathbb{N}$ be fixed and such that $G_m(\alpha) \supset \overline{Q}$ (i.e., m is large enough). Then there exists $n_0 := n_0(m)$ such that $\Omega(\alpha_n) \supset G_m(\alpha)$ $\forall n \geq n_0$. Since $\widetilde{u}_n = u_n$ in $G_m(\alpha)$ for $n \geq n_0$ we have

$$\begin{aligned} \|u_n - u\|^2_{1,Q} \leq \|u_n - u\|^2_{1,G_m(\alpha)} &\leq \|u_n\|^2_{1,\Omega_n} - 2(u_n, u)_{1,G_m(\alpha)} + \|u\|^2_{1,G_m(\alpha)} \\ &= \int_{\widehat{\Omega}} \chi_n f \widetilde{u}_n \, dx - 2(u_n, u)_{1,G_m(\alpha)} + \|u\|^2_{1,G_m(\alpha)} \end{aligned}$$

using the definition of $(\mathcal{P}(\alpha_n))$. Therefore

$$0 \leq \limsup_{n\to\infty} \|u_n - u\|^2_{1,Q} \leq \int_{\widehat{\Omega}} \chi f u \, dx - \|u\|^2_{1,G_m(\alpha)} = \|u\|^2_{1,\Omega(\alpha)} - \|u\|^2_{1,G_m(\alpha)} \tag{2.108}$$

holds for any $m \in \mathbb{N}$ sufficiently large. Letting $m \to \infty$ on the right-hand side of (2.108) we arrive at (2.106). $\square$

REMARK 2.16. The statement of Lemma 2.17 can be equivalently expressed as follows:

$$u_n \to u(\alpha) \quad \text{in } H^1_{\text{loc}}(\Omega(\alpha)), \qquad n \to \infty.$$

REMARK 2.17. Let us now consider a nonhomogeneous Neumann boundary condition $\partial u/\partial \nu = g$ on $\partial\Omega(\alpha)$, where $g \in L^2(\partial\Omega(\alpha))$ for any $\alpha \in \widetilde{U}^{ad}$. The weak formulation of this problem reads as follows:

$$\left\{\begin{array}{l} \text{Find } u := u(\alpha) \in H^1(\Omega(\alpha)) \quad \text{such that} \\ \displaystyle\int_{\Omega(\alpha)} (\nabla u \cdot \nabla v + uv)\, dx = \int_{\Omega(\alpha)} fv\, dx + \int_{\partial\Omega(\alpha)} gv\, ds \quad \forall v \in H^1(\Omega(\alpha)). \end{array}\right. \tag{$\mathcal{P}(\alpha)$}$$

The presence of the curvilinear integral on the right-hand side of $(\mathcal{P}(\alpha))$ complicates the existence analysis of the respective optimal shape design problem. Uniform convergence of boundaries, ensured by the definition of $\widetilde{\mathcal{O}}$, is not sufficient to claim that $\int_{\partial\Omega(\alpha_n)} gv\, ds \to \int_{\partial\Omega(\alpha)} gv\, ds$ so that assumption $(\mathcal{B}1)$ may not be satisfied. To overcome this difficulty one can either increase the regularity of α or reformulate the Neumann boundary condition avoiding the integrals over $\partial\Omega(\alpha)$. We shall describe in brief the latter approach. Suppose that the function g is of the form $g = \partial G/\partial s$, where $G \in H^2(\widehat{\Omega})$ is a given function and $\partial/\partial s$ stands for the derivative along $\partial\Omega(\alpha)$. From Green's formula it follows that

$$\int_{\Omega(\alpha)} \operatorname{curl} G \cdot \nabla v\, dx = \int_{\partial\Omega(\alpha)} \frac{\partial G}{\partial s} v\, ds = \int_{\partial\Omega(\alpha)} gv\, ds$$

holds for any $v \in H^1(\Omega(\alpha))$, where $\operatorname{curl} G := (\partial G/\partial x_2, -\partial G/\partial x_1)$. Thus, the nonhomogeneous Neumann boundary value problem takes the following form:

$$\left\{\begin{array}{l} \text{Find } u := u(\alpha) \in H^1(\Omega(\alpha)) \quad \text{such that} \\ \displaystyle\int_{\Omega(\alpha)} (\nabla u \cdot \nabla v + uv)\, dx = \int_{\Omega(\alpha)} fv\, dx + \int_{\Omega(\alpha)} \operatorname{curl} G \cdot \nabla v\, dx \quad \forall v \in H^1(\Omega(\alpha)). \end{array}\right. \tag{$\mathcal{P}_G(\alpha)$}$$

Since this formulation contains only integrals over $\Omega(\alpha)$, assumption $(\mathcal{B}1)$ can be verified *without changing* the system $\widetilde{\mathcal{O}}$.

We now pass to an approximation of $(\mathbb{P})$. First we specify constraints, defining U^{ad} (see (2.97)).

CONVENTION: *In all examples here and in the next subsections only the volume of $\Omega(\alpha)$ will be prescribed:*

$$U^{ad} = \{\alpha \in \widetilde{U}^{ad} \mid \operatorname{meas} \Omega(\alpha) = \gamma\},$$

where $\gamma > 0$ is a given constant.

The family $\mathcal{O}$ characterized by U^{ad} will be approximated by the same discrete systems $\{\mathcal{O}_\varkappa\}$ and $\{\mathcal{O}_{\varkappa h}\}$ as in Section 2.2 (see (2.49), (2.51)). Also all the other symbols introduced there keep the same meaning with the following minor change concerning the definition of $V_h(s_\varkappa)$, $s_\varkappa \in U_\varkappa^{ad}$:

$$V_h(s_\varkappa) = \left\{ v_h \in C(\overline{\Omega_h(s_\varkappa)}) \mid v_{h|T} \in P_1(T)\ \forall T \in \mathcal{T}(h, s_\varkappa) \right\},$$

where $\{\mathcal{T}(h, s_\varkappa)\}$, $s_\varkappa \in U_\varkappa^{ad}$, is a family of triangulations of $\overline{\Omega(r_h s_\varkappa)}$ satisfying $(\mathcal{T}1)$–$(\mathcal{T}3)$ of Section 2.2. Recall that $\Omega_h(s_\varkappa)$ stands for $\Omega(r_h s_\varkappa)$ with the given triangulation $\mathcal{T}(h, s_\varkappa)$.

For any $s_\varkappa \in U_\varkappa^{ad}$ we define the following:

$$\left\{ \begin{array}{l} \text{Find } u_h := u_h(s_\varkappa) \in V_h(s_\varkappa) \text{ such that} \\ \displaystyle\int_{\Omega_h(s_\varkappa)} (\nabla u_h \cdot \nabla v_h + u_h v_h)\, dx = \int_{\Omega_h(s_\varkappa)} f v_h\, dx \quad \forall v_h \in V_h(s_\varkappa). \end{array} \right. \tag{$\mathcal{P}_h(s_\varkappa)$}$$

For $h, \varkappa > 0$ fixed, the approximation of $(\mathbb{P})$ reads as follows:

$$\left\{ \begin{array}{l} \text{Find } s_\varkappa^* \in U_\varkappa^{ad} \text{ such that} \\ J(u_h(s_\varkappa^*)) \le J(u_h(s_\varkappa)) \quad \forall s_\varkappa \in U_\varkappa^{ad}, \end{array} \right. \tag{$\mathbb{P}_{h\varkappa}$}$$

where $u_h(s_\varkappa) \in V_h(s_\varkappa)$ solves $(\mathcal{P}_h(s_\varkappa))$. Arguing just as in Theorem 2.6 of Section 2.2 one can prove that $(\mathbb{P}_{h\varkappa})$ has at least one solution $s_\varkappa^*$ for any $h, \varkappa > 0$.

The convergence analysis will be based again on the abstract convergence theory presented in Section 2.4. As in Section 2.4 we shall consider the mesh parameter h to be a function of $\varkappa$; i.e., $h := h(\varkappa)$ such that

$$h \to 0+ \iff \varkappa \to 0+.$$

We shall verify assumptions $(\mathcal{B}1)_\varkappa$ and $(\mathcal{B}2)_\varkappa$ of Section 2.4. Assumption $(\mathcal{B}1)_\varkappa$ follows from (iii) and (iv) of Lemma 2.10.

Lemma 2.18. *(Verification of $(\mathcal{B}2)_\varkappa$.) For any sequence $\{(s_\varkappa, u_h(s_\varkappa))\}$, where $s_\varkappa \in U_\varkappa^{ad}$ and $u_h(s_\varkappa) \in V_h(s_\varkappa)$ solves $(\mathcal{P}_h(s_\varkappa))$, $h \to 0+$, there exist its subsequence and elements $\alpha \in U^{ad}$, $u \in H^1(\widehat{\Omega})$ such that*

$$s_\varkappa \rightrightarrows \alpha, \quad r_h s_\varkappa \rightrightarrows \alpha \quad \text{in } [0,1]; \tag{2.109}$$

$$\widetilde{u}_h(s_\varkappa) \rightharpoonup u \quad \text{in } H^1(\widehat{\Omega}), \quad h, \varkappa \to 0+. \tag{2.110}$$

In addition, $u(\alpha) := u_{|\Omega(\alpha)}$ solves $(\mathcal{P}(\alpha))$.

Proof. The existence of $\{s_\varkappa\}$, $\{r_h s_\varkappa\}$, $h, \varkappa \to 0+$, satisfying (2.109) for some $\alpha \in U^{ad}$ follows from (ii) and (iv) of Lemma 2.10. Since the system $\{\mathcal{O}_{\varkappa h}\}$, $\varkappa, h > 0$, possesses the uniform ε-cone property with the same $\varepsilon > 0$ as $\widetilde{\mathcal{O}}$ (i.e., independent of $h, \varkappa > 0$) we know that

$$\exists c = \text{const.} > 0 : \quad \|\widetilde{u}_h(s_\varkappa)\|_{1,\widehat{\Omega}} \le c \quad \forall h, \varkappa > 0. \tag{2.111}$$

Therefore (2.110) holds for an appropriate subsequence of $\{\widetilde{u}_h(s_\varkappa)\}$ and for an element $u \in H^1(\widehat{\Omega})$. Let us prove that u is such that $u(\alpha) := u_{|\Omega(\alpha)}$ solves $(\mathcal{P}(\alpha))$.

Let $\Xi(r_h s_\varkappa) = \widehat{\Omega} \setminus \overline{\Omega}(r_h s_\varkappa)$ and construct another family $\{\widehat{\mathcal{T}}(h, s_\varkappa)\}$ of triangulations of $\overline{\Xi(r_h s_\varkappa)}$ satisfying $(\mathcal{T}1)$–$(\mathcal{T}3)$ of Section 2.2. The union of $\mathcal{T}(h, s_\varkappa)$ and $\widehat{\mathcal{T}}(h, s_\varkappa)$ defines a regular triangulation of $\overline{\widehat{\Omega}}$. Let $v \in C^\infty(\overline{\widehat{\Omega}})$ be given and $\pi_h v$ be the piecewise linear Lagrange interpolant of v on $\mathcal{T}(h, s_\varkappa) \cup \widehat{\mathcal{T}}(h, s_\varkappa)$. From the classical approximation results and the regularity assumptions on $\{\mathcal{T}(h, s_\varkappa)\}$ and $\{\widehat{\mathcal{T}}(h, s_\varkappa)\}$ it follows that

$$\|v - \pi_h v\|_{1,\widehat{\Omega}} \le ch \to 0, \ h \to 0+, \tag{2.112}$$

with a constant $c > 0$ that does not depend on $h, \varkappa > 0$.

The definition of $(\mathcal{P}_h(s_\varkappa))$ yields

$$\int_{\widehat{\Omega}} \chi_{h\varkappa}(\nabla \widetilde{u}_h \cdot \nabla \pi_h v + \widetilde{u}_h \pi_h v)\, dx = \int_{\widehat{\Omega}} \chi_{h\varkappa} f \pi_h v\, dx, \tag{2.113}$$

where $\chi_{h\varkappa}$ is the characteristic function of $\Omega_h(s_\varkappa)$. Letting $h, \varkappa \to 0+$ in (2.113) and using (2.110), (2.112), and the fact that $\chi_{h\varkappa} \to \chi$ in $L^2(\widehat{\Omega})$, $\varkappa, h \to 0+$, where χ is the characteristic function of $\Omega(\alpha)$, we arrive at

$$\int_{\widehat{\Omega}} \chi(\nabla u \cdot \nabla v + uv)\, dx = \int_{\widehat{\Omega}} \chi f v\, dx. \tag{2.114}$$

Since $C^\infty(\overline{\widehat{\Omega}})$ is dense in $H^1(\widehat{\Omega})$, (2.114) holds for any $v \in H^1(\widehat{\Omega})$; i.e., $u(\alpha)$ solves $(\mathcal{P}(\alpha))$. $\square$

Remark 2.18. Using the same approach as in Lemma 2.17 one can show that

$$u_h(s_\varkappa) \to u(\alpha) \quad \text{in } H^1_{\text{loc}}(\Omega(\alpha)), \quad h, \varkappa \to 0+.$$

Since assumption $(\mathcal{B}3)_\varkappa$ is also satisfied for J given by (2.99), Theorem 2.11 applies. We obtain the following result.

Theorem 2.14. *For any sequence $\{(s^*_\varkappa, u_h(s^*_\varkappa))\}$ of optimal pairs of $(\mathbb{P}_{h\varkappa})$, $h \to 0+$, there exists its subsequence such that*

$$\begin{cases} s^*_\varkappa \rightrightarrows \alpha^* & \text{in } [0,1], \\ \widetilde{u}_h(s^*_\varkappa) \rightharpoonup u^* & \text{in } H^1(\widehat{\Omega}), \quad h, \varkappa \to 0+. \end{cases} \tag{2.115}$$

In addition, $(\alpha^*, u^*{}_{|\Omega(\alpha^*)})$ is an optimal pair of $(\mathbb{P})$. Any accumulation point of $\{(s^*_\varkappa, u_h(s^*_\varkappa))\}$ in the sense of (2.115) *possesses this property.*

Remark 2.19. We also have that

$$u_h(s^*_\varkappa) \to u(\alpha^*) \quad \text{in } H^1_{loc}(\Omega(\alpha^*)), \quad h, \varkappa \to 0+,$$

where $u(\alpha^*) := u^*{}_{|\Omega(\alpha^*)}$ solves $(\mathcal{P}(\alpha^*))$.

2.5.3 Shape optimization for state problems involving mixed boundary conditions

We have so far confined ourselves to state problems with boundary conditions of one type, prescribed on the *whole* boundary. In this section we shall analyze a more general situation, when different types of boundary conditions are considered on boundaries of designed domains. In addition, differential operators, defining PDEs, will not contain an absolute term. We pay special attention to this case because the proof of the uniform boundedness of solutions with respect to domains, which is a key point of the existence analysis, is no longer so straightforward.

Let U^{ad}, $\widetilde{U}^{ad}$, $\mathcal{O}$, and $\widetilde{\mathcal{O}}$ be the same as in Subsection 2.5.2; i.e., the shape of any $\Omega(\alpha) \in \widetilde{\mathcal{O}}$, $\alpha \in \widetilde{U}^{ad}$, is illustrated in Figure 2.2. In any $\Omega(\alpha) \in \widetilde{\mathcal{O}}$ we shall consider the following Dirichlet–Neumann boundary value problem:

$$\begin{cases} -\Delta u(\alpha) = f & \text{in } \Omega(\alpha), \\ u(\alpha) = 0 & \text{on } \Gamma_1(\alpha), \\ \dfrac{\partial u(\alpha)}{\partial \nu} = 0 & \text{on } \Gamma_2 = \partial\Omega(\alpha) \setminus \overline{\Gamma_1(\alpha)}, \end{cases} \tag{$\mathcal{P}'(\alpha)$}$$

where $f \in L^2(\widehat{\Omega})$ and $\Gamma_1(\alpha) = \{(x_1, 0) \mid x_1 \in]0, \alpha(0)[\}$ is the bottom of $\Omega(\alpha)$.

The weak formulation of $(\mathcal{P}'(\alpha))$ is defined as follows:

$$\begin{cases} \text{Find } u := u(\alpha) \in V(\alpha) \text{ such that} \\ \displaystyle\int_{\Omega(\alpha)} \nabla u \cdot \nabla v \, dx = \int_{\Omega(\alpha)} f v \, dx \quad \forall v \in V(\alpha), \end{cases} \tag{$\mathcal{P}(\alpha)$}$$

where

$$V(\alpha) = \{v \in H^1(\Omega(\alpha)) \mid v = 0 \text{ on } \Gamma_1(\alpha)\}. \tag{2.116}$$

Let the cost functional be given by (2.99) and consider the respective optimal shape design problem $(\mathbb{P})$ on $\mathcal{O}$ with the mixed Dirichlet–Neumann boundary value problem $(\mathcal{P}(\alpha))$ as the state equation.

The first step of the existence analysis consists of proving that the solutions $u(\alpha)$ to $(\mathcal{P}(\alpha))$ are bounded uniformly with respect to $\alpha \in \widetilde{U}^{ad}$:

$$\exists\, c = \text{const.} > 0 : \quad \|u(\alpha)\|_{1,\Omega(\alpha)} \le c \quad \forall \alpha \in \widetilde{U}^{ad}.$$

Inserting $v := u(\alpha)$ into $(\mathcal{P}(\alpha))$ we obtain

$$|u(\alpha)|^2_{1,\Omega(\alpha)} = \int_{\Omega(\alpha)} f u(\alpha) \, dx \le \|f\|_{0,\widehat{\Omega}} \|u(\alpha)\|_{1,\Omega(\alpha)}.$$

From the generalized Friedrichs inequality in $V(\alpha)$ we have

$$\beta \|u(\alpha)\|^2_{1,\Omega(\alpha)} \le |u(\alpha)|^2_{1,\Omega(\alpha)} \le \|f\|_{0,\widehat{\Omega}} \|u(\alpha)\|_{1,\Omega(\alpha)}, \tag{2.117}$$

where β is a positive constant. From this, however, one *cannot claim* that $\{\|u(\alpha)\|_{1,\Omega(\alpha)}\}$ is bounded! The reason is very simple: the constant β in (2.117) *might depend* on $\alpha \in \widetilde{U}^{ad}$. Therefore we first prove the following lemma.

Lemma 2.19. *There exists a constant $\beta > 0$ such that*

$$|v|_{1,\Omega(\alpha)} \geq \beta \|v\|_{1,\Omega(\alpha)} \quad \forall v \in V(\alpha),\ \forall \alpha \in \widetilde{U}^{ad}, \tag{2.118}$$

where $\widetilde{U}^{ad}$, $V(\alpha)$ are given by (2.96), (2.116), *respectively.*

Proof. (By contradiction.) Let, for any $k \in \mathbb{N}$, there exist $v_k \in V(\alpha_k)$, $\alpha_k \in \widetilde{U}^{ad}$, such that

$$|v_k|_{1,\Omega(\alpha_k)} \leq \frac{1}{k}\|v_k\|_{1,\Omega(\alpha_k)}. \tag{2.119}$$

We may assume that $\|v_k\|_{1,\Omega(\alpha_k)} = 1\ \forall k \in \mathbb{N}$ and, in addition, passing to a subsequence of $\{\alpha_k\}$ if necessary, that

$$\alpha_k \rightrightarrows \alpha \quad \text{in } [0,1], \quad k \to \infty, \tag{2.120}$$

for some $\alpha \in \widetilde{U}^{ad}$. Let $\widetilde{v}_k$ be the uniform extension of v_k from $\Omega(\alpha_k)$ to $\widehat{\Omega}$. Since the sequence $\{\widetilde{v}_k\}$ is bounded in $H^1(\widehat{\Omega})$ we may also assume that

$$\widetilde{v}_k \rightharpoonup \overline{v} \quad \text{in } H^1(\widehat{\Omega}), \quad k \to \infty. \tag{2.121}$$

Next we shall show that

$$|\overline{v}|_{1,\Omega(\alpha)} \leq \liminf_{k\to\infty} |v_k|_{1,\Omega(\alpha_k)} = 0. \tag{2.122}$$

Indeed, for any $m \in \mathbb{N}$ we define the set $G_m(\alpha)$ by (2.107). From (2.121) we also have that

$$\widetilde{v}_k \rightharpoonup \overline{v} \quad \text{in } H^1(G_m(\alpha)), \quad k \to \infty,$$

for any $m \in \mathbb{N}$ fixed so that

$$|\overline{v}|_{1,G_m(\alpha)} \leq \liminf_{k\to\infty} |v_k|_{1,G_m(\alpha)} \leq \liminf_{k\to\infty} |v_k|_{1,\Omega(\alpha_k)}, \tag{2.123}$$

making use of weak lower semicontinuity of the seminorm $|\cdot|_{1,G_m(\alpha)}$. Letting $m \to \infty$ in (2.123) and using (2.119) we arrive at (2.122). From (2.122) it follows that $\overline{v}$ is constant in $\Omega(\alpha)$ and, since $\overline{v} = 0$ on $\Gamma_1(\alpha)$, we have $\overline{v} = 0$ in $\Omega(\alpha)$. On the other hand,

$$\|v_k\|^2_{1,\Omega(\alpha_k)} = |v_k|^2_{1,\Omega(\alpha_k)} + \|v_k\|^2_{0,\Omega(\alpha_k)} = 1 \quad \forall k \in \mathbb{N}.$$

From (2.119) we see that for $k_0 \in \mathbb{N}$ large enough,

$$\|v_k\|^2_{0,\Omega(\alpha_k)} \geq \frac{1}{2}, \quad k \geq k_0.$$

But at the same time

$$\|\overline{v}\|_{0,\Omega(\alpha)} = \lim_{k\to\infty} \|v_k\|_{0,\Omega(\alpha_k)} \geq \frac{1}{2},$$

which contradicts $\overline{v} = 0$ in $\Omega(\alpha)$. $\square$

Remark 2.20. A similar approach can be used to prove the uniform property of other equivalent norms with respect to a class of domains satisfying the uniform ε-cone property (see [Has02]).

With this result at our disposal it is now easy to prove the following result.

Lemma 2.20. *(Verification of ($\mathcal{B}1$).) For any sequence $\{(\alpha_n, u_n)\}$, where $\alpha_n \in \widetilde{U}^{ad}$ and $u_n := u(\alpha_n) \in V(\alpha_n)$ is a solution of $(\mathcal{P}(\alpha_n))$, $n \to \infty$, there exist its subsequence and elements $\alpha \in \widetilde{U}^{ad}$, $u \in H^1(\widehat{\Omega})$ such that*

$$\begin{aligned} \alpha_n &\rightrightarrows \alpha \quad \text{in } [0,1], \\ \widetilde{u}_n &\rightharpoonup u \quad \text{in } H^1(\widehat{\Omega}), \qquad n \to \infty. \end{aligned}$$

In addition, $u(\alpha) := u_{|\Omega(\alpha)}$ solves $(\mathcal{P}(\alpha))$.

Proof. We may assume that $\alpha_n \rightrightarrows \alpha$ in $[0, 1]$ and $\alpha \in \widetilde{U}^{ad}$. From Lemma 2.19 and (2.117) it follows that $\{\|u_n\|_{1,\Omega(\alpha_n)}\}$ is bounded. Passing again to a new subsequence if necessary we have that

$$\widetilde{u}_n \rightharpoonup u \quad \text{in } H^1(\widehat{\Omega}).$$

Let us show that $u(\alpha) := u_{|\Omega(\alpha)}$ solves $(\mathcal{P}(\alpha))$. It is readily seen that $u = 0$ on $\Gamma_1(\alpha)$ so that $u(\alpha) \in V(\alpha)$. Let $v \in V(\alpha)$ be given and $\widetilde{v}$ be its uniform extension to $\widehat{\Omega}$. Then one can find a sequence $\{v_j\}$, $v_j \in C^\infty(\overline{\widehat{\Omega}})$, such that

$$\delta(j) := \text{dist}(\text{supp } v_j, \Gamma_1(\alpha)) > 0 \quad \forall j \in \mathbb{N}$$

and

$$v_j \to \widetilde{v} \quad \text{in } H^1(\widehat{\Omega}), \quad j \to \infty. \tag{2.124}$$

In other words, any function $\widetilde{v} \in H^1(\widehat{\Omega})$ such that $\widetilde{v}_{|\Omega(\alpha)} \in V(\alpha)$ can be approximated by functions from $C^\infty(\overline{\widehat{\Omega}})$ vanishing in the vicinity of $\Gamma_1(\alpha)$.

Let $j \in \mathbb{N}$ be fixed. Since $\alpha_n \rightrightarrows \alpha$ in $[0, 1]$ we see that $v_{j|\Omega(\alpha_n)} \in V(\alpha_n)$, provided that n is large enough. This restriction can be used as a test function in $(\mathcal{P}(\alpha_n))$ written in the form

$$\int_{\widehat{\Omega}} \chi_n \nabla \widetilde{u}_n \cdot \nabla v_j \, dx = \int_{\widehat{\Omega}} \chi_n f v_j \, dx, \tag{2.125}$$

where χ_n is the characteristic function of $\Omega(\alpha_n)$. Passing to the limit first with $n \to \infty$, then with $j \to \infty$, in (2.125) (using (2.124)), we obtain

$$\int_{\widehat{\Omega}} \chi \nabla u \cdot \nabla \widetilde{v} \, dx = \int_{\widehat{\Omega}} \chi f \widetilde{v} \, dx,$$

where χ is the characteristic function of the limit domain $\Omega(\alpha)$. Thus, $u(\alpha)$ solves $(\mathcal{P}(\alpha))$. $\square$

Remark 2.21. Using the same approach as in Lemma 2.17 one can show that

$$u_n \to u(\alpha) \quad \text{in } H^1_{loc}(\Omega(\alpha)), \qquad n \to \infty,$$

where $u_n, u(\alpha)$ solve $(\mathcal{P}(\alpha_n))$, $(\mathcal{P}(\alpha))$, respectively.

Since the cost functional J is also continuous, as follows from Lemma 2.16, the statement of Theorem 2.10 applies: problem $(\mathbb{P})$ has a solution.

The discretization and convergence analysis for $(\mathbb{P})$ follow the same steps as the Neumann state problems in the last subsection with appropriate modifications reflecting the mixed Dirichlet–Neumann boundary conditions prescribed on $\partial\Omega(\alpha)$. It is left as an exercise to verify that assumptions $(\mathcal{B}1)_\varkappa$–$(\mathcal{B}3)_\varkappa$ of Section 2.4 are satisfied. Therefore solutions to discrete optimal shape design problems are close in the sense of subsequences to solutions of the original, continuous setting in the sense of Theorem 2.11.

Remark 2.22. If the boundary $\partial\Omega(\alpha)$ were decomposed into $\Gamma_1(\alpha)$ and $\Gamma_2(\alpha)$ in a more general way than considered here, yet another assumption would be necessary: the decomposition of $\partial\Omega(\alpha)$ into $\Gamma_1(\alpha)$ and $\Gamma_2(\alpha)$ has to depend *continuously* on $\alpha \in \widetilde{U}^{ad}$, and the triangulations used for the construction of $V_h(s_\varkappa)$ have to satisfy assumption $(\mathcal{T}4)$ of Section 2.2.

2.5.4 Shape optimization of systems governed by variational inequalities

In the mechanics of solids we often meet problems whose mathematical models lead to variational inequalities. Contact problems of deformable bodies are examples of this type. Shape optimization of structures governed by variational inequalities is an important part of the topic. It turns out that there is yet another feature making these problems more involved: namely, the mapping that associates with any shape Ω a solution of the respective state inequality is *not differentiable* in general. This fact considerably restricts the use of classical, gradient type minimization methods. This subsection treats the case of scalar variational inequalities used as state problems. Shape optimization in contact problems will be briefly mentioned in the next section.

Let $\widetilde{U}^{ad}$, U^{ad} be defined by (2.96), (2.97), respectively, and the families $\widetilde{\mathcal{O}}$, $\mathcal{O}$ be the same as before; i.e., shapes $\Omega(\alpha) \in \widetilde{\mathcal{O}}$, $\alpha \in \widetilde{U}^{ad}$, are as in Figure 2.2. The boundary $\partial\Omega(\alpha)$ consists of two parts: $\Gamma(\alpha)$ is the graph of a function $\alpha \in \widetilde{U}^{ad}$ and $\Gamma_1(\alpha) = \partial\Omega(\alpha) \setminus \overline{\Gamma(\alpha)}$. In any $\Omega(\alpha)$, $\alpha \in \widetilde{U}^{ad}$, we define the following unilateral boundary value problem:

$$\begin{cases} -\Delta u(\alpha) = f & \text{in } \Omega(\alpha), \\ u(\alpha) = 0 & \text{on } \Gamma_1(\alpha), \\ u(\alpha) \geq 0, \ \dfrac{\partial u(\alpha)}{\partial \nu} \geq 0, \ u(\alpha)\dfrac{\partial u(\alpha)}{\partial \nu} = 0 & \text{on } \Gamma(\alpha), \end{cases} \tag{$\mathcal{P}(\alpha)'$}$$

where $f \in L^2(\widehat{\Omega})$.

To give a weak form of $(\mathcal{P}(\alpha)')$ we introduce the space

$$V(\alpha) = \{v \in H^1(\Omega(\alpha)) \mid v = 0 \text{ on } \Gamma_1(\alpha)\}$$

and its closed, convex subset

$$K(\alpha) = \{v \in V(\alpha) \mid v \geq 0 \text{ on } \Gamma(\alpha)\}.$$

The weak form of $(\mathcal{P}(\alpha)')$ reads as follows:

$$\begin{cases} \text{Find } u := u(\alpha) \in K(\alpha) \text{ such that} \\ \displaystyle\int_{\Omega(\alpha)} \nabla u \cdot \nabla(v-u)\,dx \geq \int_{\Omega(\alpha)} f(v-u)\,dx \quad \forall v \in K(\alpha). \end{cases} \tag{$\mathcal{P}(\alpha)$}$$

It is well known that $(\mathcal{P}(\alpha))$ has a unique solution $u(\alpha)$ for any $\alpha \in \widetilde{U}^{ad}$ (see Appendix A).

Remark 2.23. Since the bilinear form $a(u, v) = \int_{\Omega(\alpha)} \nabla u \cdot \nabla v\,dx$ is *symmetric*, problem $(\mathcal{P}(\alpha))$ can be equivalently formulated as the following minimization problem:

$$\begin{cases} \text{Find } u(\alpha) \in K(\alpha) \text{ such that} \\ E_\alpha(u(\alpha)) \leq E_\alpha(v) \quad \forall v \in K(\alpha), \end{cases} \tag{$\widetilde{\mathcal{P}}(\alpha)$}$$

where

$$E_\alpha(v) = \frac{1}{2}\int_{\Omega(\alpha)} \nabla v \cdot \nabla v\,dx - \int_{\Omega(\alpha)} f v\,dx.$$

In other words, $u(\alpha)$ solves $(\mathcal{P}(\alpha))$ iff $u(\alpha)$ is a solution of $(\widetilde{\mathcal{P}}(\alpha))$. Formulation $(\widetilde{\mathcal{P}}(\alpha))$ is important from a computational point of view.

We shall analyze the following optimal shape design problem:

$$\begin{cases} \text{Find } \alpha^* \in U^{ad} \text{ such that} \\ J(u(\alpha^*)) \leq J(u(\alpha)) \quad \forall \alpha \in U^{ad}, \end{cases} \tag{$\mathbb{P}$}$$

where J is given by (2.99) and $u(\alpha)$ solves $(\mathcal{P}(\alpha))$ (or $(\widetilde{\mathcal{P}}(\alpha))$).

To prove the existence of solutions to $(\mathbb{P})$ we need the following continuity type result for the trace mapping.

Lemma 2.21. *Let $\{(\alpha_n, y_n)\}$, $\alpha_n \in \widetilde{U}^{ad}$, $y_n \in H^1(\widehat{\Omega})$, be such that*

$$\alpha_n \rightrightarrows \alpha \quad \text{in } [0,1],$$
$$y_n \rightharpoonup y \quad \text{in } H^1(\widehat{\Omega}), \quad n \to \infty,$$

for some $\alpha \in \widetilde{U}^{ad}$ and $y \in H^1(\widehat{\Omega})$. Then

$$z_n \to z \quad \text{in } L^2(]0,1[), \quad n \to \infty, \tag{2.126}$$

where $z_n := y_{n|\Gamma(\alpha_n)} \circ \alpha_n$ and $z := y_{|\Gamma(\alpha)} \circ \alpha$.

Proof. It holds that

$$\int_0^1 |z_n - z|^2\,dx_2 \leq 2\left\{\int_0^1 |y_n \circ \alpha_n - y_n \circ \alpha|^2\,dx_2 + \int_0^1 |y_n \circ \alpha - y \circ \alpha|^2\,dx_2\right\}. \tag{2.127}$$

Since $|\alpha'| \le L_0$ a.e. in $]0,1[$ for any $\alpha \in \widetilde{U}^{ad}$, the second integral on the right of (2.127) can be bounded from above by the curvilinear integral over $\Gamma(\alpha)$:

$$\int_0^1 |y_n \circ \alpha - y \circ \alpha|^2 \, dx_2 \le \int_{\Gamma(\alpha)} |y_n - y|^2 \, ds \to 0, \quad n \to \infty, \tag{2.128}$$

making use of the compactness of the embedding of $H^1(\Omega(\alpha))$ into $L^2(\Gamma(\alpha))$ (see Appendix A) and weak convergence of $\{y_n\}$ to y in $H^1(\widehat{\Omega})$. Further,

$$|y_n \circ \alpha_n - y_n \circ \alpha|^2 = \left| \int_{\alpha_n}^{\alpha} \frac{\partial y_n}{\partial x_1} \, dx_1 \right|^2 \le \max_{x_2 \in [0,1]} |\alpha_n(x_2) - \alpha(x_2)| \int_0^{\alpha_{\max}} \left| \frac{\partial y_n}{\partial x_1} \right|^2 dx_1.$$

Integrating this inequality over $[0, 1]$ we obtain

$$\int_0^1 |y_n \circ \alpha_n - y_n \circ \alpha|^2 \, dx_2 \le \|\alpha_n - \alpha\|_{C([0,1])} \|y_n\|^2_{1,\widehat{\Omega}} \to 0+$$

by virtue of our assumptions. From this and (2.128) we arrive at the assertion of the lemma. $\square$

COROLLARY 2.1. *Let $\{(\alpha_n, y_n)\}$ be a sequence from Lemma 2.21 and, in addition, let $y_{n|\Omega(\alpha_n)} \in K(\alpha_n)$. Then the limit y of $\{y_n\}$ is such that $y_{|\Omega(\alpha)} \in K(\alpha)$. Indeed, from (2.126) it follows that there is a subsequence of $\{z_n\}$ tending to z a.e. in $]0,1[$. If $y_n \in K(\alpha_n)$, then all z_n are nonnegative in $]0,1[$ and so is the limit z. Hence $y_{|\Omega(\alpha)} \in K(\alpha)$.*

We shall now verify assumptions ($\mathcal{B}1$) and ($\mathcal{B}2$) of Section 2.4, guaranteeing the existence of solutions to ($\mathbb{P}$).

LEMMA 2.22. (*Verification of* ($\mathcal{B}1$).) *For any sequence $\{(\alpha_n, u_n)\}$, where $\alpha_n \in \widetilde{U}^{ad}$ and $u_n := u(\alpha_n) \in K(\alpha_n)$ is a solution of $(\mathcal{P}(\alpha_n))$, $n \to \infty$, there exist its subsequence and elements $\alpha \in \widetilde{U}^{ad}$, $u \in H^1(\widehat{\Omega})$ such that*

$$\begin{aligned} \alpha_n &\rightrightarrows \alpha \quad \text{in } [0,1], \\ \widetilde{u}_n &\rightharpoonup u \quad \text{in } H^1(\widehat{\Omega}), \quad n \to \infty. \end{aligned}$$

In addition, $u(\alpha) := u_{|\Omega(\alpha)}$ solves $(\mathcal{P}(\alpha))$.

Proof. As usual we may suppose that

$$\alpha_n \rightrightarrows \alpha \quad \text{in } [0,1] \tag{2.129}$$

and $\alpha \in \widetilde{U}^{ad}$. Inserting $v := 0 \in K(\alpha_n)$ $\forall n \in \mathbb{N}$ into $(\mathcal{P}(\alpha_n))$ we obtain

$$|u_n|^2_{1,\Omega(\alpha_n)} \le (f, u_n)_{0,\Omega(\alpha_n)} \le \|f\|_{0,\widehat{\Omega}} \|u_n\|_{1,\Omega(\alpha_n)}.$$

This and the generalized Friedrichs inequality with a constant $\beta > 0$ independent of $\alpha \in \widetilde{U}^{ad}$ (see (2.118)) imply the boundedness of $\{\|u_n\|_{1,\Omega(\alpha_n)}\}$ and hence also of $\{\|\widetilde{u}_n\|_{1,\widehat{\Omega}}\}$. One can pass to a subsequence of $\{\widetilde{u}_n\}$ weakly converging to an element $u \in H^1(\widehat{\Omega})$:

$$\widetilde{u}_n \rightharpoonup u \quad \text{in } H^1(\widehat{\Omega}). \tag{2.130}$$

We want to prove that $u(\alpha) := u_{|\Omega(\alpha)}$ solves $(\mathcal{P}(\alpha))$. First of all $u(\alpha) \in K(\alpha)$, as follows from (2.129), (2.130), and Corollary 2.1.

Let $v \in K(\alpha)$ be given. It is easy to show that there exists an extension v^* of v from $\Omega(\alpha)$ to $\widehat{\Omega}$ such that $v^* \in H_0^1(\widehat{\Omega})$. Since the trace of v^* on $\partial\widehat{\Omega} \cup \Gamma(\alpha)$ is nonnegative, one can construct a nonnegative extension φ of v^* from $\partial\widehat{\Omega} \cup \Gamma(\alpha)$ into $\widehat{\Omega}$; i.e., a function $\varphi \in H_0^1(\widehat{\Omega})$ exists such that $\varphi \geq 0$ a.e. in $\widehat{\Omega}$ and $\varphi_{|\partial\widehat{\Omega}\cup\Gamma(\alpha)} = v^*{}_{|\partial\widehat{\Omega}\cup\Gamma(\alpha)}$ (see [Neč67, p. 103]). This makes it possible to express v^* as the sum

$$v^* = \varphi + w,$$

where $w \in H_0^1(\widehat{\Omega})$ and $w = 0$ on $\Gamma(\alpha)$ or, equivalently, $w_{|\Omega(\alpha)} \in H_0^1(\Omega(\alpha))$, $w_{|\Xi(\alpha)} \in H_0^1(\Xi(\alpha))$, where $\Xi(\alpha) := \widehat{\Omega} \setminus \overline{\Omega(\alpha)}$. The function w can be approximated by a sequence $\{w_j\}$, $w_j \in C_0^\infty(\widehat{\Omega})$, such that

$$w_j \to w \quad \text{in } H_0^1(\widehat{\Omega}),\ j \to \infty, \tag{2.131}$$

$$w_{j|\Omega(\alpha)} \in C_0^\infty(\Omega(\alpha)),\ w_{j|\Xi(\alpha)} \in C_0^\infty(\Xi(\alpha)) \quad \text{for any } j \in \mathbb{N}. \tag{2.132}$$

Let $v_j := \varphi + w_j$. Then

$$v_j \to v^* \quad \text{in } H_0^1(\widehat{\Omega}), \quad j \to \infty, \tag{2.133}$$

by virtue of (2.131). Let $j \in \mathbb{N}$ be fixed. Then from (2.129), (2.132), and the definition of v_j we see that

$$v_{j|\Gamma(\alpha_n)} = \varphi_{|\Gamma(\alpha_n)} \geq 0$$

holds for any n large enough (say $n \geq n_0 := n_0(j)$) meaning that $v_{j|\Omega(\alpha_n)} \in K(\alpha_n)\ \forall n \geq n_0$. This enables us to substitute v_j into $(\mathcal{P}(\alpha_n))$, $n \geq n_0$:

$$\int_{\widehat{\Omega}} \chi_n \nabla\widetilde{u}_n \cdot \nabla(v_j - \widetilde{u}_n)\,dx \geq \int_{\widehat{\Omega}} \chi_n f(v_j - \widetilde{u}_n)\,dx,$$

where χ_n is the characteristic function of $\Omega(\alpha_n)$. Hence

$$\limsup_{n\to\infty} \int_{\widehat{\Omega}} \chi_n \nabla\widetilde{u}_n \cdot \nabla(v_j - \widetilde{u}_n)\,dx \geq \limsup_{n\to\infty} \int_{\widehat{\Omega}} \chi_n f(v_j - \widetilde{u}_n)\,dx. \tag{2.134}$$

It has already been proven that

$$\lim_{n\to\infty} \int_{\widehat{\Omega}} \chi_n f(v_j - \widetilde{u}_n)\,dx = \int_{\widehat{\Omega}} \chi f(v_j - u)\,dx, \tag{2.135}$$

$$\lim_{n\to\infty} \int_{\widehat{\Omega}} \chi_n \nabla\widetilde{u}_n \cdot \nabla v_j\,dx = \int_{\widehat{\Omega}} \chi \nabla u \cdot \nabla v_j\,dx, \tag{2.136}$$

$$\liminf_{n\to\infty} \int_{\widehat{\Omega}} \chi_n \nabla\widetilde{u}_n \cdot \nabla\widetilde{u}_n\,dx \geq \int_{\widehat{\Omega}} \chi \nabla u \cdot \nabla u\,dx, \tag{2.137}$$

where χ stands for the characteristic function of $\Omega(\alpha)$ (for the proof of (2.137) see the inequality in (2.122)). Using (2.135)–(2.137) in (2.134) we obtain that

$$\int_{\widehat{\Omega}} \chi \nabla u \cdot \nabla(v_j - u)\,dx \geq \int_{\widehat{\Omega}} \chi f(v_j - u)\,dx$$

holds for any $j \in \mathbb{N}$. The limit passage for $j \to \infty$ gives

$$\int_{\widehat{\Omega}} \chi \nabla u \cdot \nabla(v^* - u)\, dx \geq \int_{\widehat{\Omega}} \chi f(v^* - u)\, dx,$$

taking into account (2.133). Since $v := v^*{}_{|\Omega(\alpha)} \in K(\alpha)$ is arbitrary, we may conclude that $u(\alpha)$ solves $(\mathcal{P}(\alpha))$. $\square$

Remark 2.24. Adapting the proof of Lemma 2.17 one can show that

$$u_n \to u(\alpha) \quad \text{in } H^1_{loc}(\Omega(\alpha)), \quad n \to \infty.$$

Since the cost functional J satisfies $(\mathcal{B}2)$ we have proved the following result.

Theorem 2.15. *Problem* $(\mathbb{P})$ *has a solution.*

We now turn to an approximation of $(\mathbb{P})$. As in the previous subsections we shall consider the constant area constraint in the definition of $\mathcal{O}$.

We first introduce two discrete systems $\{\mathcal{O}_\varkappa\}$ and $\{\mathcal{O}_{\varkappa h}\}$, $\varkappa, h \to 0+$, by (2.49) and (2.51), respectively, keeping the meaning of the other notation of Section 2.2. Let $\{\mathcal{T}(h, s_\varkappa)\}$, $s_\varkappa \in U^{ad}_\varkappa$, be a family of triangulations of $\overline{\Omega(r_h s_\varkappa)}$ satisfying assumptions $(\mathcal{T}1)$–$(\mathcal{T}3)$ of Section 2.2. Recall that $\Omega_h(s_\varkappa)$ is a polygonal approximation of $\Omega(s_\varkappa)$, $s_\varkappa \in U^{ad}_\varkappa$, with the given triangulation $\mathcal{T}(h, s_\varkappa)$. The curved part $\Gamma(s_\varkappa)$ of the boundary of $\Omega(s_\varkappa)$ is locally represented by piecewise quadratic Bézier functions, while the moving part $\Gamma(r_h s_\varkappa)$ of the boundary of $\Omega_h(s_\varkappa)$ is given by the graph of the piecewise linear Lagrange interpolant $r_h s_\varkappa$ of $s_\varkappa$. On any $\Omega_h(s_\varkappa)$ we define the following finite element space:

$$\begin{aligned} V_h(s_\varkappa) = \{v_h \in C(\overline{\Omega_h(s_\varkappa)})\ |v_{h|T} \in P_1(T)\ \forall T \in \mathcal{T}(h, s_\varkappa), \\ v_h = 0 \text{ on } \Gamma_1(s_\varkappa) = \partial\Omega_h(s_\varkappa) \setminus \overline{\Gamma(r_h s_\varkappa)}\} \end{aligned}$$

and its closed, convex subset

$$K_h(s_\varkappa) = \{v_h \in V_h(s_\varkappa) \mid v_h(A) \geq 0\ \forall A \in \mathcal{N}_h\},$$

where $\mathcal{N}_h$ is the set of all vertices of $T \in \mathcal{T}(h, s_\varkappa)$ from the *interior* of $\Gamma(r_h s_\varkappa)$.

Remark 2.25. Since $r_h s_\varkappa$ is piecewise linear in $[0, 1]$, any $v_h \in K_h(s_\varkappa)$ satisfies the unilateral boundary condition not only at all $A \in \mathcal{N}_h$ but also along the whole $\overline{\Gamma(r_h s_\varkappa)}$, meaning that $K_h(s_\varkappa) \subset K(r_h s_\varkappa)$.

On any $\Omega_h(s_\varkappa) \in \mathcal{O}_{\varkappa h}$ we define the discretization of the state problem as follows:

$$\left\{ \begin{aligned} &\text{Find } u_h := u_h(s_\varkappa) \in K_h(s_\varkappa) \text{ such that} \\ &\int_{\Omega_h(s_\varkappa)} \nabla u_h \cdot \nabla(v_h - u_h)\, dx \geq \int_{\Omega_h(s_\varkappa)} f(v_h - u_h)\, dx \quad \forall v_h \in K_h(s_\varkappa). \end{aligned} \right. \qquad (\mathcal{P}_h(s_\varkappa))$$

Finally, the approximation of $(\mathbb{P})$ reads as follows:

$$\begin{cases} \text{Find } s_\varkappa^* \in U_\varkappa^{ad} \text{ such that} \\ J(u_h(s_\varkappa^*)) \leq J(u_h(s_\varkappa)) \quad \forall s_\varkappa \in U_\varkappa^{ad}, \end{cases} \tag{$\mathbb{P}_{h\varkappa}$}$$

with $u_h(s_\varkappa) \in K_h(s_\varkappa)$ solving $(\mathcal{P}_h(s_\varkappa))$.

By virtue of assumptions $(\mathcal{T}1)$–$(\mathcal{T}3)$ it is easy to prove that $(\mathbb{P}_{h\varkappa})$ has a solution for any $h, \varkappa > 0$. Next we show that solutions to $(\mathbb{P}_{h\varkappa})$ and $(\mathbb{P})$ are close in the sense of subsequences in the sense of Theorem 2.11. To this end we shall verify assumptions $(\mathcal{B}1)_\varkappa$–$(\mathcal{B}3)_\varkappa$ of Section 2.4. The validity of $(\mathcal{B}1)_\varkappa$ and $(\mathcal{B}3)_\varkappa$ has been already established so that it remains to prove a lemma.

Lemma 2.23. *(Verification of $(\mathcal{B}2)_\varkappa$.) For any sequence $\{(s_\varkappa, u_h(s_\varkappa))\}$, where $s_\varkappa \in U_\varkappa^{ad}$ and $u_h(s_\varkappa) \in K_h(s_\varkappa)$ solves $(\mathcal{P}_h(s_\varkappa))$, $h \to 0+$, there exist its subsequence and elements $\alpha \in U^{ad}$, $u \in H^1(\widehat{\Omega})$ such that*

$$s_\varkappa \rightrightarrows \alpha, \quad r_h s_\varkappa \rightrightarrows \alpha \quad \text{in } [0,1], \tag{2.138}$$

$$\widetilde{u}_h(s_\varkappa) \rightharpoonup u \quad \text{in } H^1(\widehat{\Omega}), \quad h, \varkappa \to 0+. \tag{2.139}$$

In addition, $u(\alpha) := u_{|\Omega(\alpha)}$ solves $(\mathcal{P}(\alpha))$.

Proof. Equation (2.138) follows from (ii) and (iv) of Lemma 2.10. The boundedness of $\{\|u_h(s_\varkappa)\|_{1,\Omega_h(s_\varkappa)}\}$ implying the boundedness of $\{\|\widetilde{u}_h(s_\varkappa)\|_{1,\widehat{\Omega}}\}$ can be proven just as in Lemma 2.22. Thus one can pass to a subsequence of $\{\widetilde{u}_h(s_\varkappa)\}$ such that (2.139) holds for some $u \in H^1(\widehat{\Omega})$. From (2.138), (2.139), Corollary 2.1, and the fact that $u_h(s_\varkappa) \in K_h(s_\varkappa) \subset K(r_h s_\varkappa)$ we have that $u(\alpha) := u_{|\Omega(\alpha)} \in K(\alpha)$. Next we show that $u(\alpha)$ solves $(\mathcal{P}(\alpha))$.

Let $v^* \in H_0^1(\widehat{\Omega})$ be such that $v := v^*{}_{|\Omega(\alpha)} \in K(\alpha)$. From the proof of Lemma 2.22 we know that v^* can be approximated by a sequence $\{v_j\}$, $v_j \in H_0^1(\widehat{\Omega})$, of the form

$$v_j = \varphi + w_j,$$

where $\varphi \in H_0^1(\widehat{\Omega})$, $\varphi \geq 0$ a.e. in $\widehat{\Omega}$, and $\{w_j\}$ satisfies (2.131) and (2.132). The function φ also can be approached by another sequence $\{\varphi_j\}$ such that $\varphi_j \in C_0^\infty(\widehat{\Omega})$, $\varphi_j \geq 0$ in $\widehat{\Omega}$, for any $j \in \mathbb{N}$:

$$\varphi_j \to \varphi \quad \text{in } H_0^1(\widehat{\Omega}), \; j \to \infty. \tag{2.140}$$

Define $v_j^* := \varphi_j + w_j$. Then $v_j^* \in C_0^\infty(\widehat{\Omega})$ for any $j \in \mathbb{N}$ and

$$v_j^* \to v^* \quad \text{in } H_0^1(\widehat{\Omega}), \quad j \to \infty, \tag{2.141}$$

as follows from (2.131) and (2.140).

Let $\Xi(r_h s_\varkappa) := \widehat{\Omega} \setminus \overline{\Omega(r_h s_\varkappa)}$ and $\{\widehat{\mathcal{T}}(h, s_\varkappa)\}$ be a family of triangulations of $\overline{\Xi(r_h s_\varkappa)}$ satisfying $(\mathcal{T}1)$–$(\mathcal{T}3)$. The union of $\mathcal{T}(h, s_\varkappa)$ and $\widehat{\mathcal{T}}(h, s_\varkappa)$ defines a regular triangulation of $\overline{\widehat{\Omega}}$. Since $v_j^* \in C_0^\infty(\widehat{\Omega})$, one can construct its piecewise linear Lagrange interpolant $\pi_h v_j^*$ on $\mathcal{T}(h, s_\varkappa) \cup \widehat{\mathcal{T}}(h, s_\varkappa)$. The classical approximation result says that

$$\|v_j^* - \pi_h v_j^*\|_{1,\widehat{\Omega}} \leq ch\|v_j^*\|_{2,\widehat{\Omega}} \quad \forall j \in \mathbb{N}, \tag{2.142}$$

where c is a positive constant that does not depend on j and h.

Fix $j \in \mathbb{N}$. Then in view of (2.132), (2.138), and the nonnegativeness of φ_j in $\widehat{\Omega}$ we see that $v^*_{j|\Omega_h(s_\varkappa)} \in K(r_h s_\varkappa)$ and consequently $\pi_h v^*_{j|\Omega_h(s_\varkappa)} \in K_h(s_\varkappa)$ provided that h, $\varkappa$ are small enough. Thus one can use $\pi_h v^*_{j|\Omega_h(s_\varkappa)}$ as a test function in $(\mathcal{P}_h(s_\varkappa))$:

$$\int_{\widehat{\Omega}} \chi_{h\varkappa} \nabla \widetilde{u}_h \cdot \nabla(\pi_h v_j^* - \widetilde{u}_h)\, dx \geq \int_{\widehat{\Omega}} \chi_{h\varkappa} f(\pi_h v_j^* - \widetilde{u}_h)\, dx \tag{2.143}$$

holds for any $h, \varkappa > 0$ sufficiently small and any $j \in \mathbb{N}$. Here $\chi_{h\varkappa}$ stands for the characteristic function of $\Omega_h(s_\varkappa)$. We now proceed as in the proof of Lemma 2.22. Applying $\limsup_{h,\varkappa\to 0+}$ to both sides of (2.143) we arrive at

$$\int_{\widehat{\Omega}} \chi \nabla u \cdot \nabla(v_j^* - u)\, dx \geq \int_{\widehat{\Omega}} \chi f(v_j^* - u)\, dx \quad \forall j \in \mathbb{N}, \tag{2.144}$$

where χ is the characteristic function of $\Omega(\alpha)$, taking into account (2.139) and (2.142). Letting $j \to \infty$ in (2.144) and using (2.141) we finally obtain

$$\int_{\widehat{\Omega}} \chi \nabla u \cdot \nabla(v^* - u)\, dx \geq \int_{\widehat{\Omega}} \chi f(v^* - u)\, dx.$$

In other words, $u(\alpha)$ solves $(\mathcal{P}(\alpha))$. □

Remark 2.26. It can be shown again that

$$u_h(s_\varkappa) \to u(\alpha) \quad \text{in } H^1_{loc}(\Omega(\alpha)), \quad h, \varkappa \to 0+.$$

Since $(\mathcal{B}1)_\varkappa$–$(\mathcal{B}3)_\varkappa$ of Section 2.4 are satisfied we arrive at the following.

Theorem 2.16. *For any sequence $\{(s^*_\varkappa, u_h(s^*_\varkappa))\}$ of optimal pairs of $(\mathbb{P}_{h\varkappa})$, $h \to 0+$, there exists its subsequence such that*

$$\begin{cases} s^*_\varkappa \rightrightarrows \alpha^* & \text{in } [0,1], \\ \widetilde{u}_h(s^*_\varkappa) \rightharpoonup u^* & \text{in } H^1(\widehat{\Omega}), \quad h, \varkappa \to 0+. \end{cases} \tag{2.145}$$

In addition, $(\alpha^*, u^*_{|\Omega(\alpha^*)})$ is an optimal pair of $(\mathbb{P})$. Any accumulation point of $\{(s^*_\varkappa, \widetilde{u}_h(s^*_\varkappa))\}$ in the sense of (2.145) *possesses this property.*

In the remainder of this subsection we present the algebraic form of $(\mathbb{P}_{h\varkappa})$ for $h, \varkappa > 0$ fixed. Let $\mathcal{T}_D$ and $\mathcal{T}_S$ be the isomorphisms between $U^{ad}_\varkappa$ and $\mathcal{U}$ (the set of discrete design variables defined by (2.48)) and $V_h(s_\varkappa)$ and $\mathbb{R}^n$ ($n = \dim V_h(s_\varkappa)$), respectively (see

(2.21), (2.22)). As in Remark 2.23 the approximate solution $u_h(s_{\varkappa})$ of $(\mathcal{P}_h(s_{\varkappa}))$ can be characterized as a minimizer of $E_{r_h s_{\varkappa}}$ over $K_h(s_{\varkappa})$:

$$\begin{cases} \text{Find } u_h(s_{\varkappa}) \in K_h(s_{\varkappa}) \text{ such that} \\ E_{r_h s_{\varkappa}}(u_h(s_{\varkappa})) \le E_{r_h s_{\varkappa}}(v_h) \quad \forall v_h \in K_h(s_{\varkappa}), \end{cases} \tag{$\widetilde{\mathcal{P}}_h(s_{\varkappa})$}$$

where

$$E_{r_h s_{\varkappa}}(v_h) = \frac{1}{2}\int_{\Omega_h(s_{\varkappa})} \nabla v_h \cdot \nabla v_h \, dx - \int_{\Omega_h(s_{\varkappa})} f v_h \, dx.$$

The algebraic form of $(\widetilde{\mathcal{P}}_h(s_{\varkappa}))$ leads to the following *quadratic programming problem*:

$$\begin{cases} \text{Find } \boldsymbol{q}(\boldsymbol{\alpha}) \in \mathcal{K} \text{ such that} \\ \mathcal{E}(\boldsymbol{\alpha}, \boldsymbol{q}(\boldsymbol{\alpha})) \le \mathcal{E}(\boldsymbol{\alpha}, \boldsymbol{q}) \quad \forall \boldsymbol{q} \in \mathcal{K}, \end{cases} \tag{$\mathcal{P}(\boldsymbol{\alpha})$}$$

where

$$\mathcal{E}(\boldsymbol{\alpha}, \boldsymbol{q}) = \frac{1}{2}\boldsymbol{q}^{\mathrm{T}}\boldsymbol{K}(\boldsymbol{\alpha})\boldsymbol{q} - \boldsymbol{q}^{\mathrm{T}}\boldsymbol{f}(\boldsymbol{\alpha})$$

is the quadratic function defined by the stiffness matrix $\boldsymbol{K}(\boldsymbol{\alpha})$ and the force vector $\boldsymbol{f}(\boldsymbol{\alpha})$ both depending on the discrete design variable $\boldsymbol{\alpha} \in \mathcal{U}$ and

$$\mathcal{K} = \{\boldsymbol{x} \in \mathbb{R}^n \mid \mathcal{T}_S^{-1}\boldsymbol{x} \in K_h(s_{\varkappa})\}.$$

It is readily seen that $\mathcal{K}$ is a closed, convex subset of $\mathbb{R}^n$ defined by

$$\mathcal{K} = \{\boldsymbol{x} \in \mathbb{R}^n \mid x_{j_i} \ge 0 \ \forall\, j_i \in \mathcal{I}\},$$

where $\mathcal{I}$ contains the indices of the constrained components of $\boldsymbol{x}$ corresponding to the nodal values of $u_h(s_{\varkappa})$ at the points of $\mathcal{N}_h$:

$$j_i \in \mathcal{I} \iff \exists\, A_i \in \mathcal{N}_h : \ x_{j_i} = u_h(s_{\varkappa})(A_i).$$

Observe that $\mathcal{K}$ does *not* depend explicitly on $\boldsymbol{\alpha} \in \mathcal{U}$. Optimization problem $(\mathbb{P}_{h\varkappa})$ leads to the following *nonlinear mathematical programming problem*:

$$\begin{cases} \text{Find } \boldsymbol{\alpha}^* \in \mathcal{U} \text{ such that} \\ \mathcal{J}(\boldsymbol{q}(\boldsymbol{\alpha}^*)) \le \mathcal{J}(\boldsymbol{q}(\boldsymbol{\alpha})) \quad \forall \boldsymbol{\alpha} \in \mathcal{U}, \end{cases} \tag{$\mathbb{P}_d$}$$

where $\boldsymbol{q}(\boldsymbol{\alpha}) \in \mathcal{K}$ solves $(\mathcal{P}(\boldsymbol{\alpha}))$ and $\mathcal{J}$ is the algebraic representation of J. As we have already mentioned at the beginning of this subsection, problem $(\mathbb{P}_d)$ is more involved than the ones analyzed up to now due to the fact that the state problem is given by the variational inequality $(\mathcal{P}(\boldsymbol{\alpha}))$.

2.5.5 Shape optimization in linear elasticity and contact problems

The previous subsections were devoted to shape optimization of structures governed by *scalar* state problems. We saw that the way to prove the existence and convergence results depends on the type of boundary conditions in the state problem. We discussed in detail four typical cases: the Dirichlet and Neumann boundary conditions prescribed on the *whole* boundary, the mixed Dirichlet–Neumann boundary condition, and finally the unilateral boundary conditions on the designed part of the boundary. These techniques can be used (eventually and with minor modifications) in more general state problems.

This subsection deals with shape optimization in the mechanics of solids. We restrict ourselves to the case of linearly elastic structures. Before we formulate the problem, recall in brief the basic notions of linear elasticity that will be used (for details we refer to [NH81]).

CONVENTION: *In what follows Einstein's summation convention is used: any term in which the same index appears twice indicates summation with respect to this index over the range from* 1 *to the spatial dimension* (2 *or* 3)*. For example,*

$$u_i v_i := \sum_{i=1}^{3} u_i v_i, \quad \frac{\partial \tau_{ij}}{\partial x_j} := \sum_{j=1}^{3} \frac{\partial \tau_{ij}}{\partial x_j}.$$

Let a body occupying a bounded domain $\Omega \subset \mathbb{R}^3$ be loaded by *body forces* of density $f = (f_1, f_2, f_3)$ and *surface tractions* of density $P = (P_1, P_2, P_3)$ on a portion Γ_P of the boundary of Ω. On the remaining part $\Gamma_u = \partial\Omega \setminus \overline{\Gamma}_P$, the body is fixed. We want to find an equilibrium state of Ω. This state is characterized by a symmetric *stress tensor* $(\tau_{ij})_{i,j=1}^3$, $\tau_{ij} = \tau_{ji}$, defined in Ω, in equilibrium with f and P, i.e., satisfying

$$\frac{\partial \tau_{ij}}{\partial x_j} + f_i = 0 \quad \text{in } \Omega, \quad i = 1, 2, 3; \tag{2.146}$$

$$\tau_{ij} \nu_j = P_i \quad \text{on } \Gamma_P, \quad i = 1, 2, 3. \tag{2.147}$$

The deformation of Ω is characterized by a *displacement vector* $u = (u_1, u_2, u_3)$ and the respective *linearized strain tensor* $\varepsilon(u) = (\varepsilon_{ij}(u))_{i,j=1}^3$, where $\varepsilon_{ij}(u) = \frac{1}{2}\left(\partial u_i/\partial x_j + \partial u_j/\partial x_i\right)$.

The stress-strain relation is given by a linear *Hooke's law*:

$$\tau_{ij} = c_{ijkl}\varepsilon_{kl}, \quad i, j, k, l = 1, 2, 3, \tag{2.148}$$

with $c_{ijkl} \in L^\infty(\Omega)$ $\forall i, j, k, l$ satisfying the following *symmetry* and *ellipticity* conditions:

$$c_{ijkl}(x) = c_{jikl}(x) = c_{klij}(x) \quad \text{for almost all (a.a.) } x \in \Omega, \tag{2.149}$$

$$\exists q = \text{const.} > 0 : \quad c_{ijkl}(x)\xi_{ij}\xi_{kl} \geq q\xi_{ij}\xi_{ij} \quad \text{for a.a. } x \in \Omega, \tag{2.150}$$

where $(\xi_{ij})_{i,j=1}^3$ is an arbitrary 3×3 symmetric matrix.

We seek a displacement vector u such that (2.146) and (2.147) are satisfied with $\tau := \tau(u)$, where $\tau(u)$ is related to $\varepsilon(u)$ through (2.148). In addition, the displacement vector u has to satisfy the homogeneous boundary condition on Γ_u:

$$u_i = 0 \quad \text{on } \Gamma_u, \quad i = 1, 2, 3. \tag{2.151}$$

To derive the weak formulation of this problem we use the following Green's formula:

$$\int_\Omega \tau_{ij}\varepsilon_{ij}(v)\, dx = -\int_\Omega \frac{\partial \tau_{ij}}{\partial x_j} v_i\, dx + \int_{\partial\Omega} \tau_{ij}\nu_j v_i\, ds, \tag{2.152}$$

which is valid for any τ and v smooth enough.

Let $\mathbb{V}(\Omega)$ be the space of admissible displacements

$$\mathbb{V}(\Omega) = \left\{ v \in (H^1(\Omega))^3 \mid v = 0 \text{ on } \Gamma_u \right\}. \tag{2.153}$$

From (2.146), (2.147), and the definition of $\mathbb{V}(\Omega)$ it is readily seen that a solution u (if any) can be characterized as an element of $\mathbb{V}(\Omega)$ satisfying the following integral identity:

$$\int_\Omega \tau_{ij}(u)\varepsilon_{ij}(v)\, dx = \int_\Omega f_i v_i \, dx + \int_{\Gamma_P} P_i v_i \, ds \quad \forall v \in \mathbb{V}(\Omega), \tag{2.154}$$

which will be used as a basis for the weak formulation of our problem.

Thus the *weak formulation* of the linear elasticity problem reads as follows:

$$\begin{cases} \text{Find } u \in \mathbb{V}(\Omega) \text{ such that} \\ a(u, v) = L(v) \quad \forall v \in \mathbb{V}(\Omega), \end{cases} \tag{$\mathcal{P}_1$}$$

where the bilinear form a and the linear form L are defined by the left and right sides of (2.154), respectively.

Next we shall suppose that $f \in (L^2(\Omega))^3,\ P \in (L^2(\Gamma_P))^3$, and Γ_u is *nonempty* and open in $\partial\Omega$. From (2.148) and (2.150) we see that

$$a(v, v) \geq q \int_\Omega \varepsilon_{ij}(v)\varepsilon_{ij}(v)\, dx = q\|\varepsilon(v)\|_{0,\Omega}^2 \quad \forall v \in \mathbb{V}(\Omega). \tag{2.155}$$

The most important assumption of the Lax–Milgram lemma, namely the $\mathbb{V}(\Omega)$-ellipticity of the bilinear form a, follows from Korn's inequality:

$$\exists \beta = \text{const.} > 0 : \quad \|\varepsilon(v)\|_{0,\Omega}^2 \geq \beta \|v\|_{1,\Omega}^2, \tag{2.156}$$

which holds for any $v \in \mathbb{V}(\Omega)$. Therefore $(\mathcal{P}_1)$ has a unique solution.

We now turn to shape optimization. To simplify our presentation we restrict ourselves to *plane* elasticity problems in what follows (due to symmetry conditions that are frequently present, such a dimensional reduction is possible).

Shapes of admissible domains will again be parametrized by means of Lipschitz continuous functions from $\widetilde{U}^{ad}$, defined by (2.96) (see also Figure 2.2). On any $\Omega(\alpha),\ \alpha \in \widetilde{U}^{ad}$, we consider the following linear elasticity problem:

$$\begin{cases} \text{Find } u := u(\alpha) \in \mathbb{V}(\alpha) \quad \text{such that} \\ a_\alpha(u, v) = L_\alpha(v) \quad \forall v \in \mathbb{V}(\alpha), \end{cases} \tag{$\mathcal{P}_1(\alpha)$}$$

where

$$a_\alpha(u, v) = \int_{\Omega(\alpha)} \tau_{ij}(u)\varepsilon_{ij}(v)\, dx, \tag{2.157}$$

$$L_\alpha(v) = \int_{\Omega(\alpha)} f_i v_i \, dx + \int_{\Gamma_P(\alpha)} P_i v_i \, ds, \tag{2.158}$$

and $\mathbb{V}(\alpha)$ is the space of admissible displacements defined by (2.153) on $\Omega(\alpha)$. As in Remark 2.22 we shall suppose that $\Gamma_u(\alpha),\ \Gamma_P(\alpha)$ depend continuously on $\alpha \in \widetilde{U}^{ad}$ and, in addition, there exists $\delta > 0$ such that the one-dimensional Lebesgue measure meas $_1\Gamma_u(\alpha) \geq \delta > 0$ for any $\alpha \in \widetilde{U}^{ad}$. Finally, let

$$f \in (L^2(\widehat{\Omega}))^2, \tag{2.159}$$

$$\exists c = \text{const.} > 0 : \quad \|P\|_{0,\Gamma_P(\alpha)} \leq c \quad \forall \alpha \in \widetilde{U}^{ad}, \tag{2.160}$$

and (2.149), (2.150) be satisfied in $\widehat{\Omega} \supset \Omega(\alpha)\ \forall \alpha \in \widetilde{U}^{ad}$.

One of the most typical problems we meet in shape optimization of deformable structures can be formulated as follows: Determine the shape of the structure of a prescribed volume exhibiting the highest stiffness. The mathematical formulation of this problem is as follows:

$$\begin{cases} \text{Find } \alpha^* \in U^{ad} \quad \text{such that} \\ J(\alpha^*, u(\alpha^*)) \leq J(\alpha, u(\alpha)) \quad \forall \alpha \in U^{ad}, \end{cases} \tag{$\mathbb{P}_1$}$$

where

$$J(\alpha, y) = \int_{\Omega(\alpha)} f_i y_i \, dx + \int_{\Gamma_P(\alpha)} P_i y_i \, ds \tag{2.161}$$

is the compliance functional, $u(\alpha) \in \mathbb{V}(\alpha)$ solves $(\mathcal{P}_1(\alpha))$, and U^{ad} is defined by (2.97) involving the constant volume constraint.

Before we prove the existence of solutions to $(\mathbb{P}_1)$ we shall specify the location of $\Gamma_u(\alpha)$ and $\Gamma_P(\alpha)$. *In what follows we shall suppose that $\Gamma_u(\alpha)$ is represented by the graph of $\alpha \in \widetilde{U}^{ad}$; i.e., $\Gamma_u(\alpha) = \Gamma(\alpha)$ and $\Gamma_P(\alpha) = \partial\Omega(\alpha) \setminus \overline{\Gamma(\alpha)}$.* We shall verify assumptions $(\mathcal{B}1)$ and $(\mathcal{B}2)$ of Section 2.4.

Lemma 2.24. *(Verification of $(\mathcal{B}1)$.) Let $\Gamma_u(\alpha)$, $\Gamma_P(\alpha)$ be as above and $\{(\alpha_n, u_n)\}$ be an arbitrary sequence, where $\alpha_n \in \widetilde{U}^{ad}$ and $u_n := u(\alpha_n) \in \mathbb{V}(\alpha_n)$ solves $(\mathcal{P}_1(\alpha_n))$, $n \to \infty$. Then one can find its subsequence and elements $\alpha \in \widetilde{U}^{ad}$, $u \in (H^1(\widehat{\Omega}))^2$ such that*

$$\begin{cases} \alpha_n \rightrightarrows \alpha & \text{in } [0,1], \\ \widetilde{u}_n \to u & \text{in } (H^1(\widehat{\Omega}))^2, \quad n \to \infty, \end{cases} \tag{2.162}$$

where $\widetilde{u}_n$ denotes the extension of u_n from $\Omega(\alpha_n)$ to $\widehat{\Omega}$ by zero. In addition, $u(\alpha) := u_{|\Omega(\alpha)}$ solves $(\mathcal{P}_1(\alpha))$.

Proof. From the definition of $(\mathcal{P}_1(\alpha_n))$, (2.155), (2.159), and (2.160) we see that

$$q\|\varepsilon(u_n)\|^2_{0,\Omega(\alpha)} \leq a_{\alpha_n}(u_n, u_n) = L_{\alpha_n}(u_n) \leq c\|u_n\|_{1,\Omega(\alpha_n)} \tag{2.163}$$

holds for any $n \in \mathbb{N}$ with a positive constant c that does not depend on $\alpha \in \widetilde{U}^{ad}$. Next we use a nontrivial result saying that the constant β of Korn's inequality (2.156) is uniform with respect to a class of domains possessing the uniform ε-cone property (see [Nit81]). In particular, this constant can be chosen independently of $\alpha \in \widetilde{U}^{ad}$. From this and (2.163) we see that there exists a positive constant c such that

$$\|u_n\|_{1,\Omega(\alpha_n)} \leq c \quad \forall n \in \mathbb{N}. \tag{2.164}$$

The rest of the proof is now obvious. Since $u_n = 0$ on $\Gamma(\alpha_n)$, one can use its zero extension $\widetilde{u}_n$ from $\Omega(\alpha_n)$ to $\widehat{\Omega}$, satisfying (2.164), as well. Thus there exists a subsequence of $\{(\alpha_n, \widetilde{u}_n)\}$ satisfying $(2.162)_1$ for some $\alpha \in U^{ad}$ and converging weakly to an element u in $(H^1(\widehat{\Omega}))^2$. The fact that $u = 0$ in $\widehat{\Omega} \setminus \overline{\Omega(\alpha)}$, implying that $u(\alpha) := u_{|\Omega(\alpha)} \in \mathbb{V}(\alpha)$, can be verified just as in Lemma 2.5. To show that $u(\alpha)$ solves $(\mathcal{P}_1(\alpha))$, fix $v \in \mathbb{V}(\alpha)$. Then one can find a sequence $\{v_j\}$ with $v_j \in (C^\infty(\overline{\widehat{\Omega}}))^2$ vanishing in a vicinity of $\Gamma(\alpha)$ and such that

$$v_j \to v \quad \text{in } (H^1(\widehat{\Omega}))^2, \quad j \to \infty. \tag{2.165}$$

For $j \in \mathbb{N}$ fixed, the function $v_{j|\Omega(\alpha_n)}$ belongs to $\mathbb{V}(\alpha_n)$ for n large enough. This means that it can be used as a test function in $(\mathcal{P}_1(\alpha_n))$. Passing to the limit first with $n \to \infty$, then with $j \to \infty$, we obtain

$$\lim_{j\to\infty}\left(\lim_{n\to\infty} a_{\alpha_n}(u_n, v_j)\right) = a_\alpha(u(\alpha), v),$$

$$\lim_{j\to\infty}\left(\lim_{n\to\infty} L_{\alpha_n}(v_j)\right) = L_\alpha(v),$$

making use of $(2.162)_1$, (2.165), and weak convergence of $\{\widetilde{u}_n\}$ to u. From this we conclude that $u(\alpha)$ solves $(\mathcal{P}_1(\alpha))$. Strong convergence in $(2.162)_2$ can be shown in the standard way. $\square$

Remark 2.27. The linear term L_α also involves the integral over $\Gamma_P(\alpha)$. We have mentioned in Subsection 2.5.2 that uniform convergence of $\{\alpha_n\}$ to α generally does not ensure convergence of the respective sequence of the curvilinear integrals. In our particular case, however, the situation is simpler, since the integration over $\Gamma_P(\alpha)$ is carried out on the straight line segments and only the integrals along the top and bottom of $\Omega(\alpha)$ depend on α. These integrals are of the form

$$\int_0^{\alpha_n(0)} P_i v_i \, dx_1, \qquad \int_0^{\alpha_n(1)} P_i v_i \, dx_1,$$

and clearly uniform convergence of $\{\alpha_n\}$ to α is more than enough to pass to the limit with $n \to \infty$. If the surface tractions P were prescribed on a *curved* part, it would be possible to proceed as in Remark 2.17, avoiding the use of more regular design variables α.

It is left as an easy exercise to show that J is *continuous* in the following sense:

$$\left.\begin{array}{lll} \alpha_n \rightrightarrows \alpha & \text{in } [0,1], & \alpha_n, \alpha \in \widetilde{U}^{ad} \\ y_n \rightharpoonup y & \text{in } (H^1(\widehat{\Omega}))^2, & y_n, y \in (H^1(\widehat{\Omega}))^2 \end{array}\right\} \implies \lim_{n\to\infty} J(\alpha_n, y_{n|\Omega(\alpha_n)}) = J(\alpha, y_{|\Omega(\alpha)}). \tag{2.166}$$

From this and Lemma 2.24 we arrive at the following result.

Theorem 2.17. *Problem* $(\mathbb{P}_1)$ *has a solution.*

We now describe very briefly the discretization of $(\mathbb{P}_1)$. The discrete families of admissible domains $\mathcal{O}_\varkappa$, $\mathcal{O}_{\varkappa h}$ will be defined as in Section 2.2. The space of admissible displacements will be discretized by continuous, piecewise linear *vector* functions on a triangulation $\mathcal{T}(h, s_\varkappa)$ of $\overline{\Omega(r_h s_\varkappa)}$, $s_\varkappa \in U_\varkappa^{ad}$:

$$\mathbb{V}_h(s_\varkappa) = \{v_h \in (C(\overline{\Omega_h(s_\varkappa)}))^2 \mid v_{h|T} \in (P_1(T))^2 \quad \forall T \in \mathcal{T}(h, s_\varkappa), \\ v_h = 0 \quad \text{on } \Gamma_u(r_h s_\varkappa)\}. \tag{2.167}$$

Recall that $\Gamma_u(r_h s_\varkappa) = \Gamma(r_h s_\varkappa)$ is the graph of the piecewise linear interpolant of $s_\varkappa$.

In any $\Omega_h(s_\varkappa) \in \mathcal{O}_{\varkappa h}$ we define the following discrete linear elasticity problem:

$$\begin{cases} \text{Find } u_h := u_h(s_\varkappa) \in \mathbb{V}_h(s_\varkappa) \text{ such that} \\ a_{r_h s_\varkappa}(u_h, v_h) = L_{r_h s_\varkappa}(v_h) \quad \forall v_h \in \mathbb{V}_h(s_\varkappa), \end{cases} \tag{$\mathcal{P}_{1h}(s_\varkappa)$}$$

where

$$a_{r_h s_\varkappa}(u_h, v_h) = \int_{\Omega_h(s_\varkappa)} \tau_{ij}(u_h)\varepsilon_{ij}(v_h)\, dx,$$

$$L_{r_h s_\varkappa}(v_h) = \int_{\Omega_h(s_\varkappa)} f_i v_{hi}\, dx + \int_{\Gamma_P(r_h s_\varkappa)} P_i v_{hi}\, ds \quad (\Gamma_P(r_h s_\varkappa) = \partial\Omega_h(s_\varkappa) \setminus \overline{\Gamma_u(r_h s_\varkappa)}).$$

For $h, \varkappa > 0$ fixed, the discretization of $(\mathbb{P}_1)$ is as follows:

$$\begin{cases} \text{Find } s_\varkappa^* \in U_\varkappa^{ad} \quad \text{such that} \\ J(s_\varkappa^*, u_h(s_\varkappa^*)) \le J(s_\varkappa, u_h(s_\varkappa)) \quad \forall s_\varkappa \in U_\varkappa^{ad}, \end{cases} \tag{$\mathbb{P}_{h\varkappa}^1$}$$

where

$$J(s_\varkappa, y_h) = \int_{\Omega_h(s_\varkappa)} f_i y_{hi}\, dx + \int_{\Gamma_P(r_h s_\varkappa)} P_i y_{hi}\, ds, \quad y_h \in \mathbb{V}_h(s_\varkappa).$$

The convergence analysis for this problem can be done in a similar fashion as in the previous parts of the book (see Problem 2.18).

We now turn to an important class of problems called *contact shape optimization*, i.e., optimization of structures assembling several deformable bodies in mutual contact. For the sake of simplicity, we restrict ourselves to contact problems between one elastic body and a rigid foundation—the case known in literature as the *Signorini* problem.

Consider a body Ω loaded by body forces of density f and surface tractions of density P on a part $\Gamma_P \subset \partial\Omega$ and fixed along Γ_u, respectively. In addition, the body will be supported along a portion Γ_C of the boundary by a rigid foundation S, limiting its deformation (see Figure 2.5).

The body deforms in such a way that it still remains in the complement of S. Why is this problem more complicated than the classical elasticity problem discussed above? The deformation of Ω depends not only on the given forces but also on contact pressures occurring on Γ_C. Unfortunately, these pressures *are not known* a priori; they are unknowns of the problem. To make the form of contact conditions simpler, we confine ourselves to a two-dimensional case and, in addition, shapes of bodies will be as shown in Figure 2.6: the rigid foundation supporting $\Omega(\alpha)$ is represented by the half-plane $S = \{x = (x_1, x_2) \in \mathbb{R}^2 \mid x_2 \le 0\}$ and the contact part Γ_C, usually the main object of optimization by the bottom of $\Omega(\alpha)$. We shall suppose that Γ_C is represented by the graph of functions $\alpha \in \widetilde{U}^{ad}$, where

$$\widetilde{U}^{ad} = \{\alpha \in C^{0,1}([0,1]) \mid 0 \le \alpha \le \alpha_{\max} \text{ in } [0,1], \quad |\alpha'| \le L_0 \text{ a.e. in }]0,1[\} \tag{2.168}$$

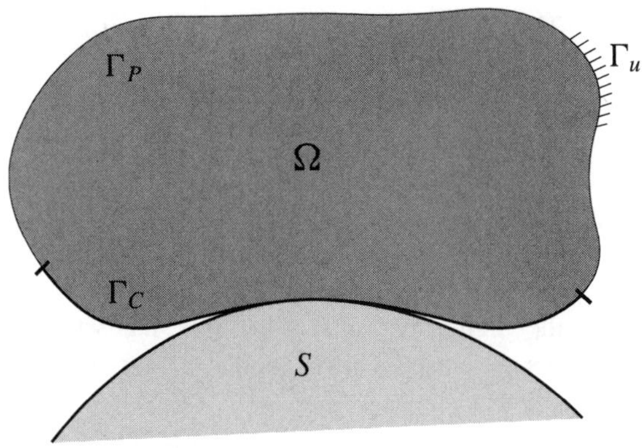

Figure 2.5. *Deformable body on a rigid support.*

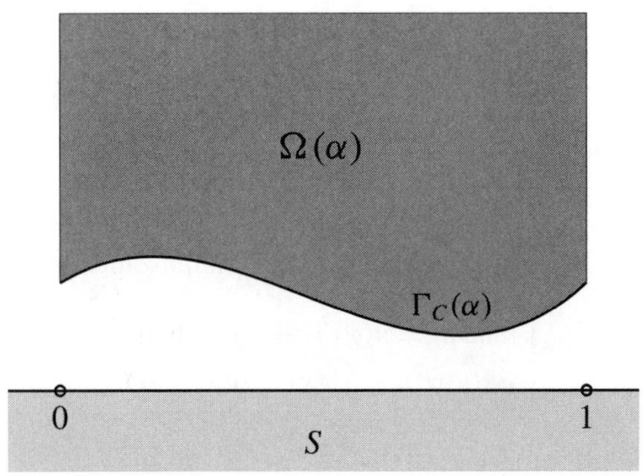

Figure 2.6. *Shape of admissible domains.*

for some positive constants $\alpha_{\max}$, L_0. The family $\widetilde{\mathcal{O}}$ consists of domains

$$\Omega(\alpha) = \{(x_1, x_2) \in \mathbb{R}^2 \mid \alpha(x_1) < x_2 < \overline{\gamma},\ 0 < x_1 < 1\}, \quad \overline{\gamma} > 0 \text{ given}, \tag{2.169}$$

where $\alpha \in \widetilde{U}^{ad}$ and $\alpha_{\max} \leq \overline{\gamma}/2$.

The contact conditions on $\Gamma_C := \Gamma_C(\alpha)$, $\alpha \in \widetilde{U}^{ad}$, are expressed as follows:

$$u_2(x) \geq -\alpha(x_1), \quad T_2(x) \geq 0 \quad \forall x = (x_1, \alpha(x_1)),\ x_1 \in]0, 1[, \tag{2.170}$$

$$(u_2(x) + \alpha(x_1))\, T_2(x) = 0, \quad \forall x = (x_1, \alpha(x_1)),\ x_1 \in]0, 1[, \tag{2.171}$$

$$T_1(x) = 0, \quad \forall x \in \Gamma_C(\alpha), \tag{2.172}$$

where $T_i(x) = \tau_{ij}(u(x))\nu_j$, $x \in \partial\Omega(\alpha)$, stands for the ith component ($i = 1, 2$) of the stress vector T.

Condition (2.170) says that $\Omega(\alpha)$ cannot penetrate into S and only pressures may occur on $\Gamma_C(\alpha)$. If there is no contact at a point $x \in \Gamma_C(\alpha)$, then $T_2(x) = 0$ (no pressure).

This is expressed by (2.171). Assuming perfectly smooth surfaces of $\Omega(\alpha)$ and S, the influence of friction can be neglected (see (2.172)).

By a *classical* solution of the Signorini problem we mean any displacement field $u = (u_1, u_2)$ satisfying the analogy of (2.146), (2.147) for plane problems with $\tau := \tau(u)$, where $\tau(u)$ is related to the strain tensor $\varepsilon(u)$ by means of the linear Hooke's law, as well as the homogeneous conditions on $\Gamma_u(\alpha)$ and the contact conditions (2.170)–(2.172) on $\Gamma_C(\alpha)$. As before we shall suppose that both $\Gamma_u(\alpha)$ and $\Gamma_P(\alpha)$ depend continuously on $\alpha \in \widetilde{U}^{ad}$ and $\text{meas}_1 \Gamma_u(\alpha) \geq \delta > 0$ for some $\delta > 0$ and all $\alpha \in \widetilde{U}^{ad}$.

The weak form of the Signorini problem in $\Omega(\alpha)$ leads to an elliptic variational inequality. We first introduce the closed, convex, and nonempty subset of $\mathbb{V}(\alpha)$:

$$\mathbb{K}(\alpha) = \{v = (v_1, v_2) \in \mathbb{V}(\alpha) \mid v_2(x_1, \alpha(x_1)) \geq -\alpha(x_1) \text{ for a.a. } x_1 \in]0, 1[\}. \tag{2.173}$$

By a *weak* solution of the Signorini problem we mean any function u that is the solution of the following:

$$\begin{cases} \text{Find } u := u(\alpha) \in \mathbb{K}(\alpha) \text{ such that} \\ a_\alpha(u, v - u) \geq L_\alpha(v - u) \quad \forall v \in \mathbb{K}(\alpha), \end{cases} \tag{$\mathcal{P}_2(\alpha)$}$$

where a_α, L_α are defined by (2.157), (2.158), respectively. It is well known that $(\mathcal{P}_2(\alpha))$ has a unique solution $u(\alpha)$ for any $\alpha \in \widetilde{U}^{ad}$ (see [HHN96]). It is left as an exercise to show that the weak and classical formulations are formally equivalent. Since the bilinear form a_α is *symmetric*, $(\mathcal{P}_2(\alpha))$ is equivalent to the following minimization problem:

$$\begin{cases} \text{Find } u := u(\alpha) \in \mathbb{K}(\alpha) \text{ such that} \\ E_\alpha(u(\alpha)) \leq E_\alpha(v) \quad \forall v \in \mathbb{K}(\alpha), \end{cases} \tag{$\widetilde{\mathcal{P}}_2(\alpha)$}$$

where

$$E_\alpha(v) = \frac{1}{2} a_\alpha(v, v) - L_\alpha(v)$$

is the total potential energy functional.

Next we shall analyze the following optimal shape design problem:

$$\begin{cases} \text{Find } \alpha^* \in U^{ad} \text{ such that} \\ J(\alpha^*, u(\alpha^*)) \leq J(\alpha, u(\alpha)) \quad \forall \alpha \in U^{ad}, \end{cases} \tag{$\mathbb{P}_2$}$$

where

$$U^{ad} = \left\{ \alpha \in \widetilde{U}^{ad} \text{given by (2.168)} \mid \int_0^1 \alpha(x_1)\, dx_1 = \gamma \right\}, \quad \gamma > 0 \text{ given}, \tag{2.174}$$

and

$$J(\alpha, u(\alpha)) := E_\alpha(u(\alpha)) \quad \forall \alpha \in U^{ad}; \tag{2.175}$$

i.e., the value of the cost functional equals the value of the total potential energy in the equilibrium state $u(\alpha) \in \mathbb{K}(\alpha)$, the solution of $(\mathcal{P}_2(\alpha))$. As we shall see in Chapter 3, shape optimization with such a choice of J has an important feature: namely the resulting contact stresses are *evenly* distributed along the *optimal* contact part.

The existence analysis for $(\mathbb{P}_2)$ is based on the following auxiliary result, whose proof can be found in [HN96] (see also Problem 2.19).

Lemma 2.25. *Let $\alpha_n \rightrightarrows \alpha$ in $[0,1]$, $\alpha_n, \alpha \in \widetilde{U}^{ad}$, and $v \in \mathbb{K}(\alpha)$ be given. Then there exists a sequence $\{v_j\}$, $v_j \in (H^1(\widehat{\Omega}))^2$, such that*

$$\begin{cases} v_{j|\Omega(\alpha_n)} \in \mathbb{K}(\alpha_n) & \forall j \in \mathbb{N}\ \forall n \geq n_0(j) \in \mathbb{N}, \\ v_j \to \widetilde{v} \quad \text{in } (H^1(\widehat{\Omega}))^2, & j \to \infty, \end{cases} \tag{2.176}$$

where $\widetilde{v} = p_{\Omega(\alpha)} v$ is the uniform extension of v from $\Omega(\alpha)$ to $\widehat{\Omega}$.

With this lemma in hand one can easily prove the following one.

Lemma 2.26. *(Verification of $(\mathcal{B}1)$.) For any sequence $\{(\alpha_n, u_n)\}$, where $\alpha_n \in \widetilde{U}^{ad}$ and $u_n := u(\alpha_n) \in \mathbb{K}(\alpha_n)$ is a solution of $(\mathcal{P}_2(\alpha_n))$, $n \to \infty$, there exist its subsequence and elements $\alpha \in \widetilde{U}^{ad}$, $u \in (H^1(\widehat{\Omega}))^2$ such that*

$$\begin{cases} \alpha_n \rightrightarrows \alpha & \text{in } [0,1], \\ \widetilde{u}_n \rightharpoonup u & \text{in } (H^1(\widehat{\Omega}))^2, \quad n \to \infty, \end{cases} \tag{2.177}$$

where $\widetilde{u}_n = p_{\Omega(\alpha_n)} u_n$. In addition, $u(\alpha) := u_{|\Omega(\alpha)}$ solves $(\mathcal{P}_2(\alpha))$.

Proof. (Sketch.) Inserting $v := 0$ into $(\mathcal{P}_2(\alpha_n))$ we obtain

$$a_{\alpha_n}(u_n, u_n) \leq L_{\alpha_n}(u_n) \leq c\|u_n\|_{1,\Omega(\alpha_n)} \quad \forall n \in \mathbb{N}, \tag{2.178}$$

making use of (2.159) and (2.160). Exactly as in the linear elasticity problem one obtains the boundedness of $\{\|u_n\|_{1,\Omega(\alpha_n)}\}$ and hence of $\{\|\widetilde{u}_n\|_{1,\widehat{\Omega}}\}$.

We may assume that (2.177) holds for an appropriate subsequence and elements $\alpha \in \widetilde{U}^{ad}$, $u \in (H^1(\widehat{\Omega}))^2$. Set $u(\alpha) := u_{|\Omega(\alpha)}$. To prove that $u(\alpha) \in \mathbb{K}(\alpha)$, it is sufficient to verify that $u_2(x_1, \alpha(x_1)) \geq -\alpha(x_1)$ a.e. in $]0,1[$. Indeed, let $z_n := u_{n2} \circ \alpha_n + \alpha_n$, $z := u_2 \circ \alpha + \alpha$ be the functions belonging to $L^2(]0,1[)$. As in Lemma 2.21 one can show that

$$z_n \to z \quad \text{in } L^2(]0,1[), \quad n \to \infty.$$

Since z_n are nonnegative for any $n \in \mathbb{N}$, so is the limit z, implying that $u(\alpha) \in \mathbb{K}(\alpha)$. It remains to prove that $u(\alpha)$ solves $(\mathcal{P}_2(\alpha))$. Fix $v \in \mathbb{K}(\alpha)$. From Lemma 2.25 the existence of $\{v_j\}$ satisfying (2.176) follows. In particular, for $j \in \mathbb{N}$ fixed, $v_{j|\Omega(\alpha_n)}$ can be used as a test function in $(\mathcal{P}_2(\alpha_n))$ provided that $n \geq n_0(j)$:

$$a_{\alpha_n}(u_n, v_j - u_n) \geq L_{\alpha_n}(v_j - u_n).$$

It is now easy to show that the limit passage for $n \to \infty$ and then for $j \to \infty$ in the previous inequality yields

$$a_\alpha(u(\alpha), v - u(\alpha)) \geq L_\alpha(v - u(\alpha)),$$

making use of (2.176) and (2.177). From this we conclude that $u(\alpha)$ solves $(\mathcal{P}_2(\alpha))$. $\square$

We now prove the following.

Lemma 2.27. (*Verification of* ($\mathcal{B}2$).) *The cost functional* J*, defined by* (2.175)*, is lower semicontinuous in the following sense:*

$$\left.\begin{array}{llll} \alpha_n \rightrightarrows \alpha & in\ [0,1], & \alpha_n, \alpha \in \widetilde{U}^{ad} \\ y_n \rightharpoonup y & in\ (H^1(\widehat{\Omega}))^2, & y_n, y \in (H^1(\widehat{\Omega}))^2 \end{array}\right\} \Longrightarrow \liminf_{n\to\infty} J(\alpha_n, y_{n|\Omega(\alpha_n)}) \geq J(\alpha, y_{|\Omega(\alpha)}).$$

Proof. From the definition of J it follows that

$$J(\alpha, y) = E_\alpha(y) = \frac{1}{2}a_\alpha(y, y) - L_\alpha(y).$$

Since the bilinear form a_α is symmetric for any $\alpha \in \widetilde{U}^{ad}$, it holds that the mapping $y \mapsto \sqrt{a_\alpha(y, y)}$, $y \in \mathbb{V}(\alpha)$, is a norm in $\mathbb{V}(\alpha)$. Arguing as in the proof of Lemma 2.19 one can show that

$$\liminf_{n\to\infty} a_{\alpha_n}(y_{n|\Omega(\alpha_n)}, y_{n|\Omega(\alpha_n)}) \geq a_\alpha(y_{|\Omega(\alpha)}, y_{|\Omega(\alpha)}).$$

Since also

$$\lim_{n\to\infty} L_{\alpha_n}(y_{n|\Omega(\alpha_n)}) = L_\alpha(y_{|\Omega(\alpha)}),$$

we arrive at the assertion. $\square$

As a consequence of Lemmas 2.26 and 2.27 we obtain the following existence result.

Theorem 2.18. *Problem* ($\mathbb{P}_2$) *has a solution.*

In the remainder of this subsection we briefly describe the discretization of ($\mathbb{P}_2$). Recall that the geometry of admissible domains is now given by Figure 2.6. The only difference in contact shape optimization is the fact that the state problem is given by a variational inequality. Let $\Delta_h : 0 = b_0 < b_1 < \cdots < b_{\bar{d}(h)} = 1$ be a partition of $[0, 1]$ and, as before, denote by $r_h s_\varkappa$ the piecewise linear interpolant of $s_\varkappa \in U^{ad}_\varkappa$ (given by (2.50)) on Δ_h defining the computational domain $\Omega(r_h s_\varkappa)$. Let $\mathcal{T}(h, s_\varkappa)$ be a triangulation of $\overline{\Omega(r_h s_\varkappa)}$ and $\mathcal{M}_h = \{A_i\}_{i=0}^{\bar{d}(h)}$, $A_i = (b_i, r_h s_\varkappa(b_i)) = (b_i, s_\varkappa(b_i))$, be the system of all nodes of $\mathcal{T}(h, s_\varkappa)$ lying on $\overline{\Gamma_C(r_h s_\varkappa)}$. By $\mathcal{N}_h \subseteq \mathcal{M}_h$ we denote the system of all the *contact nodes*; i.e., $A_i \in \mathcal{N}_h$ iff $A_i \in \overline{\Gamma_C(r_h s_\varkappa)} \setminus \overline{\Gamma_u(r_h s_\varkappa)}$ (observe that $\mathcal{N}_h$ is a proper subset of $\mathcal{M}_h$ provided $\overline{\Gamma_C(r_h s_\varkappa)} \cap \overline{\Gamma_u(r_h s_\varkappa)} \neq \emptyset$). We set

$$\mathbb{K}_h(s_\varkappa) = \{v_h = (v_{h1}, v_{h2}) \in \mathbb{V}_h(s_\varkappa) \mid v_{h2}(A_i) \geq -s_\varkappa(b_i) \quad \forall A_i \in \mathcal{N}_h\}, \tag{2.179}$$

where $\mathbb{V}_h(s_\varkappa)$ is defined by (2.167).

Remark 2.28. As in Remark 2.25 one can show that $\mathbb{K}_h(s_\varkappa) \subset \mathbb{K}(r_h s_\varkappa)$; i.e., functions from $\mathbb{K}_h(s_\varkappa)$ satisfy the unilateral condition not only at the contact nodes but also along the whole piecewise linear contact boundary $\Gamma_C(r_h s_\varkappa)$.

On any $\Omega_h(s_\varkappa) \in \mathcal{O}_{\varkappa h}$ we define the discrete state inequality as follows:

$$\begin{cases} \text{Find } u_h := u_h(s_\varkappa) \in \mathbb{K}_h(s_\varkappa) \quad \text{such that} \\ a_{r_h s_\varkappa}(u_h, v_h - u_h) \geq L_{r_h s_\varkappa}(v_h - u_h) \quad \forall v_h \in \mathbb{K}_h(s_\varkappa). \end{cases} \tag{$\mathcal{P}_{2h}(s_\varkappa)$}$$

Remark 2.29. In analogy to the continuous setting the solution $u_h(s_\varkappa)$ of $(\mathcal{P}_{2h}(s_\varkappa))$ can be characterized as follows:

$$\begin{cases} \text{Find } u_h(s_\varkappa) \in \mathbb{K}_h(s_\varkappa) \quad \text{such that} \\ E_{r_h s_\varkappa}(u_h(s_\varkappa)) \leq E_{r_h s_\varkappa}(v_h) \quad \forall v_h \in \mathbb{K}_h(s_\varkappa), \end{cases} \tag{$\widetilde{\mathcal{P}}_{2h}(s_\varkappa)$}$$

where

$$E_{r_h s_\varkappa}(v_h) = \frac{1}{2} a_{r_h s_\varkappa}(v_h, v_h) - L_{r_h s_\varkappa}(v_h).$$

The discretization of $(\mathbb{P}_2)$ is now obvious. For $h, \varkappa > 0$ fixed we define the following:

$$\begin{cases} \text{Find } s_\varkappa^* \in U_\varkappa^{ad} \quad \text{such that} \\ J(s_\varkappa^*, u_h(s_\varkappa^*)) \leq J(s_\varkappa, u_h(s_\varkappa)) \quad \forall s_\varkappa \in U_\varkappa^{ad}, \end{cases} \tag{$\mathbb{P}_{h\varkappa}^2$}$$

where $u_h(s_\varkappa) \in \mathbb{K}_h(s_\varkappa)$ solves $(\mathcal{P}_{2h}(s_\varkappa))$ and

$$J(s_\varkappa, u_h(s_\varkappa)) := E_{r_h s_\varkappa}(u_h(s_\varkappa)).$$

To establish a relation between $(\mathbb{P}_2)$ and $(\mathbb{P}_{h\varkappa}^2)$ for $h, \varkappa \to 0+$ we have to verify the assumptions of Theorem 2.11.

Lemma 2.28. *(Verification of $(\mathcal{B}2)_\varkappa$.) For any sequence $\{(s_\varkappa, u_h(s_\varkappa))\}$, where $s_\varkappa \in U_\varkappa^{ad}$ and $u_h(s_\varkappa) \in \mathbb{K}_h(s_\varkappa)$ solves $(\mathcal{P}_{2h}(s_\varkappa))$, $h \to 0+$, there exist its subsequence and elements $\alpha \in U^{ad}$, $u \in (H^1(\widehat{\Omega}))^2$ such that*

$$\begin{cases} s_\varkappa \rightrightarrows \alpha, \quad r_h s_\varkappa \rightrightarrows \alpha \quad \text{in } [0,1], \\ \widetilde{u}_h(s_\varkappa) \rightharpoonup u \quad \text{in } (H^1(\widehat{\Omega}))^2, \quad h, \varkappa \to 0+, \end{cases} \tag{2.180}$$

where $\widetilde{u}_h(s_\varkappa) = p_{\Omega(r_h s_\varkappa)} u_h(s_\varkappa)$ is the uniform extension of $u_h(s_\varkappa)$ from $\Omega(r_h s_\varkappa)$ to $\widehat{\Omega}$. In addition, $u(\alpha) := u_{|\Omega(\alpha)}$ solves $(\mathcal{P}_2(\alpha))$.

Proof. (Sketch.) The existence of a subsequence and of elements $\alpha \in U^{ad}$, $u \in (H^1(\widehat{\Omega}))^2$ satisfying (2.180) is obvious. Also, the fact that $u(\alpha) := u_{|\Omega(\alpha)} \in \mathbb{K}(\alpha)$ can be shown as in Lemma 2.26. Next we prove that $u(\alpha)$ solves $(\mathcal{P}_2(\alpha))$. From Lemma 2.25 we know that any function $v \in \mathbb{K}(\alpha)$ can be approximated by a sequence $\{v_j\}$, $v_j \in (H^1(\widehat{\Omega}))^2$, satisfying (2.176) with $\{r_h s_\varkappa\}$ instead of $\{\alpha_n\}$. It is possible to show (see [HNT86]) that the functions v_j can be chosen even more regular, namely $v_j \in (C^\infty(\overline{\widehat{\Omega}}))^2$ $\forall j \in \mathbb{N}$ and $v_{j|\Omega(r_h s_\varkappa)} \in \mathbb{K}(r_h s_\varkappa)$ provided that $h, \varkappa$ are small enough, still preserving $(2.176)_2$. Let $\{\widehat{\mathcal{T}}(h, s_\varkappa)\}$ be a system of triangulations of $\widehat{\Omega} \setminus \overline{\Omega(r_h s_\varkappa)}$ satisfying $(\mathcal{T}1)$–$(\mathcal{T}3)$ and let $\pi_h v_j$

be the piecewise linear Lagrange interpolant of v_j on $\mathcal{T}(h, s_\varkappa) \cup \widehat{\mathcal{T}}(h, s_\varkappa)$. Fix $j \in \mathbb{N}$. Then

$$\pi_h v_j \to v_j \quad \text{in } (H^1(\widehat{\Omega}))^2, \quad h \to 0+,$$

and $\pi_h v_{j|\Omega_h(s_\varkappa)} \in \mathbb{K}_h(s_\varkappa)$ for $h, \varkappa$ small enough. Therefore $\pi_h v_j$ can be used as a test function in $(\mathcal{P}_{2h}(s_\varkappa))$. The rest of the proof proceeds in the standard way: we pass to the limit in $(\mathcal{P}_{2h}(s_\varkappa))$ first with $h \to 0+$, then with $j \to \infty$. $\square$

Lemma 2.29. *(Verification of $(\mathcal{B}3)_\varkappa$.) Let $\{(s_\varkappa, u_h(s_\varkappa))\}$, where $s_\varkappa \in U^{ad}_\varkappa$ and $u_h(s_\varkappa) \in \mathbb{K}_h(s_\varkappa)$ solves $(\mathcal{P}_{2h}(s_\varkappa))$, $h \to 0+$, be such that* (2.180) *holds for some $\alpha \in U^{ad}$ and $u \in (H^1(\widehat{\Omega}))^2$. Then*

$$\lim_{h,\varkappa \to 0+} J(s_\varkappa, u_h(s_\varkappa)) = J(\alpha, u_{|\Omega(\alpha)}).$$

Proof. From the proof of the previous lemma we know that any function from $\mathbb{K}(\alpha)$, in particular the solution $u(\alpha)$ of $(\mathcal{P}_2(\alpha))$, can be approximated by a sequence $\{u_j\}$, $u_j \in (C^\infty(\overline{\widehat{\Omega}}))^2$, $u_{j|\Omega(r_h s_\varkappa)} \in \mathbb{K}(r_h s_\varkappa)$, $h, \varkappa$ small enough, such that $(2.176)_2$ holds with $\widetilde{v} := \widetilde{u}(\alpha)$. Let $\pi_h u_j$ be the piecewise linear Lagrange interpolant of u_j constructed in the previous lemma. We know that for any $j \in \mathbb{N}$ fixed

$$\begin{cases} \pi_h u_j \to u_j \quad \text{in } (H^1(\widehat{\Omega}))^2, \quad h \to 0+, \\ \pi_h u_{j|\Omega_h(s_\varkappa)} \in \mathbb{K}_h(s_\varkappa) \end{cases} \tag{2.181}$$

for $h, \varkappa$ small enough. From the definition of J, Remark 2.29, and (2.181) it follows that

$$J(s_\varkappa, u_h(s_\varkappa)) = E_{r_h s_\varkappa}(u_h(s_\varkappa)) \le E_{r_h s_\varkappa}(\pi_h u_j).$$

From this and (2.181) we easily obtain

$$\begin{aligned} \limsup_{h,\varkappa\to 0+} J(s_\varkappa, u_h(s_\varkappa)) &\le \limsup_{j\to\infty} \left(\limsup_{h,\varkappa\to 0+} E_{r_h s_\varkappa}(\pi_h u_j) \right) \\ &= \lim_{j\to\infty} \left(\lim_{h,\varkappa\to 0+} E_{r_h s_\varkappa}(\pi_h u_j) \right) = E_\alpha(u(\alpha)) \end{aligned} \tag{2.182}$$

(for the sake of simplicity of notation we write $\pi_h u_j$ instead of $\pi_h u_{j|\Omega(r_h s_\varkappa)}$). On the other hand, from Lemma 2.27, it follows that

$$\liminf_{h,\varkappa\to 0+} J(s_\varkappa, u_h(s_\varkappa)) \ge J(\alpha, u(\alpha)) = E_\alpha(u(\alpha)).$$

Combining this and (2.182) we see that

$$\lim_{h,\varkappa\to 0+} J(s_\varkappa, u_h(s_\varkappa)) = J(\alpha, u(\alpha)). \quad \square$$

Since all the convergence assumptions are satisfied, we have proved that $(\mathbb{P}_2)$ and $(\mathbb{P}^2_{h\varkappa})$, $h \to 0+$, are close in the sense of subsequences in the sense of Theorem 2.11.

The matrix formulation of $(\widetilde{\mathcal{P}}_{2h}(s_\varkappa))$ for $h, \varkappa > 0$ fixed leads again to a quadratic programming problem similar to the one presented in the previous subsection. However, there is one distinction in comparison with the scalar case: the closed, convex subset of $\mathbb{R}^n$ ($n = \dim \mathbb{V}_h(s_\varkappa)$), which is isomorphic with $\mathbb{K}_h(s_\varkappa)$, now *depends explicitly* on the discrete design variable $\boldsymbol{\alpha} \in \mathcal{U}$. Indeed, the function $s_\varkappa \in U_\varkappa^{ad}$ defining the shape of $\Omega(s_\varkappa)$ is uniquely determined by a discrete design variable $\boldsymbol{\alpha} \in \mathcal{U}$ (see (2.48) and (2.50)). To stress this fact let us write $s_\varkappa(\boldsymbol{\alpha})$ instead of $s_\varkappa$. Then

$$\mathcal{K}(\boldsymbol{\alpha}) = \{\boldsymbol{x} \in \mathbb{R}^n \mid x_{j_i} \geq -s_\varkappa(b_i, \boldsymbol{\alpha}) \quad \forall j_i \in \mathcal{I}\}, \tag{2.183}$$

where $\mathcal{I}$ is the set of indices of all (constrained) x_2 components of $u_h(s_\varkappa)$ at the nodes of $\mathcal{N}_h$:

$$j_i \in \mathcal{I} \iff \exists A_i \in \mathcal{N}_h : \quad x_{j_i} = u_{h2}(A_i),\ A_i = (b_i, s_\varkappa(b_i, \boldsymbol{\alpha})).$$

2.5.6 Shape optimization in fluid mechanics

In the third part of this book we shall present several examples of shape optimization in the field of fluid mechanics with applications in practice. The aim of this subsection is to show how the techniques developed so far can be used in such problems. For the sake of simplicity we restrict ourselves to *two-dimensional stationary problems* governed by the Stokes equations and to a cost functional J depending *only on the velocity*. For more details on the formulation and the numerical realization of problems in fluid mechanics we refer to [GR79].

We first introduce the state problem. Let an incompressible Newtonian fluid occupy a domain $\Omega \subset \mathbb{R}^2$. We want to find the velocity field $u = (u_1, u_2)$ and the pressure p of the fluid satisfying the Stokes equations

$$\begin{cases} -\mu \Delta u_i + \dfrac{\partial p}{\partial x_i} = f_i & \text{in } \Omega, \quad i = 1, 2, \\ \operatorname{div} u = 0 & \text{in } \Omega, \\ u_i = 0 & \text{on } \partial\Omega, \quad i = 1, 2, \end{cases} \tag{2.184}$$

where $\mu > 0$ is a viscosity parameter and $f \in (L^2_{loc}(\mathbb{R}^2))^2$ is the density of external forces.

For simplicity let $\mu = 1$. To get a weak form of (2.184) we introduce the space

$$\mathbb{V}(\Omega) = \{v \in (H_0^1(\Omega))^2 \mid \operatorname{div} v = 0 \text{ in } \Omega\}. \tag{2.185}$$

We start with the so-called velocity formulation. A velocity u is said to be a *weak* solution of the Stokes problem iff

$$u \in \mathbb{V}(\Omega) : \quad \int_\Omega \nabla u : \nabla v \, dx = \int_\Omega f \cdot v \, dx \quad \forall v \in \mathbb{V}(\Omega), \tag{$\mathcal{P}$}$$

where $\nabla u : \nabla v := \nabla u_i \cdot \nabla v_i$ and $f \cdot v := f_i v_i$ using again the summation convention. The pressure p disappears from this formulation owing to the definition of the space $\mathbb{V}(\Omega)$ in which the incompressibility constraint $\operatorname{div} v = 0$ in Ω is satisfied a priori. On the other hand this condition can be viewed to be an additional constraint that can be released by means of a Lagrange multiplier technique. In this way we arrive at the *velocity-pressure formulation*, frequently used in computations.

A couple $(u, p) \in (H_0^1(\Omega))^2 \times L_0^2(\Omega)$ is called a *weak solution of the velocity-pressure formulation* of the Stokes equations iff

$$\begin{cases} \displaystyle\int_\Omega \nabla u : \nabla v \, dx - \int_\Omega p \operatorname{div} v \, dx = \int_\Omega f \cdot v \, dx \quad \forall v \in (H_0^1(\Omega))^2, \\ \displaystyle\int_\Omega q \operatorname{div} u \, dx = 0 \quad \forall q \in L_0^2(\Omega), \end{cases} \tag{$\widetilde{\mathcal{P}}$}$$

where $L_0^2(\Omega)$ is the subspace of functions from $L^2(\Omega)$ that are orthogonal to P_0, i.e., to all constants. The pressure p can be interpreted as the Lagrange multiplier associated with the constraint $\operatorname{div} u = 0$ in Ω. It is well known that $(\mathcal{P})$ and $(\widetilde{\mathcal{P}})$ are equivalent in the following sense: if $u \in \mathbb{V}(\Omega)$ solves $(\mathcal{P})$, then there exists $p \in L_0^2(\Omega)$ such that the pair (u, p) solves $(\widetilde{\mathcal{P}})$. On the contrary, if (u, p) solves $(\widetilde{\mathcal{P}})$ then the first component u solves $(\mathcal{P})$. In addition, $(\mathcal{P})$ and $(\widetilde{\mathcal{P}})$ have unique solutions u and (u, p), respectively.

In what follows we shall consider an optimal design problem with the Stokes equations as the state problem. As before we restrict ourselves to domains depicted in Figure 2.2; i.e., the designed part is again parametrized by functions $\alpha \in \widetilde{U}^{ad}$ given by (2.96). On any $\Omega(\alpha)$, $\alpha \in \widetilde{U}^{ad}$, the Stokes problem will be defined. One of the typical shape optimization problems in fluid mechanics consists of finding a profile of the structure that minimizes the dissipated energy. The mathematical formulation of this problem is as follows:

$$\begin{cases} \text{Find } \alpha^* \in U^{ad} \quad \text{such that} \\ J(\alpha^*, u(\alpha^*)) \leq J(\alpha, u(\alpha)) \quad \forall \alpha \in U^{ad}, \end{cases} \tag{$\mathbb{P}$}$$

where

$$J(\alpha, v) = \frac{1}{2} \int_{\Omega(\alpha)} \varepsilon_{ij}(v) \varepsilon_{ij}(v) \, dx, \qquad \varepsilon_{ij}(v) = \frac{1}{2} \left(\frac{\partial v_i}{\partial x_j} + \frac{\partial v_j}{\partial x_i} \right); \tag{2.186}$$

$u(\alpha)$ solves the Stokes equations in $\Omega(\alpha)$; and U^{ad} is defined by (2.97). Below we prove the existence of a solution to $(\mathbb{P})$. To this end we shall verify assumptions $(\mathcal{B}1)$ and $(\mathcal{B}2)$ of Section 2.4. Since the cost functional J involves only the velocity field u, we shall use the weak velocity formulation of the Stokes problem in $\Omega(\alpha)$:

$$\begin{cases} \text{Find } u := u(\alpha) \in \mathbb{V}(\alpha) \quad \text{such that} \\ \displaystyle\int_{\Omega(\alpha)} \nabla u : \nabla v \, dx = \int_{\Omega(\alpha)} f \cdot v \, dx \quad \forall v \in \mathbb{V}(\alpha), \end{cases} \tag{$\mathcal{P}(\alpha)$}$$

where $\mathbb{V}(\alpha)$ is the space defined by (2.185) in $\Omega := \Omega(\alpha)$, $\alpha \in \widetilde{U}^{ad}$.

Convention: *Throughout this subsection, the symbol* $\tilde{}$ *above the function stands for its extension by zero outside of the domain of its definition.*

Lemma 2.30. *(Verification of $(\mathcal{B}1)$.) For any sequence $\{(\alpha_n, u_n)\}$, where $\alpha_n \in \widetilde{U}^{ad}$ and $u_n := u(\alpha_n) \in \mathbb{V}(\alpha_n)$ solves $(\mathcal{P}(\alpha_n))$, $n \to \infty$, there exist its subsequence and elements $\alpha \in \widetilde{U}^{ad}$, $u \in (H_0^1(\widehat{\Omega}))^2$ such that*

$$\begin{cases} \alpha_n \rightrightarrows \alpha & \text{in } [0, 1], \\ \widetilde{u}_n \to u & \text{in } (H_0^1(\widehat{\Omega}))^2, \quad n \to \infty. \end{cases} \tag{2.187}$$

In addition, $u(\alpha) := u_{|\Omega(\alpha)}$ solves $(\mathcal{P}(\alpha))$.

Proof. The proof proceeds in the same way as the proof of Lemma 2.5 using that $\mathbb{V}(\alpha) \cap (C_0^\infty(\Omega(\alpha)))^2$ is dense in $\mathbb{V}(\alpha)$ for any $\alpha \in \widetilde{U}^{ad}$ (for the proof see [Tem77]) and the fact that if $v \in \mathbb{V}(\alpha)$, then its zero extension $\widetilde{v}$ belongs to $\mathbb{V}(\widehat{\Omega}) = \{\varphi \in (H_0^1(\widehat{\Omega}))^2 \mid \operatorname{div}\varphi = 0 \text{ in } \widehat{\Omega}\}$. Only one more step has to be proven, namely that $u(\alpha)$ is divergence free in $\Omega(\alpha)$, or

$$\int_{\Omega(\alpha)} u(\alpha) \cdot \nabla\varphi \, dx = 0 \quad \forall \varphi \in C_0^\infty(\Omega(\alpha)). \tag{2.188}$$

Let $\varphi \in C_0^\infty(\Omega(\alpha))$ be fixed. Then $\widetilde{\varphi}_{|\Omega(\alpha_n)} \in C_0^\infty(\Omega(\alpha_n))$ for any $n \geq n_0$, where $n_0 \in \mathbb{N}$ is large enough. Since $u_n \in \mathbb{V}(\alpha_n)$ we have that

$$\int_{\Omega(\alpha_n)} u_n \cdot \nabla\widetilde{\varphi} \, dx = 0 \quad \forall n \geq n_0.$$

Passing to the limit with $n \to \infty$ and using (2.187) we arrive at (2.188). $\square$

It is readily seen that assumption $(\mathcal{B}2)$ is also satisfied for the cost functional J defined by (2.186). Therefore from Theorem 2.10 we obtain the following.

Theorem 2.19. *Problem* $(\mathbb{P})$ *has a solution.*

Let us pass to an approximation of $(\mathbb{P})$. As far as the approximation of the geometry is concerned, we use the same systems $\mathcal{O}_\varkappa$ and $\mathcal{O}_{\varkappa h}$ as in Section 2.2. We will focus on the discretization of the state problem. The incompressibility condition is the main difficulty we face in the numerical realization of the Stokes equations. There are several ways to treat this condition. Here we briefly mention three of them: (i) *the stream function formulation,* (ii) *mixed finite element formulations,* and (iii) *penalization techniques.*

The stream function formulation relies on the fact that any function $v \in \mathbb{V}(\alpha)$ in (simply connected) $\Omega(\alpha)$ can be written in the form $v = \operatorname{curl}\psi$, where the so-called *stream function* ψ belongs to $H_0^2(\Omega(\alpha))$. Inserting $u := \operatorname{curl}\varphi$, $v := \operatorname{curl}\psi$ into $(\mathcal{P}(\alpha))$, $\varphi, \psi \in H_0^2(\Omega(\alpha))$, we derive the stream function formulation of the Stokes problem:

$$\left\{\begin{array}{l} \text{Find } \varphi := \varphi(\alpha) \in H_0^2(\Omega(\alpha)) \quad \text{such that} \\ \displaystyle\int_{\Omega(\alpha)} \Delta\varphi\Delta\psi \, dx = \int_{\Omega(\alpha)} f \cdot \operatorname{curl}\psi \, dx \quad \forall \psi \in H_0^2(\Omega(\alpha)), \end{array}\right. \tag{$\mathcal{P}_{\mathrm{str}}(\alpha)$}$$

which will be the basis of a finite element approximation.

Remark 2.30. In the derivation of $(\mathcal{P}_{\mathrm{str}}(\alpha))$ we used the fact that

$$\int_{\Omega(\alpha)} \frac{\partial^2\varphi}{\partial x_1^2}\frac{\partial^2\psi}{\partial x_2^2} \, dx = \int_{\Omega(\alpha)} \frac{\partial^2\varphi}{\partial x_1\partial x_2}\frac{\partial^2\psi}{\partial x_1\partial x_2} \, dx \quad \forall \varphi, \psi \in H_0^2(\Omega(\alpha)).$$

Let $W_h(s_\varkappa) \subset H_0^2(\Omega(r_h s_\varkappa))$ be a finite element space constructed by means of C^1-finite elements over a triangulation $\mathcal{T}(h, s_\varkappa)$ of $\overline{\Omega(r_h s_\varkappa)}$, $s_\varkappa \in U_\varkappa^{ad}$ (we may use, for example, polynomials of the fifth degree on triangles with 21 degrees of freedom; for details we refer to [Cia02]).

The discretization of the stream function formulation is now defined in the standard way:

$$\begin{cases} \text{Find } \varphi_h := \varphi_h(s_\varkappa) \in W_h(s_\varkappa) \quad \text{such that} \\ \displaystyle\int_{\Omega_h(s_\varkappa)} \Delta\varphi_h \Delta\psi_h \, dx = \int_{\Omega_h(s_\varkappa)} f \cdot \operatorname{curl} \psi_h \, dx \quad \forall \psi_h \in W_h(s_\varkappa). \end{cases} \qquad (\mathcal{P}_{\mathrm{str},h}(s_\varkappa))$$

With $\varphi_h(s_\varkappa)$ at hand, the function $u_h(s_\varkappa) := \operatorname{curl} \varphi_h(s_\varkappa)$ approximates the velocity field. Observe that the incompressibility condition $\operatorname{div} u_h(s_\varkappa) = 0$ is satisfied *exactly* in $\Omega_h(s_\varkappa)$ in this case.

The discretization of $(\mathbb{P})$ now reads as follows:

$$\begin{cases} \text{Find } s^*_\varkappa \in U^{ad}_\varkappa \quad \text{such that} \\ J(s^*_\varkappa, \operatorname{curl} \varphi_h(s^*_\varkappa)) \le J(s_\varkappa, \operatorname{curl} \varphi_h(s_\varkappa)) \quad \forall s_\varkappa \in U^{ad}_\varkappa, \end{cases} \qquad (\mathbb{P}_{h\varkappa})$$

with $\varphi_h(s_\varkappa) \in W_h(s_\varkappa)$ being the solution of $(\mathcal{P}_{\mathrm{str},h}(s_\varkappa))$.

We want to prove that $(\mathbb{P})$ and $(\mathbb{P}_{h\varkappa})$ are close as $h, \varkappa \to 0+$ in the sense of Theorem 2.11. We first shall verify assumption $(\mathcal{B}2)_\varkappa$ of Section 2.4.

Lemma 2.31. *(Verification of $(\mathcal{B}2)_\varkappa$.) For any sequence $\{(s_\varkappa, \varphi_h(s_\varkappa))\}$, where $s_\varkappa \in U^{ad}_\varkappa$ and $\varphi_h := \varphi_h(s_\varkappa) \in W_h(s_\varkappa)$ is the solution of $(\mathcal{P}_{\mathrm{str},h}(s_\varkappa))$, $h \to 0+$, there exist its subsequence and elements $\alpha \in U^{ad}$, $\varphi \in H^2_0(\widehat{\Omega})$ such that*

$$\begin{cases} s_\varkappa \rightrightarrows \alpha \quad \text{in } [0,1], \\ \widetilde{\varphi}_h(s_\varkappa) \to \varphi \quad \text{in } H^2_0(\widehat{\Omega}), \quad h, \varkappa \to 0+. \end{cases} \tag{2.189}$$

In addition, $\varphi(\alpha) := \varphi_{|\Omega(\alpha)}$ solves $(\mathcal{P}_{\mathrm{str}}(\alpha))$.

Proof. Inserting $\psi_h := \varphi_h$ into $(\mathcal{P}_{\mathrm{str},h}(s_\varkappa))$ we obtain (see Remark 2.30)

$$|\varphi_h|^2_{2,\Omega_h(s_\varkappa)} = \|\Delta\varphi_h\|^2_{0,\Omega_h(s_\varkappa)} \le \|f\|_{0,\widehat{\Omega}} \|\varphi_h\|_{2,\Omega_h(s_\varkappa)}. \tag{2.190}$$

Since the extended function $\widetilde{\varphi}_h$ belongs to $H^2_0(\widehat{\Omega})$, (2.190) and the Friedrichs inequality in $H^2_0(\widehat{\Omega})$ yield

$$c\|\widetilde{\varphi}_h\|^2_{2,\widehat{\Omega}} \le |\widetilde{\varphi}_h|^2_{2,\widehat{\Omega}} = |\varphi_h|^2_{2,\Omega_h(s_\varkappa)} \le \|f\|_{0,\widehat{\Omega}} \|\widetilde{\varphi}_h\|_{2,\widehat{\Omega}},$$

where $c > 0$ does not depend on $h, \varkappa > 0$, implying the boundedness of $\{\|\widetilde{\varphi}_h\|_{2,\widehat{\Omega}}\}$. Thus one can pass to a subsequence of $\{(s_\varkappa, \widetilde{\varphi}_h)\}$ satisfying $(2.189)_1$ for some $\alpha \in U^{ad}$ and converging weakly to an element $\varphi \in H^2_0(\widehat{\Omega})$. To prove that $\varphi(\alpha) := \varphi_{|\Omega(\alpha)}$ belongs to $H^2_0(\Omega(\alpha))$ we have to show that $\varphi(\alpha) = \partial\varphi(\alpha)/\partial\nu = 0$ on $\Gamma(\alpha)$. But this follows from the fact that $\varphi = 0$ in $\widehat{\Omega} \setminus \overline{\Omega(\alpha)}$ (see the proof of Lemma 2.5). It remains to prove that $\varphi(\alpha)$ solves $(\mathcal{P}_{\mathrm{str}}(\alpha))$.

Let $\{\widehat{\mathcal{T}}(h, s_\varkappa)\}$ be a regular family of triangulations of $\widehat{\Omega} \setminus \overline{\Omega(r_h s_\varkappa)}$ satisfying $(\mathcal{T}1)$–$(\mathcal{T}3)$ of Section 2.2 and $W_h(\widehat{\Omega})$ be the finite element space over $\mathcal{T}(h, s_\varkappa) \cup \widehat{\mathcal{T}}(h, s_\varkappa)$ using the same type of C^1-finite elements as above. Let $\psi \in C^\infty_0(\Omega(\alpha))$ be given and denote by $\pi_h\widetilde{\psi} \in W_h(\widehat{\Omega})$ the piecewise Hermite interpolant of $\widetilde{\psi}$. Then

$$\pi_h\widetilde{\psi} \to \widetilde{\psi} \quad \text{in } H^2_0(\widehat{\Omega}), \quad h \to 0+. \tag{2.191}$$

In addition, the restriction $\pi_h \widetilde{\psi}_{|\Omega_h(s_\varkappa)}$ belongs to $W_h(s_\varkappa)$ as follows from $(2.189)_1$ for $h, \varkappa > 0$ small enough. Thus it can be used as a test function in $(\mathcal{P}_{\mathrm{str},h}(s_\varkappa))$:

$$\int_{\Omega_h(s_\varkappa)} \Delta\varphi_h \Delta(\pi_h \widetilde{\psi})\, dx = \int_{\Omega_h(s_\varkappa)} f \cdot \operatorname{curl}(\pi_h \widetilde{\psi})\, dx$$

or, equivalently,

$$\int_{\widehat{\Omega}} \chi_{h\varkappa} \Delta\widetilde{\varphi}_h \Delta(\pi_h \widetilde{\psi})\, dx = \int_{\widehat{\Omega}} \chi_{h\varkappa} f \cdot \operatorname{curl}(\pi_h \widetilde{\psi})\, dx,$$

where $\chi_{h\varkappa}$ is the characteristic function of $\Omega_h(s_\varkappa)$. Letting $h, \varkappa \to 0+$ we arrive at

$$\int_{\widehat{\Omega}} \chi \Delta\varphi \Delta\widetilde{\psi}\, dx = \int_{\Omega(\alpha)} \chi f \cdot \operatorname{curl} \widetilde{\psi}\, dx, \tag{2.192}$$

where χ is the characteristic function of $\Omega(\alpha)$, making use of $(2.189)_1$, (2.191), and weak convergence of a subsequence of $\{\widetilde{\varphi}_h(s_\varkappa)\}$ to φ. Since $C_0^\infty(\Omega(\alpha))$ is dense in $H_0^2(\Omega(\alpha))$, (2.192) holds for any $\psi \in H_0^2(\Omega(\alpha))$; i.e., $\varphi_{|\Omega(\alpha)}$ solves $(\mathcal{P}_{\mathrm{str}}(\alpha))$. The proof of strong convergence of $\{\widetilde{\varphi}_h(s_\varkappa)\}$ to φ in the $H_0^2(\widehat{\Omega})$-norm is obvious:

$$|\widetilde{\varphi}_h|^2_{2,\widehat{\Omega}} = (f, \operatorname{curl} \widetilde{\varphi}_h(s_\varkappa))_{0,\widehat{\Omega}} \to (f, \operatorname{curl} \varphi)_{0,\widehat{\Omega}} = |\varphi|^2_{2,\widehat{\Omega}}. \qquad \square$$

Since assumption $(\mathcal{B}3)_\varkappa$ is also satisfied for the cost functional J, we arrive at the following result.

Theorem 2.20. *Let $\{(s^*_\varkappa, \varphi_h(s^*_\varkappa))\}$ be a sequence of optimal pairs of $(\mathbb{P}_{h\varkappa})$, $h \to 0+$. Then one can find its subsequence such that*

$$\begin{cases} s^*_\varkappa \rightrightarrows \alpha^* & \text{in } [0,1], \\ \widetilde{\varphi}_h(s^*_\varkappa) \to \varphi^* & \text{in } H_0^2(\widehat{\Omega}), \quad h, \varkappa \to 0+. \end{cases} \tag{2.193}$$

In addition, $(\alpha^*, \varphi^*{}_{|\Omega(\alpha^*)})$ is an optimal pair of $(\mathbb{P})$. Any accumulation point of $\{(s^*_\varkappa, \widetilde{\varphi}_h(s^*_\varkappa))\}$ in the sense of (2.193) possesses this property.

Due to their complexity, C^1-finite elements used for the approximation of the stream function formulation represent the main drawback of the previous approach. For this reason, mixed finite element techniques based on the velocity-pressure formulation are nowadays more popular. For more details on mixed variational formulations see also Chapter 6.

On any $\Omega(\alpha) \in \widetilde{\mathcal{O}}$ we shall consider the velocity-pressure formulation of the Stokes problem:

$$\left\{\begin{aligned} &\text{Find } (u(\alpha), p(\alpha)) \in (H_0^1(\Omega(\alpha)))^2 \times L_0^2(\Omega(\alpha)) \quad \text{such that} \\ &\int_{\Omega(\alpha)} \nabla u(\alpha) : \nabla v\, dx - \int_{\Omega(\alpha)} p(\alpha) \operatorname{div} v\, dx \\ &\qquad = \int_{\Omega(\alpha)} f \cdot v\, dx \qquad \forall v \in (H_0^1(\Omega(\alpha)))^2, \\ &\int_{\Omega(\alpha)} q \operatorname{div} u(\alpha)\, dx = 0 \qquad \forall q \in L_0^2(\Omega(\alpha)). \end{aligned}\right. \tag{$\widetilde{\mathcal{P}}(\alpha)$}$$

As we have already mentioned we are not now restricted to divergence free velocities.

Let $\mathbb{X}_h(s_\varkappa) \subset \left(H_0^1(\Omega(r_h s_\varkappa))\right)^2$, $M_h(s_\varkappa) \subset L_0^2(\Omega(r_h s_\varkappa))$ be finite element spaces constructed by means of triangular elements in $\Omega_h(s_\varkappa)$. By a *mixed finite element approximation* of the Stokes equations we mean a problem of finding a pair $(u_h(s_\varkappa), p_h(s_\varkappa)) \in \mathbb{X}_h(s_\varkappa) \times M_h(s_\varkappa)$ satisfying

$$\begin{cases} \displaystyle\int_{\Omega_h(s_\varkappa)} \nabla u_h(s_\varkappa) : \nabla v_h \, dx - \int_{\Omega_h(s_\varkappa)} p_h(s_\varkappa) \mathrm{div}\, v_h \, dx \\ \qquad\qquad \displaystyle = \int_{\Omega_h(s_\varkappa)} f \cdot v_h \, dx \quad \forall v_h \in \mathbb{X}_h(s_\varkappa), \\ \displaystyle\int_{\Omega_h(s_\varkappa)} q_h \mathrm{div}\, u_h(s_\varkappa) \, dx = 0 \qquad \forall q_h \in M_h(s_\varkappa). \end{cases} \tag{$\widetilde{\mathcal{P}}_h(s_\varkappa)$}$$

The existence and uniqueness of a solution to $(\widetilde{\mathcal{P}}_h(s_\varkappa))$ is ensured provided that the following condition is satisfied:

$$\int_{\Omega_h(s_\varkappa)} q_h \mathrm{div}\, v_h \, dx = 0 \quad \forall v_h \in \mathbb{X}_h(s_\varkappa) \implies q_h = 0 \quad \text{in } \Omega_h(s_\varkappa). \tag{2.194}$$

This condition *restricts* the choice of $\mathbb{X}_h(s_\varkappa)$ and $M_h(s_\varkappa)$ used in $(\widetilde{\mathcal{P}}_h(s_\varkappa))$. For pairs satisfying (2.194) we refer to [GR79] and [BF91]. Contrary to the stream function formulation, the divergence free constraint in $(\widetilde{\mathcal{P}}_h(s_\varkappa))$ is satisfied only in the following *approximate sense*:

$$\int_{\Omega_h(s_\varkappa)} q_h \mathrm{div}\, v_h \, dx = 0 \qquad \forall q_h \in M_h(s_\varkappa).$$

On the other hand, mixed finite element methods enable us to approximate *simultaneously* and *independently* the velocity u and the pressure p. The convergence analysis for $(\mathbb{P}_{h\varkappa})$, $h, \varkappa \to 0+$, using mixed finite element formulations of the Stokes state equations can again be carried out by adapting the abstract theory of Section 2.4.

A penalty technique is another way to treat the incompressibility condition. We shall illustrate this approach in the continuous setting of the Stokes equations.

It is known that the velocity formulation $(\mathcal{P}(\alpha))$ is equivalent to the following minimization problem:

$$\begin{cases} \text{Find } u(\alpha) \in \mathbb{V}(\alpha) \quad \text{such that} \\ E_\alpha(u(\alpha)) \leq E_\alpha(v) \quad \forall v \in \mathbb{V}(\alpha), \end{cases} \tag{$\mathcal{P}'(\alpha)$}$$

where

$$E_\alpha(v) = \frac{1}{2} \int_{\Omega(\alpha)} \nabla v : \nabla v \, dx - \int_{\Omega(\alpha)} f \cdot v \, dx, \quad \alpha \in \widetilde{U}^{ad}, \ v \in (H_0^1(\Omega(\alpha)))^2.$$

The divergence free condition appearing in the definition of $\mathbb{V}(\alpha)$ represents a constraint that can be involved in the problem by means of a suitable *penalty term*.

Let $\varepsilon > 0$ be a penalty parameter destined to tend to zero and define the augmented functional $E_{\alpha,\varepsilon}$:

$$E_{\alpha,\varepsilon}(v) := E_\alpha(v) + \frac{1}{2\varepsilon} \int_{\Omega(\alpha)} (\mathrm{div}\, v)^2 \, dx, \quad \alpha \in \widetilde{U}^{ad}, \quad v \in (H_0^1(\Omega(\alpha)))^2.$$

The *penalized form* of the Stokes equations in $\Omega(\alpha)$ is defined as follows:

$$\begin{cases} \text{Find } u_\varepsilon(\alpha) \in (H_0^1(\Omega(\alpha)))^2 \quad \text{such that} \\ E_{\alpha,\varepsilon}(u_\varepsilon(\alpha)) \le E_{\alpha,\varepsilon}(v) \quad \forall v \in (H_0^1(\Omega(\alpha)))^2 \end{cases} \tag{$\mathcal{P}'_\varepsilon(\alpha)$}$$

or, equivalently:

$$\begin{cases} \text{Find } u_\varepsilon := u_\varepsilon(\alpha) \in (H_0^1(\Omega(\alpha)))^2 \quad \text{such that} \\ \displaystyle\int_{\Omega(\alpha)} \nabla u_\varepsilon : \nabla v\, dx + \frac{1}{\varepsilon}\int_{\Omega(\alpha)} \operatorname{div} u_\varepsilon \operatorname{div} v\, dx = \int_{\Omega(\alpha)} f \cdot v\, dx \;\; \forall v \in (H_0^1(\Omega(\alpha)))^2. \end{cases} \tag{$\mathcal{P}_\varepsilon(\alpha)$}$$

Notice that $(\mathcal{P}'_\varepsilon(\alpha))$ is already the *unconstrained* minimization problem, which approximates $(\mathcal{P}(\alpha))$ in the following sense (see [Lio69]):

$$u_\varepsilon(\alpha) \to u(\alpha) \quad \text{in } (H_0^1(\Omega(\alpha)))^2, \quad \varepsilon \to 0+. \tag{2.195}$$

For any $\varepsilon > 0$ fixed, we shall define the following optimal shape design problem:

$$\begin{cases} \text{Find } \alpha_\varepsilon^* \in U^{ad} \quad \text{such that} \\ J(\alpha_\varepsilon^*, u_\varepsilon(\alpha_\varepsilon^*)) \le J(\alpha, u_\varepsilon(\alpha)) \quad \forall \alpha \in U^{ad}, \end{cases} \tag{$\mathbb{P}_\varepsilon$}$$

where $u_\varepsilon(\alpha) \in (H_0^1(\Omega(\alpha)))^2$ is the solution to $(\mathcal{P}_\varepsilon(\alpha))$ and J is defined by (2.186). Instead of the original state problem $(\mathcal{P}(\alpha))$ we use its penalty approximation $(\mathcal{P}_\varepsilon(\alpha))$ whose quality depends on, among other factors, the magnitude of ε. The optimal shape in $(\mathbb{P}_\varepsilon)$ depends on ε. This is why we use the notation α_ε^*. A natural question arises: What happens when $\varepsilon \to 0+$? Are solutions to $(\mathbb{P})$ and $(\mathbb{P}_\varepsilon)$ close for $\varepsilon \to 0+$ and, if yes, in which sense? We shall study this in more detail.

First of all, using the abstract theory of Section 2.4, one can easily show that $(\mathbb{P}_\varepsilon)$ has at least one solution α_ε^* for any $\varepsilon > 0$.

The following lemma plays a crucial role in the convergence analysis.

Lemma 2.32. *For any sequence $\{(\alpha_\varepsilon, u_\varepsilon(\alpha_\varepsilon))\}$, where $\alpha_\varepsilon \in \widetilde{U}^{ad}$ and $u_\varepsilon(\alpha_\varepsilon) \in (H_0^1(\Omega(\alpha_\varepsilon)))^2$ is the solution of $(\mathcal{P}_\varepsilon(\alpha_\varepsilon))$, $\varepsilon \to 0+$, there exist its subsequence and elements $\alpha \in \widetilde{U}^{ad}$, $u \in (H_0^1(\widehat{\Omega}))^2$ such that*

$$\begin{cases} \alpha_\varepsilon \rightrightarrows \alpha \quad \text{in } [0,1], \\ \widetilde{u}_\varepsilon(\alpha_\varepsilon) \to u \quad \text{in } (H_0^1(\widehat{\Omega}))^2, \quad \varepsilon \to 0+. \end{cases} \tag{2.196}$$

In addition, $u(\alpha) := u_{|\Omega(\alpha)}$ solves $(\mathcal{P}(\alpha))$.

Proof. The definition of $(\mathcal{P}_\varepsilon(\alpha_\varepsilon))$ yields

$$\int_{\Omega(\alpha_\varepsilon)} \nabla u_\varepsilon : \nabla u_\varepsilon\, dx + \frac{1}{\varepsilon}\int_{\Omega(\alpha_\varepsilon)} (\operatorname{div} u_\varepsilon)^2\, dx = \int_{\Omega(\alpha_\varepsilon)} f \cdot u_\varepsilon\, dx. \tag{2.197}$$

From this and the Friedrichs inequality in $(H_0^1(\widehat{\Omega}))^2$ it easily follows that $\{\|\widetilde{u}_\varepsilon\|_{1,\widehat{\Omega}}\}$ is bounded. Therefore (2.196) holds for an appropriate subsequence of $\{(\alpha_\varepsilon, u_\varepsilon(\alpha_\varepsilon))\}$ and for some $\alpha \in \widetilde{U}^{ad}$ and $u \in (H_0^1(\widehat{\Omega}))^2$. It is easy to prove that $u = 0$ in $\widehat{\Omega} \setminus \overline{\Omega(\alpha)}$, meaning that $u(\alpha) := u_{|\Omega(\alpha)} \in (H_0^1(\Omega(\alpha)))^2$.

Next we prove that $u(\alpha)$ is divergence free in $\Omega(\alpha)$; i.e., $u(\alpha) \in \mathbb{V}(\alpha)$. To this end it will be sufficient to show that

$$\int_{\Omega(\alpha)} q \operatorname{div} u(\alpha)\, dx = 0 \quad \forall q \in L(\alpha) \tag{2.198}$$

for a dense subset $L(\alpha)$ of $L_0^2(\Omega(\alpha))$. Indeed, if (2.198) were satisfied for $q \in L(\alpha)$, then it also would be satisfied for any q belonging to the direct sum $P_0 \oplus L(\alpha)$, which is dense in $L^2(\Omega(\alpha))$, since

$$\int_{\Omega(\alpha)} \operatorname{div} v\, dx = 0 \quad \forall v \in (H_0^1(\Omega(\alpha)))^2,$$

implying $\operatorname{div} u(\alpha) = 0$ in $\Omega(\alpha)$. Below we show that

$$L(\alpha) = \{q \in L_0^2(\Omega(\alpha)) \mid \exists v \in (C_0^\infty(\Omega(\alpha)))^2 : q = \operatorname{div} v\} \tag{2.199}$$

possesses this property. We first prove that such an $L(\alpha)$ is dense in $L_0^2(\Omega(\alpha))$.

Let $q \in L_0^2(\Omega(\alpha))$ and $\eta > 0$ be given. Then there exists a function $v \in (H_0^1(\Omega(\alpha)))^2$ (see [GR79]) such that

$$\operatorname{div} v = q \quad \text{in } \Omega(\alpha). \tag{2.200}$$

From the density of $(C_0^\infty(\Omega(\alpha)))^2$ in $(H_0^1(\Omega(\alpha)))^2$, the existence of $w \in (C_0^\infty(\Omega(\alpha)))^2$ such that

$$\|v - w\|_{1,\Omega(\alpha)} < \eta$$

follows. From this and (2.200) we see that

$$\|q - \overline{q}\|_{0,\Omega(\alpha)} < \eta,$$

where $\overline{q} = \operatorname{div} w$ in $\Omega(\alpha)$, proving the required density result.

Let $w \in (C_0^\infty(\Omega(\alpha)))^2$ be given. Then $\widetilde{w}_{|\Omega(\alpha_\varepsilon)} \in (C_0^\infty(\Omega(\alpha_\varepsilon)))^2$ for any $\varepsilon \le \varepsilon_0$ where $\varepsilon_0 > 0$ is small enough. The definition of $(\mathcal{P}_\varepsilon(\alpha_\varepsilon))$, $\varepsilon \le \varepsilon_0$, yields

$$\int_{\Omega(\alpha_\varepsilon)} \operatorname{div} u_\varepsilon \operatorname{div} \widetilde{w}\, dx = \varepsilon\left(\int_{\Omega(\alpha_\varepsilon)} f \cdot \widetilde{w}\, dx - \int_{\Omega(\alpha_\varepsilon)} \nabla u_\varepsilon : \nabla \widetilde{w}\, dx\right). \tag{2.201}$$

Letting $\varepsilon \to 0+$ in (2.201) and using the boundedness of $\{\|u_\varepsilon\|_{1,\Omega(\alpha_\varepsilon)}\}$ we obtain

$$\lim_{\varepsilon \to 0+} \int_{\Omega(\alpha_\varepsilon)} \operatorname{div} u_\varepsilon \operatorname{div} \widetilde{w}\, dx = 0.$$

On the other hand it is easy to show that

$$0 = \lim_{\varepsilon \to 0+} \int_{\Omega(\alpha_\varepsilon)} \operatorname{div} u_\varepsilon \operatorname{div} \widetilde{w}\, dx = \int_{\Omega(\alpha)} \operatorname{div} u(\alpha)\, q\, dx, \tag{2.202}$$

proving (2.198) for any element $q = \operatorname{div} \widetilde{w} \in L(\alpha)$ given by (2.199). It remains to show that $u(\alpha)$ solves $(\mathcal{P}(\alpha))$.

Let $w \in (C_0^\infty(\Omega(\alpha)))^2$, $\operatorname{div} w = 0$ in $\Omega(\alpha)$, be given. Then $\widetilde{w}_{|\Omega(\alpha_\varepsilon)} \in \mathbb{V}(\alpha_\varepsilon)$ for any $\varepsilon \le \varepsilon_0$ with $\varepsilon_0 > 0$ small enough. From the definition of $(\mathcal{P}_\varepsilon(\alpha_\varepsilon))$ it follows that

$$\int_{\Omega(\alpha_\varepsilon)} \nabla u_\varepsilon : \nabla \widetilde{w}\, dx = \int_{\Omega(\alpha_\varepsilon)} f \cdot \widetilde{w}\, dx \tag{2.203}$$

holds for any $\varepsilon \le \varepsilon_0$. The limit passage $\varepsilon \to 0+$ in (2.203) yields

$$\int_{\Omega(\alpha)} \nabla u(\alpha) : \nabla \widetilde{w}\, dx = \int_{\Omega(\alpha)} f \cdot \widetilde{w}\, dx. \tag{2.204}$$

This equality holds for any $w \in \mathbb{V}(\alpha) \cap (C_0^\infty(\Omega(\alpha)))^2$ that is dense in $\mathbb{V}(\alpha)$. Therefore (2.204) holds for any $w \in \mathbb{V}(\alpha)$; i.e., $u(\alpha)$ solves $(\mathcal{P}(\alpha))$. Let us show now that $\{\widetilde{u}_\varepsilon(\alpha_\varepsilon)\}$ tends strongly to u in $(H_0^1(\widehat{\Omega}))^2$. From the definition of $(\mathcal{P}_\varepsilon(\alpha_\varepsilon))$ it follows that

$$|\widetilde{u}_\varepsilon|^2_{1,\widehat{\Omega}} \le \int_{\Omega(\alpha_\varepsilon)} \nabla u_\varepsilon : \nabla u_\varepsilon\, dx + \frac{1}{\varepsilon}\int_{\Omega(\alpha_\varepsilon)} (\operatorname{div} u_\varepsilon)^2\, dx = \int_{\Omega(\alpha_\varepsilon)} f \cdot u_\varepsilon\, dx$$

so that

$$\begin{aligned}\limsup_{\varepsilon\to 0+} |\widetilde{u}_\varepsilon|^2_{1,\widehat{\Omega}} \le \lim_{\varepsilon\to 0+} \int_{\Omega(\alpha_\varepsilon)} f \cdot u_\varepsilon\, dx &= \int_{\Omega(\alpha)} f \cdot u(\alpha)\, dx \\ &= \int_{\Omega(\alpha)} \nabla u(\alpha) : \nabla u(\alpha)\, dx = |u|^2_{1,\widehat{\Omega}},\end{aligned} \tag{2.205}$$

making use of the fact that $u(\alpha) = u_{|\Omega(\alpha)}$ solves $(\mathcal{P}(\alpha))$. On the other hand,

$$\liminf_{\varepsilon\to 0+} |\widetilde{u}_\varepsilon|^2_{1,\widehat{\Omega}} \ge |u|^2_{1,\widehat{\Omega}},$$

as follows from the weak convergence of $\{\widetilde{u}_\varepsilon\}$ to u. From this and (2.205) we see that

$$\lim_{\varepsilon\to 0+} |\widetilde{u}_\varepsilon|_{1,\widehat{\Omega}} = |u|_{1,\widehat{\Omega}},$$

implying strong convergence, taking into account that $|\cdot|_{1,\widehat{\Omega}}$ is a norm in $H_0^1(\widehat{\Omega})$. □

On the basis of the previous lemma we are ready to analyze the relation between solutions to $(\mathbb{P})$ and $(\mathbb{P}_\varepsilon)$ as $\varepsilon \to 0+$.

Theorem 2.21. *Let $\{(\alpha_\varepsilon^*, u_\varepsilon(\alpha_\varepsilon^*))\}$ be a sequence of optimal pairs of $(\mathbb{P}_\varepsilon)$, $\varepsilon \to 0+$. Then one can find its subsequence such that*

$$\begin{cases} \alpha_\varepsilon^* \rightrightarrows \alpha^* & \text{in } [0,1], \\ \widetilde{u}_\varepsilon(\alpha_\varepsilon^*) \to u^* & \text{in } (H_0^1(\widehat{\Omega}))^2, \quad \varepsilon \to 0+. \end{cases} \tag{2.206}$$

In addition, $(\alpha^*, u^*_{|\Omega(\alpha^*)})$ is an optimal pair of $(\mathbb{P})$. Furthermore, any accumulation point of $\{(\alpha_\varepsilon^*, u_\varepsilon(\alpha_\varepsilon^*))\}$ in the sense of (2.206) possesses this property.

Proof. From Lemma 2.32 we already know that (2.206) holds for an appropriate subsequence of $\{(\alpha_\varepsilon^*, u_\varepsilon(\alpha_\varepsilon^*))\}$ and that $u(\alpha^*) := u^*_{|\Omega(\alpha^*)}$ solves $(\mathcal{P}(\alpha^*))$. We shall show that α^*

solves ($\mathbb{P}$). Let $\overline{\alpha} \in U^{ad}$ be given and $u(\overline{\alpha}) \in \mathbb{V}(\overline{\alpha})$ be the respective solution to the Stokes problem in $\Omega(\overline{\alpha})$. From (2.195) we know that $u(\overline{\alpha})$ can be approximated by the solutions of the penalized problems $(\mathcal{P}_\varepsilon(\overline{\alpha}))$ (keeping $\overline{\alpha}$ fixed):

$$u_\varepsilon(\overline{\alpha}) \to u(\overline{\alpha}) \quad \text{in } (H_0^1(\Omega(\overline{\alpha})))^2, \quad \varepsilon \to 0+. \tag{2.207}$$

From the definition of $(\mathbb{P}_\varepsilon)$,

$$J(\alpha_\varepsilon^*, u_\varepsilon(\alpha_\varepsilon^*)) \le J(\overline{\alpha}, u_\varepsilon(\overline{\alpha})).$$

Passing to the limit with $\varepsilon \to 0+$, using (2.206), (2.207), and continuity of J, we arrive at

$$J(\alpha^*, u(\alpha^*)) \le J(\overline{\alpha}, u(\overline{\alpha})).$$

Since $\overline{\alpha} \in U^{ad}$ is arbitrary, α^* solves ($\mathbb{P}$). □

Remark 2.31. Instead of the penalty functional $j_\varepsilon(v) = (1/\varepsilon)\int_{\Omega(\alpha)} (\operatorname{div} v)^2\, dx$, which is the simplest one, more sophisticated penalizations of the Stokes problem (see [GR79]) are usually used.

We now briefly describe the discretization of ($\mathbb{P}$) using the penalized form of the Stokes equations. Let $s_\varkappa \in U_\varkappa^{ad}$ be given and $\mathbb{V}_h(s_\varkappa)$ be a finite element subspace of $(H_0^1(\Omega(r_h s_\varkappa)))^2$ made of continuous, piecewise linear vector functions over $\mathcal{T}(h, s_\varkappa)$. Finally, let the penalty parameter $\varepsilon := \varepsilon(h)$ be a function of h such that $\varepsilon(h) \to 0+$ iff $h \to 0+$. We first discretize the penalized Stokes problem as follows:

$$\left\{\begin{aligned} &\text{Find } u_{\varepsilon h} := u_{\varepsilon h}(s_\varkappa) \in \mathbb{V}_h(s_\varkappa) \quad \text{such that}\\ &\int_{\Omega_h(s_\varkappa)} \nabla u_{\varepsilon h} : \nabla v_h\, dx + \frac{1}{\varepsilon(h)}\int_{\Omega_h(s_\varkappa)} \operatorname{div} u_{\varepsilon h} \operatorname{div} v_h\, dx\\ &\qquad = \int_{\Omega_h(s_\varkappa)} f\cdot v_h\, dx \qquad \forall v_h \in \mathbb{V}_h(s_\varkappa), \end{aligned}\right. \tag{$\mathcal{P}_{\varepsilon h}(s_\varkappa)$}$$

and then we define the following discrete optimization problem:

$$\left\{\begin{aligned} &\text{Find } s_{\varepsilon\varkappa}^* \in U_\varkappa^{ad} \quad \text{such that}\\ &J(s_{\varepsilon\varkappa}^*, u_{\varepsilon h}(s_{\varepsilon\varkappa}^*)) \le J(s_\varkappa, u_{\varepsilon h}(s_\varkappa)) \quad \forall s_\varkappa \in U_\varkappa^{ad}, \end{aligned}\right. \tag{$\mathbb{P}_{\varepsilon h\varkappa}$}$$

where $u_{\varepsilon h}(s_\varkappa)$ solves $(\mathcal{P}_{\varepsilon h}(s_\varkappa))$. One can show again that problems ($\mathbb{P}$) and $(\mathbb{P}_{\varepsilon h\varkappa})$ are close in the sense of Theorem 2.11 as $h, \varkappa \to 0+$.

Remark 2.32. The penalty technique can be used with success in any optimization problem involving constraints. In particular, it can be used in the optimization of structures whose behavior is governed by variational inequalities. As we have already mentioned, these problems are not generally differentiable due to the fact that the state is represented by a variational inequality. An appropriate penalization of constraints makes it possible to approximate state inequalities by a sequence of equations and therefore to replace an originally nonsmooth problem with a sequence of smooth ones. This approach has been used, for example, in contact shape optimization (see [HN96]).

Problems

Problem 2.1. Prove Theorem 2.2.

Problem 2.2. [BG75] Prove that for any $e \in U^{ad}$ there exists $\{e_h\}$, $e_h \in U_h^{ad}$, such that $e_h \rightrightarrows e$ in $I, h \to 0+$, where U^{ad}, U_h^{ad} are defined by (2.3), (2.15), respectively.

Hint: For any $h > 0$ define the continuous, piecewise linear function e_h on Δ_h as follows:

$$e_h(a_i) = \frac{1}{h} \int_{a_{i-1/2}}^{a_{i+1/2}} \widetilde{e}\, dx, \quad i = 0, \ldots, d,$$

where $\widetilde{e}$ is the even extension of e from $[0, \ell]$ to $[-h/2, \ell + h/2]$ and $a_{i+1/2}$ stands for the midpoint of $[a_i, a_{i+1}]$ $(a_{-1/2} = -h/2,\ a_{d+1/2} = \ell + h/2)$. Show that $e_h \in U_h^{ad}$ and $\{e_h\}$ has the required convergence property.

Problem 2.3. Prove the formal equivalence between the classical formulation (2.2) of the state problem and its weak formulation $(\mathcal{P}(e))$.

Problem 2.4. Prove that the cost functionals ($e \in U^{ad}$ (see (2.3)), $y \in H_0^2(I)$)

$$J(e, y) = \frac{1}{2}\|y - y_d\|_{j,I}^2 + \frac{1}{2}\int_I e^2\, dx, \quad y_d \in H_0^j(I) \text{ given},\ j = 0, 1, 2;$$

$$J(e, y) = \frac{1}{2}\left|\frac{d^k}{dx^k} y(x_0) - c_k\right|^2 + \frac{1}{2}\int_I e^2\, dx, \quad c_k \in \mathbb{R} \text{ and } x_0 \in]0, \ell[\text{ given}, k = 0, 1,$$

satisfy (2.12).

Problem 2.5. Prove that the function $\widetilde{e}_h$ defined by (2.35) belongs to $\widetilde{U}_h^{ad}$ given by (2.32) and

$$\widetilde{e}_h \to e \quad \text{in } L^\infty(I), \quad h \to 0+,$$

for any $e \in U^{ad}$.

Problem 2.6. Prove Lemma 2.4. (Hint: use the Lebesgue-dominated convergence theorem.)

Problem 2.7. Prove Lemma 2.7.

Problem 2.8. Prove Lemma 2.8.

Problem 2.9. Prove that the following cost functionals satisfy (2.46) ($\alpha \in U^{ad}$ (see Section 2.2), $y \in H_0^1(\widehat{\Omega})$):

$$J(\alpha, y) = \frac{1}{2}\|y - z_d\|_{0,\Omega(\alpha)}^2, \quad z_d \in L^2(\widehat{\Omega}) \text{ given};$$

$$J(\alpha, y) = \frac{1}{2}\|\nabla y - z_d\|_{0,\Omega(\alpha)}^2, \quad z_d \in (L^2(\widehat{\Omega}))^2 \text{ given};$$

$$J(\alpha, y) = \frac{1}{2}\|\nabla y\|_{0,\Omega(\alpha)}^2 - (f, y)_{0,\Omega(\alpha)}.$$

Problem 2.10. Prove Lemma 2.14 under the conditions that

$$\varphi(0) \leq 0, \quad \varphi(\ell) \leq 0,$$

where φ is the function appearing in the definition of K.

Problem 2.11. Let K_h and K be the same as in Subsection 2.5.1. Let $\{v_h\}$, $v_h \in K_h$, be such that

$$v_h(x) \to v(x), \; h \to 0+, \text{ for all } x \in I,$$

where $v \in C(I)$. Prove that $v \in K$.

Problem 2.12. Prove that $(\mathcal{P}(e))$ from Subsection 2.5.1 has a unique solution u. Using integration by parts shows that u satisfies the following set of conditions:

$$\begin{aligned} &A(e)u \geq f \quad \text{in }]0, \ell[, \\ &u(0) = u'(0) = u(\ell) = u'(\ell) = 0, \quad u \geq \varphi, \quad \text{in } [0, \ell], \\ &(A(e)u - f)(u - \varphi) = 0 \quad \text{in }]0, \ell[, \end{aligned}$$

where $A(e)u := (\beta e^3 u'')''$ provided u is smooth enough.

Problem 2.13. Give the algebraic formulation of the discrete state inequality $(\mathcal{P}_h(e_h))$ from Subsection 2.5.1.

Problem 2.14. Prove Lemma 2.16.

Problem 2.15. Consider problem $(\mathbb{P})$ with nonhomogeneous Neumann state problems whose weak formulation is defined by $(\mathcal{P}_G(\alpha))$ (see Remark 2.17) and with the cost functional J:

$$J(\alpha, y) = \int_{\Omega(\alpha)} \nabla y \cdot \nabla y \, dx, \quad y \in H^1(\Omega(\alpha)), \quad \alpha \in U^{ad},$$

where U^{ad} is given by (2.97). Prove the existence of solutions to $(\mathbb{P})$.

Problem 2.16. Define the approximation of $(\mathbb{P})$ with the Dirichlet–Neumann boundary state problems and verify that assumptions $(\mathcal{B}1)_\varkappa$–$(\mathcal{B}3)_\varkappa$ of Section 2.4 are satisfied.

Problem 2.17. Prove (2.166).

Problem 2.18. Complete the convergence analysis for optimal shape design problem $(\mathbb{P}_1)$ from Subsection 2.5.5.

Problem 2.19. Prove Lemma 2.25.

Hint: let $v = (v_1, v_2) \in \mathbb{K}(\alpha)$, $\widetilde{v} := p_{\Omega(\alpha)}v = (\widetilde{v}_1, \widetilde{v}_2)$ be the extension of v from $\Omega(\alpha)$ to $\widehat{\Omega}$. Define $\psi_2(x_1, x_2) := \max\{\widetilde{v}_2(x_1, x_2), -x_2\}$, $(x_1, x_2) \in \widehat{\Omega}$, and set $\psi = (\widetilde{v}_1, \psi_2)$. Prove that $\psi_{|\Omega(\alpha)} \in \mathbb{K}(\alpha)$. Decompose $\widetilde{v}$: $\widetilde{v} = \psi + \Phi$, where $\Phi = (\Phi_1, \Phi_2)$ is such that $\Phi_{1|\Gamma_u(\alpha)} = \Phi_{2|\Gamma_u(\alpha)} = \Phi_{2|\Gamma_C(\alpha)} = 0$. Approximate Φ by a sequence $\{\Phi_j\}$, $\Phi_j =$

$(\Phi_{j1}, \Phi_{j2}) \in (C^\infty(\overline{\widehat{\Omega}}))^2$, with Φ_{j1}, Φ_{j2} vanishing in the vicinity of $\overline{\Gamma_u(\alpha)}$, $\overline{\Gamma_u(\alpha) \cup \Gamma_C(\alpha)}$, respectively, and show that $\{v_j\}$, where $v_j := \psi + \Phi_j$, satisfies (2.176) using the fact that $\psi_2(x_1, x_2) \geq -x_2$ in $\widehat{\Omega}$.

Problem 2.20. Let $\widetilde{U}^{ad}$ be given by (2.96) and consider the respective family $\widetilde{\mathcal{O}}$ of admissible domains. On any $\Omega(\alpha) \in \widetilde{\mathcal{O}}$, $\alpha \in \widetilde{U}^{ad}$, we shall consider the linear elasticity problem $(\mathcal{P}_1(\alpha))$ as in Subsection 2.5.5 with the following minor change: the surface traction P is now applied on $\Gamma_P(\alpha) = \Gamma(\alpha)$, i.e., on the designed part of the structure, while the zero displacements are prescribed on $\Gamma_u(\alpha) = \partial\Omega(\alpha) \setminus \overline{\Gamma_P(\alpha)}$. Formulate the traction conditions on $\Gamma_P(\alpha)$ by using the technique of Remark 2.17.

Problem 2.21. Prove the formal equivalence between the classical and weak formulations of the Signorini problem without friction.

Problem 2.22. Prove that optimal shape design problem $(\mathbb{P}_\varepsilon)$ from Subsection 2.5.6 using the penalized Stokes equations $(\mathcal{P}_\varepsilon(\alpha))$ has a solution for any $\varepsilon > 0$.

Problem 2.23. Consider an approximation $(\mathbb{P}_{h\varkappa})$ of $(\mathbb{P})$ using a mixed finite element approximation $(\mathcal{P}_h(s_\varkappa))$ of the Stokes problem with finite element spaces $\mathbb{X}_h(s_\varkappa)$ and $M_h(s_\varkappa)$ satisfying (2.194). Prove that $(\mathbb{P})$ and $(\mathbb{P}_{h\varkappa})$ are close in the sense of Theorem 2.11 for $h, \varkappa \to 0+$ provided that the systems $\{\mathbb{X}_h(s_\varkappa)\}$, $\{M_h(s_\varkappa)\}$ satisfy the following conditions: $\forall \alpha \in U^{ad}$ $\forall (v, p) \in (H_0^1(\Omega(\alpha)))^2 \times L_0^2(\Omega(\alpha))$ $\exists \{(s_\varkappa, v_h, p_h)\}$, $s_\varkappa \in U_\varkappa^{ad}$, $v_h \in \mathbb{X}_h(s_\varkappa)$, $p_h \in M_h(s_\varkappa)$,

$$\|\widetilde{v} - v_h\|_{1,\Omega_h(s_\varkappa)} \to 0,$$
$$\|\widetilde{p} - p_h\|_{0,\Omega_h(s_\varkappa)} \to 0 \quad \text{as } h, \varkappa \to 0+.$$

Problem 2.24. Consider the approximation $(\mathbb{P}_{\varepsilon h\varkappa})$ of $(\mathbb{P})$ defined in Subsection 2.5.6 using the discretization $(\mathcal{P}_{\varepsilon h}(s_\varkappa))$ of the penalized form of the Stokes problem. Let the penalty parameter $\varepsilon := \varepsilon(h)$ satisfy $\varepsilon(h) \to 0+$ iff $h \to 0+$. Prove that $(\mathbb{P})$ and $(\mathbb{P}_{\varepsilon h\varkappa})$ are close in the sense of Theorem 2.11 for $h, \varkappa \to 0+$ provided the family $\{\mathcal{T}(h, s_\varkappa)\}$ satisfies assumptions $(\mathcal{T}1)$–$(\mathcal{T}3)$ of Section 2.2.

Part II

Computational Aspects of Sizing and Shape Optimization

Chapter 3

Sensitivity Analysis

From the previous chapter we already know that a continuous dependence of solutions to state problems on design variations is a fundamental property ensuring the existence of optimal solutions. Continuity is important but not enough. To better understand the problem, other properties are needed. *Differentiability* is one of the most important of these. The need to deal with such information gave rise to a special discipline in optimization called *sensitivity analysis*. Sensitivity analysis develops appropriate tools and concepts enabling us to analyze the differentiability of various objects, such as solutions to state problems, cost and constraint functionals, etc., with respect to control variables, and in particular with respect to design variables in sizing and shape optimization (SSO). On the basis of these results one can derive *necessary optimality conditions* satisfied by solutions to optimal control problems. The interpretation of optimality conditions reveals important properties of optimal solutions that are not usually directly seen from the original setting of the problem. Sensitivity analysis also plays an important role in computations: it provides us with gradient information required by the gradient type methods most frequently used for the numerical minimization of discretized problems.

This chapter deals with sensitivity analysis in SSO. The basic ideas are the same. Each branch of structural optimization, however, develops its own techniques, taking into account its features. This will be seen, in particular, in the case of shape optimization, for which appropriate tools enabling us to describe the change in geometry have to be introduced. This chapter starts with sensitivity analysis in the algebraic setting of problems because of its simplicity. Here we explain common ideas. These results will then be adapted following the specific needs of SSO.

3.1 Algebraic sensitivity analysis

Let us consider a system of linear algebraic equations depending on a parameter $\boldsymbol{\alpha}$:

$$\boldsymbol{A}(\boldsymbol{\alpha})\boldsymbol{x}(\boldsymbol{\alpha}) = \boldsymbol{f}(\boldsymbol{\alpha}), \quad \boldsymbol{A}(\boldsymbol{\alpha}) \in \mathbb{R}^{n\times n}, \quad \boldsymbol{f}(\boldsymbol{\alpha}) \in \mathbb{R}^n, \qquad (\mathcal{P}(\boldsymbol{\alpha}))$$

in which $\boldsymbol{A}$ and $\boldsymbol{f}$ are *matrix* and *vector* functions, respectively, of $\boldsymbol{\alpha}$, which belongs to an *open* set $\mathcal{U} \subset \mathbb{R}^d$. Further, let $J : \mathcal{U} \times \mathbb{R}^n \to \mathbb{R}$ be a real function and define $\mathcal{J} : \mathcal{U} \to \mathbb{R}$

by

$$\mathcal{J}(\boldsymbol{\alpha}) := J(\boldsymbol{\alpha}, \boldsymbol{x}(\boldsymbol{\alpha})) \tag{3.1}$$

with $\boldsymbol{x}(\boldsymbol{\alpha}) \in \mathbb{R}^n$ a solution of $(\mathcal{P}(\boldsymbol{\alpha}))$.

Remark 3.1. In the framework of SSO, $(\mathcal{P}(\boldsymbol{\alpha}))$ represents a finite element approximation of a linear elliptic equation, $\boldsymbol{\alpha}$ is a vector of discrete design variables, $\mathcal{U}$ is isometrically isomorphic with U_h^{ad}, and J is an algebraic representation of a cost functional.

In what follows we shall analyze the differentiability of the mappings $\boldsymbol{x} : \mathcal{U} \to \mathbb{R}^n$ and $\mathcal{J} : \mathcal{U} \to \mathbb{R}$ defined by $(\mathcal{P}(\boldsymbol{\alpha}))$ and (3.1), respectively. To this end we shall suppose that

(i) $\boldsymbol{A}(\boldsymbol{\alpha})$ *is regular for any* $\boldsymbol{\alpha} \in \mathcal{U}$;
(ii) $\boldsymbol{A}, \boldsymbol{f}$ *are* C^1*-mappings in* $\mathcal{U}$: $\boldsymbol{A} \in C^1(\mathcal{U}; \mathbb{R}^{n\times n}),\ \boldsymbol{f} \in C^1(\mathcal{U}; \mathbb{R}^n)$;
(iii) $J \in C^1(\mathcal{U} \times \mathbb{R}^n)$.

Remark 3.2. From (ii) it follows that the directional derivatives of $\boldsymbol{A}$ and $\boldsymbol{f}$ exist at any point $\boldsymbol{\alpha} \in \mathcal{U}$ and in any direction $\boldsymbol{\beta} \in \mathbb{R}^d$:

$$\boldsymbol{A}'(\boldsymbol{\alpha}; \boldsymbol{\beta}) := \lim_{t\to 0+} \frac{\boldsymbol{A}(\boldsymbol{\alpha} + t\boldsymbol{\beta}) - \boldsymbol{A}(\boldsymbol{\alpha})}{t} \in \mathbb{R}^{n\times n},$$
$$\boldsymbol{f}'(\boldsymbol{\alpha}; \boldsymbol{\beta}) := \lim_{t\to 0+} \frac{\boldsymbol{f}(\boldsymbol{\alpha} + t\boldsymbol{\beta}) - \boldsymbol{f}(\boldsymbol{\alpha})}{t} \in \mathbb{R}^{n}.$$

The elements of $\boldsymbol{A}'(\boldsymbol{\alpha}; \boldsymbol{\beta})$ and $\boldsymbol{f}'(\boldsymbol{\alpha}; \boldsymbol{\beta})$ can be computed in the standard way:

$$a'_{ij}(\boldsymbol{\alpha}; \boldsymbol{\beta}) = (\nabla_{\boldsymbol{\alpha}} a_{ij}(\boldsymbol{\alpha}))^{\mathrm{T}} \boldsymbol{\beta} \quad \forall i, j = 1, \dots, n,$$
$$f'_i(\boldsymbol{\alpha}; \boldsymbol{\beta}) = (\nabla_{\boldsymbol{\alpha}} f_i(\boldsymbol{\alpha}))^{\mathrm{T}} \boldsymbol{\beta} \quad \forall i = 1, \dots, n,$$

where $\nabla_{\boldsymbol{\alpha}}$ is the gradient of a function with respect to $\boldsymbol{\alpha}$.

Convention: *To simplify notation we shall sometimes use the symbols* $\boldsymbol{A}'(\boldsymbol{\alpha})$, $\boldsymbol{f}'(\boldsymbol{\alpha})$, *etc., instead of* $\boldsymbol{A}'(\boldsymbol{\alpha}; \boldsymbol{\beta})$, $\boldsymbol{f}'(\boldsymbol{\alpha}; \boldsymbol{\beta})$, *etc. The direction of differentiation will be seen from the context.*

From (i)–(iii) *continuous differentiability* of the mappings $\boldsymbol{x}$ and $\mathcal{J}$ in $\mathcal{U}$ easily follows. Indeed, let $\boldsymbol{\alpha} \in \mathcal{U}$, $\boldsymbol{\beta} \in \mathbb{R}^d$ be given and $\boldsymbol{x}(\boldsymbol{\alpha} + t\boldsymbol{\beta}) \in \mathbb{R}^n$ be a solution to $(\mathcal{P}(\boldsymbol{\alpha} + t\boldsymbol{\beta}))$:

$$\boldsymbol{A}(\boldsymbol{\alpha} + t\boldsymbol{\beta})\boldsymbol{x}(\boldsymbol{\alpha} + t\boldsymbol{\beta}) = \boldsymbol{f}(\boldsymbol{\alpha} + t\boldsymbol{\beta}), \quad t > 0.$$

The classical implicit function theorem ensures the existence of the directional derivative $\boldsymbol{x}'(\boldsymbol{\alpha}) := \boldsymbol{x}'(\boldsymbol{\alpha}; \boldsymbol{\beta})$ for any $\boldsymbol{\beta} \in \mathbb{R}^d$:

$$\boldsymbol{x}'(\boldsymbol{\alpha}) = \boldsymbol{A}(\boldsymbol{\alpha})^{-1} \left(\boldsymbol{f}'(\boldsymbol{\alpha}) - \boldsymbol{A}'(\boldsymbol{\alpha})\boldsymbol{x}(\boldsymbol{\alpha})\right). \tag{3.2}$$

Remark 3.3. If A is symmetric for any $\boldsymbol{\alpha} \in \mathcal{U}$, then the directional derivative $\boldsymbol{x}'(\boldsymbol{\alpha})$ can be equivalently characterized as the solution of the following minimization problem:

$$\begin{cases} \text{Find } \boldsymbol{x}'(\boldsymbol{\alpha}) \in \mathbb{R}^n \quad \text{such that} \\ \mathcal{G}_{\boldsymbol{\alpha},\boldsymbol{\beta}}(\boldsymbol{x}'(\boldsymbol{\alpha})) \leq \mathcal{G}_{\boldsymbol{\alpha},\boldsymbol{\beta}}(\boldsymbol{y}) \quad \forall \boldsymbol{y} \in \mathbb{R}^n, \end{cases}$$

where

$$\mathcal{G}_{\boldsymbol{\alpha},\boldsymbol{\beta}}(\boldsymbol{y}) = \frac{1}{2}\boldsymbol{y}^{\mathrm{T}}\boldsymbol{A}(\boldsymbol{\alpha})\boldsymbol{y} + \boldsymbol{y}^{\mathrm{T}}\boldsymbol{A}'(\boldsymbol{\alpha})\boldsymbol{x}(\boldsymbol{\alpha}) - \boldsymbol{y}^{\mathrm{T}}\boldsymbol{f}'(\boldsymbol{\alpha}). \tag{3.3}$$

By virtue of (ii) the mapping $\boldsymbol{x}' : \boldsymbol{\alpha} \mapsto \boldsymbol{x}'(\boldsymbol{\alpha}; \cdot)$ is also continuous in $\mathcal{U}$ so that $\boldsymbol{x} \in C^1(\mathcal{U}, \mathbb{R}^n)$. From this and (iii) we may conclude that $\mathcal{J} \in C^1(\mathcal{U})$ and, in addition, the directional derivative $\mathcal{J}'(\boldsymbol{\alpha}; \boldsymbol{\beta})$ can be computed by using the classical chain rule of differentiation:

$$\mathcal{J}'(\boldsymbol{\alpha}; \boldsymbol{\beta}) = (\nabla_{\boldsymbol{\alpha}} J(\boldsymbol{\alpha}, \boldsymbol{x}(\boldsymbol{\alpha})))^{\mathrm{T}}\boldsymbol{\beta} + (\nabla_{\boldsymbol{x}} J(\boldsymbol{\alpha}, \boldsymbol{x}(\boldsymbol{\alpha})))^{\mathrm{T}}\boldsymbol{x}'(\boldsymbol{\alpha}, \boldsymbol{\beta}), \tag{3.4}$$

where $\nabla_{\boldsymbol{\alpha}}$, $\nabla_{\boldsymbol{x}}$ denote the partial gradients of J with respect to $\boldsymbol{\alpha} \in \mathcal{U}$, $\boldsymbol{x} \in \mathbb{R}^n$, respectively. To get full information on the gradient of $\mathcal{J}$, d linearly independent directions $\boldsymbol{\beta} \in \mathbb{R}^d$ are needed. If (3.4) were directly used for its computation, it would be necessary to solve (3.2) d times for each direction separately to get $\nabla_{\boldsymbol{\alpha}}\boldsymbol{x}(\boldsymbol{\alpha})$. Fortunately there is a way to overcome this difficulty. We first introduce the so-called adjoint state system

$$\boldsymbol{A}(\boldsymbol{\alpha})^{\mathrm{T}}\boldsymbol{p}(\boldsymbol{\alpha}) = \nabla_{\boldsymbol{x}} J(\boldsymbol{\alpha}, \boldsymbol{x}(\boldsymbol{\alpha})). \tag{3.5}$$

The vector $\boldsymbol{p}(\boldsymbol{\alpha})$ is termed the *adjoint state*. Multiplying (3.5) by $\boldsymbol{x}'(\boldsymbol{\alpha})$ we obtain

$$\boldsymbol{p}(\boldsymbol{\alpha})^{\mathrm{T}}\left(\boldsymbol{f}'(\boldsymbol{\alpha}) - \boldsymbol{A}'(\boldsymbol{\alpha})\boldsymbol{x}(\boldsymbol{\alpha})\right) = \boldsymbol{p}(\boldsymbol{\alpha})^{\mathrm{T}}\boldsymbol{A}(\boldsymbol{\alpha})\boldsymbol{x}'(\boldsymbol{\alpha}) = (\nabla_{\boldsymbol{x}} J(\boldsymbol{\alpha}, \boldsymbol{x}(\boldsymbol{\alpha})))^{\mathrm{T}}\boldsymbol{x}'(\boldsymbol{\alpha}), \tag{3.6}$$

making use of (3.2). From (3.4) and (3.6) we arrive at the final expression for $\mathcal{J}'(\boldsymbol{\alpha}; \boldsymbol{\beta})$:

$$\mathcal{J}'(\boldsymbol{\alpha}; \boldsymbol{\beta}) = (\nabla_{\boldsymbol{\alpha}} J(\boldsymbol{\alpha}, \boldsymbol{x}(\boldsymbol{\alpha})))^{\mathrm{T}}\boldsymbol{\beta} + \boldsymbol{p}(\boldsymbol{\alpha})^{\mathrm{T}}\left(\boldsymbol{f}'(\boldsymbol{\alpha}) - \boldsymbol{A}'(\boldsymbol{\alpha})\boldsymbol{x}(\boldsymbol{\alpha})\right). \tag{3.7}$$

The previous results are summarized below.

Theorem 3.1. *Let* (i)–(iii) *be satisfied. Then the solution of* ($\mathcal{P}(\boldsymbol{\alpha})$) *and the function* $\mathcal{J}$ *defined by* (3.1) *are continuously differentiable in* $\mathcal{U}$ *with respect to* $\boldsymbol{\alpha}$. *The directional derivatives* $\boldsymbol{x}'(\boldsymbol{\alpha}; \boldsymbol{\beta})$, *and* $\mathcal{J}'(\boldsymbol{\alpha}; \boldsymbol{\beta})$ *are given by* (3.2) *and* (3.7), *respectively. The adjoint state* $\boldsymbol{p}(\boldsymbol{\alpha})$ *appearing in* (3.7) *is defined by* (3.5).

The same result can be achieved by using a duality approach: state equation ($\mathcal{P}(\boldsymbol{\alpha})$) will be treated as a constraint by means of a Lagrange multiplier technique. Indeed, it is readily seen that the value $J(\boldsymbol{\alpha}, \boldsymbol{x}(\boldsymbol{\alpha}))$, where $\boldsymbol{x}(\boldsymbol{\alpha}) \in \mathbb{R}^n$ solves ($\mathcal{P}(\boldsymbol{\alpha})$), can be expressed as follows:

$$J(\boldsymbol{\alpha}, \boldsymbol{x}(\boldsymbol{\alpha})) = \inf_{\boldsymbol{x} \in \mathbb{R}^n} \sup_{\boldsymbol{p} \in \mathbb{R}^n} \mathcal{L}(\boldsymbol{\alpha}, \boldsymbol{x}, \boldsymbol{p}), \tag{3.8}$$

where $\mathcal{L} : \mathcal{U} \times \mathbb{R}^n \times \mathbb{R}^n \to \mathbb{R}$ is the Lagrangian defined by

$$\mathcal{L}(\boldsymbol{\alpha}, \boldsymbol{x}, \boldsymbol{p}) = J(\boldsymbol{\alpha}, \boldsymbol{x}) + \boldsymbol{p}^{\mathrm{T}} (\boldsymbol{f}(\boldsymbol{\alpha}) - \boldsymbol{A}(\boldsymbol{\alpha})\boldsymbol{x}) . \tag{3.9}$$

Let $(\overline{\boldsymbol{x}}, \overline{\boldsymbol{p}}) \in \mathbb{R}^n \times \mathbb{R}^n$ be such that

$$J(\boldsymbol{\alpha}, \boldsymbol{x}(\boldsymbol{\alpha})) = \inf_{\boldsymbol{x} \in \mathbb{R}^n} \sup_{\boldsymbol{p} \in \mathbb{R}^n} \mathcal{L}(\boldsymbol{\alpha}, \boldsymbol{x}, \boldsymbol{p}) = \mathcal{L}(\boldsymbol{\alpha}, \overline{\boldsymbol{x}}, \overline{\boldsymbol{p}}).$$

Then necessarily

$$\nabla_p \mathcal{L}(\boldsymbol{\alpha}, \overline{\boldsymbol{x}}, \overline{\boldsymbol{p}}) = \mathbf{0} \iff \boldsymbol{A}(\boldsymbol{\alpha})\overline{\boldsymbol{x}} = \boldsymbol{f}(\boldsymbol{\alpha})$$

and

$$\nabla_x \mathcal{L}(\boldsymbol{\alpha}, \overline{\boldsymbol{x}}, \overline{\boldsymbol{p}}) = \mathbf{0} \iff \boldsymbol{A}(\boldsymbol{\alpha})^{\mathrm{T}}\overline{\boldsymbol{p}} = \nabla_x J(\boldsymbol{\alpha}, \overline{\boldsymbol{x}});$$

i.e., $\overline{\boldsymbol{x}} = \boldsymbol{x}(\boldsymbol{\alpha})$ solves $(\mathcal{P}(\boldsymbol{\alpha}))$ and $\overline{\boldsymbol{p}} = \boldsymbol{p}(\boldsymbol{\alpha})$ is the respective adjoint state. From this, (3.1), and (3.9) we have

$$\mathcal{J}(\boldsymbol{\alpha}) = J(\boldsymbol{\alpha}, \boldsymbol{x}(\boldsymbol{\alpha})) + \boldsymbol{p}(\boldsymbol{\alpha})^{\mathrm{T}} (\boldsymbol{f}(\boldsymbol{\alpha}) - \boldsymbol{A}(\boldsymbol{\alpha})\boldsymbol{x}(\boldsymbol{\alpha})) .$$

Assuming that the adjoint state is directionally differentiable in $\mathcal{U}$, the classical chain rule of differentiation applies to the right-hand side of the previous expression:

$$\begin{aligned} \mathcal{J}'(\boldsymbol{\alpha}, \boldsymbol{\beta}) = \boldsymbol{\beta}^{\mathrm{T}} \nabla_\alpha J(\boldsymbol{\alpha}, \boldsymbol{x}(\boldsymbol{\alpha})) &+ \big(\nabla_x J(\boldsymbol{\alpha}, \boldsymbol{x}(\boldsymbol{\alpha}))\big)^{\mathrm{T}} \boldsymbol{x}'(\boldsymbol{\alpha}) \\ &+ \big(\boldsymbol{p}'(\boldsymbol{\alpha})\big)^{\mathrm{T}} \big(\boldsymbol{f}(\boldsymbol{\alpha}) - \boldsymbol{A}(\boldsymbol{\alpha})\boldsymbol{x}(\boldsymbol{\alpha})\big) \\ &+ \boldsymbol{p}(\boldsymbol{\alpha})^{\mathrm{T}} \big(\boldsymbol{f}'(\boldsymbol{\alpha}) - \boldsymbol{A}'(\boldsymbol{\alpha})\boldsymbol{x}(\boldsymbol{\alpha}) - \boldsymbol{A}(\boldsymbol{\alpha})\boldsymbol{x}'(\boldsymbol{\alpha})\big). \end{aligned} \tag{3.10}$$

The third term on the right of (3.10) disappears since $\boldsymbol{x}(\boldsymbol{\alpha})$ solves $(\mathcal{P}(\boldsymbol{\alpha}))$. The second and fourth terms can be rearranged as follows:

$$\begin{aligned} \boldsymbol{x}'(\boldsymbol{\alpha})^{\mathrm{T}} \big(\nabla_x J(\boldsymbol{\alpha}, \boldsymbol{x}(\boldsymbol{\alpha})) - \boldsymbol{A}(\boldsymbol{\alpha})^{\mathrm{T}} \boldsymbol{p}(\boldsymbol{\alpha})\big) + \boldsymbol{p}(\boldsymbol{\alpha})^{\mathrm{T}} \big(\boldsymbol{f}'(\boldsymbol{\alpha}) - \boldsymbol{A}'(\boldsymbol{\alpha})\boldsymbol{x}(\boldsymbol{\alpha})\big) \\ = \boldsymbol{p}(\boldsymbol{\alpha})^{\mathrm{T}} \big(\boldsymbol{f}'(\boldsymbol{\alpha}) - \boldsymbol{A}'(\boldsymbol{\alpha})\boldsymbol{x}(\boldsymbol{\alpha})\big) , \end{aligned}$$

taking into account (3.5). We have recovered (3.7).

Remark 3.4. It would be possible to consider nonlinear state equations of the form

$$\boldsymbol{A}(\boldsymbol{\alpha}, \boldsymbol{x}(\boldsymbol{\alpha})) = \boldsymbol{f}(\boldsymbol{\alpha}),$$

where $\boldsymbol{A} \in C^1(\mathcal{U} \times \mathbb{R}^n, \mathbb{R}^n)$, $\boldsymbol{f} \in C^1(\mathcal{U}, \mathbb{R}^n)$ are two vector functions. The previous sensitivity analysis extends straightforwardly to this case (see Problem 3.1).

We shall now pay attention to the case when $(\mathcal{P}(\boldsymbol{\alpha}))$ is given by the following algebraic inequality, depending on a parameter $\boldsymbol{\alpha} \in \mathcal{U}$:

$$\begin{cases} \text{Find } \boldsymbol{x}(\boldsymbol{\alpha}) \in \mathcal{K}(\boldsymbol{\alpha}) \text{ such that} \\ (\boldsymbol{y} - \boldsymbol{x}(\boldsymbol{\alpha}))^{\mathrm{T}} \boldsymbol{A}(\boldsymbol{\alpha})\boldsymbol{x}(\boldsymbol{\alpha}) \geq (\boldsymbol{y} - \boldsymbol{x}(\boldsymbol{\alpha}))^{\mathrm{T}} \boldsymbol{f}(\boldsymbol{\alpha}) \quad \forall \boldsymbol{y} \in \mathcal{K}(\boldsymbol{\alpha}), \end{cases} \tag{$\mathcal{P}_{in}(\boldsymbol{\alpha})$}$$

where

$$\mathcal{K}(\boldsymbol{\alpha}) = \left\{ \boldsymbol{y} \in \mathbb{R}^n \mid y_{j_i} \geq -\alpha_i \; \forall j_i \in \mathcal{I} \right\}, \quad \boldsymbol{\alpha} = (\alpha_1, \ldots, \alpha_d) \in \mathcal{U} \subset \mathbb{R}^d, \tag{3.11}$$

and $\mathcal{I}$ is a set of indices of the constrained components of $\boldsymbol{y} \in \mathbb{R}^n$. As we have already mentioned, optimization of systems governed by variational inequalities deserves special attention. In contrast to state equations, this time the control state mapping *need not be continuously differentiable*. The remainder of this section will be devoted to the sensitivity analysis of solutions to $(\mathcal{P}_{in}(\boldsymbol{\alpha}))$.

In addition to (i)–(iii), two more assumptions are supposed to be satisfied:

(iv) $\boldsymbol{A}(\boldsymbol{\alpha})$ *is symmetric for any* $\boldsymbol{\alpha} \in \mathcal{U}$;

(v) $\exists m = const. > 0: \quad \boldsymbol{y}^{\mathrm{T}}\boldsymbol{A}(\boldsymbol{\alpha})\boldsymbol{y} \geq m\|\boldsymbol{y}\|^2 \quad \forall\, \boldsymbol{y} \in \mathbb{R}^n, \quad \forall \boldsymbol{\alpha} \in \mathcal{U}.$

Remark 3.5. Assumption (iv) is not needed. It helps us to interpret the next results in a variational way. Also (v) can be weakened. It is sufficient to assume that

(v′) $\boldsymbol{y}^{\mathrm{T}}\boldsymbol{A}(\boldsymbol{\alpha})\boldsymbol{y} > 0 \quad \forall\, \boldsymbol{y} \in \mathbb{R}^n, \; \forall \boldsymbol{\alpha} \in \mathcal{U}$

(see Problem 3.2).

It is left as an exercise to show that the control state mapping defined by $(\mathcal{P}_{in}(\boldsymbol{\alpha}))$ is *Lipschitz continuous* in $\mathcal{U}$. This results from (ii) and the following.

Lemma 3.1. *Let* (v) *be satisfied. Then, for any two solutions* $\boldsymbol{x}(\boldsymbol{\alpha})$, $\boldsymbol{x}(\boldsymbol{\beta})$ *of* $(\mathcal{P}_{in}(\boldsymbol{\alpha}))$, $(\mathcal{P}_{in}(\boldsymbol{\beta}))$, *respectively, where* $\boldsymbol{\alpha}, \boldsymbol{\beta} \in \mathcal{U}$,

$$\|\boldsymbol{x}(\boldsymbol{\alpha}) - \boldsymbol{x}(\boldsymbol{\beta})\| \leq c\,\{\|\boldsymbol{A}(\boldsymbol{\alpha}) - \boldsymbol{A}(\boldsymbol{\beta})\| + \|\boldsymbol{f}(\boldsymbol{\alpha}) - \boldsymbol{f}(\boldsymbol{\beta})\| + \|\boldsymbol{\alpha} - \boldsymbol{\beta}\|\}, \tag{3.12}$$

where $c > 0$ *does not depend on* $\boldsymbol{\alpha}, \boldsymbol{\beta} \in \mathcal{U}$.

Proof. See Problem 3.3. $\square$

Remark 3.6. If (v′) were satisfied then the control state mapping would be only *locally Lipschitz continuous* in $\mathcal{U}$, having no influence on the sensitivity analysis.

From the previous lemma and the Rademacher theorem it follows that the mapping $\boldsymbol{x} : \boldsymbol{\alpha} \mapsto \boldsymbol{x}(\boldsymbol{\alpha}) \in \mathcal{K}(\boldsymbol{\alpha})$ is differentiable almost everywhere in $\mathcal{U}$. Next we prove that $\boldsymbol{x}$ is in fact *directionally differentiable* at any $\boldsymbol{\alpha} \in \mathcal{U}$ and in any direction $\boldsymbol{\beta} \in \mathbb{R}^d$ and we show how to compute $\boldsymbol{x}'(\boldsymbol{\alpha}; \boldsymbol{\beta})$.

We start with an equivalent formulation of $(\mathcal{P}(\boldsymbol{\alpha}))$. It is well known that $\boldsymbol{x}(\boldsymbol{\alpha}) \in \mathcal{K}(\boldsymbol{\alpha})$ solves $(\mathcal{P}(\boldsymbol{\alpha}))$ iff there exists a vector

$$\boldsymbol{\lambda} := \boldsymbol{\lambda}(\boldsymbol{\alpha}) = (\lambda_1(\boldsymbol{\alpha}), \ldots, \lambda_m(\boldsymbol{\alpha})) \in \mathbb{R}^m, \qquad m = \operatorname{card} \mathcal{I},$$

such that the pair $(\boldsymbol{x}(\boldsymbol{\alpha}), \boldsymbol{\lambda}(\boldsymbol{\alpha}))$ satisfies the following conditions (the summation convention is used):

$$\begin{cases} a_{ij}(\boldsymbol{\alpha})x_j(\boldsymbol{\alpha}) = f_i(\boldsymbol{\alpha}) \quad \forall i \notin \mathcal{I}, \\ a_{j_ik}(\boldsymbol{\alpha})x_k(\boldsymbol{\alpha}) = f_{j_i}(\boldsymbol{\alpha}) + \lambda_i(\boldsymbol{\alpha}) \quad \forall j_i \in \mathcal{I}, \\ \lambda_i(\boldsymbol{\alpha}) \geq 0 \quad \forall i, \quad x_{j_i}(\boldsymbol{\alpha}) + \alpha_i \geq 0 \quad \forall j_i \in \mathcal{I}, \\ \lambda_i(\boldsymbol{\alpha})(x_{j_i}(\boldsymbol{\alpha}) + \alpha_i) = 0 \quad \text{(no sum)} \quad \forall j_i \in \mathcal{I}, \end{cases} \tag{KKT($\boldsymbol{\alpha}$)}$$

known as the *Karush–Kuhn–Tucker (KKT) optimality conditions* for $(\mathcal{P}_{in}(\boldsymbol{\alpha}))$ (see [Fle87]). From Lemma 3.1 and (KKT($\boldsymbol{\alpha}$)) we see that the mapping $\boldsymbol{\lambda} : \boldsymbol{\alpha} \mapsto \boldsymbol{\lambda}(\boldsymbol{\alpha})$ is Lipschitz continuous in $\mathcal{U}$, as well.

Let $(\boldsymbol{x}(\boldsymbol{\alpha}+t\boldsymbol{\beta}), \boldsymbol{\lambda}(\boldsymbol{\alpha}+t\boldsymbol{\beta}))$ be the solution of (KKT($\boldsymbol{\alpha}+t\boldsymbol{\beta}$)), where $\boldsymbol{\alpha} \in \mathcal{U}$, $\boldsymbol{\beta} \in \mathbb{R}^d$, and $t > 0$. In view of Lipschitz continuity, the difference quotients

$$\frac{\boldsymbol{x}(\boldsymbol{\alpha}+t\boldsymbol{\beta}) - \boldsymbol{x}(\boldsymbol{\alpha})}{t}, \qquad \frac{\boldsymbol{\lambda}(\boldsymbol{\alpha}+t\boldsymbol{\beta}) - \boldsymbol{\lambda}(\boldsymbol{\alpha})}{t}$$

are bounded for $t > 0$. Thus one can find a sequence $\{t_n\}$, $t_n \to 0+$, and vectors $\dot{\boldsymbol{x}} := \dot{\boldsymbol{x}}(\{t_n\})$, $\dot{\boldsymbol{\lambda}} := \dot{\boldsymbol{\lambda}}(\{t_n\})$ such that

$$\frac{\boldsymbol{x}(\boldsymbol{\alpha}+t_n\boldsymbol{\beta}) - \boldsymbol{x}(\boldsymbol{\alpha})}{t_n} \to \dot{\boldsymbol{x}}, \quad \frac{\boldsymbol{\lambda}(\boldsymbol{\alpha}+t_n\boldsymbol{\beta}) - \boldsymbol{\lambda}(\boldsymbol{\alpha})}{t_n} \to \dot{\boldsymbol{\lambda}}, \quad t_n \to 0+. \tag{3.13}$$

We do not yet know if $\dot{\boldsymbol{x}}$ and $\dot{\boldsymbol{\lambda}}$ depend on a particular choice of $\{t_n\}$. This is why we used the symbol $\{t_n\}$ in the argument of the previous limits. From (ii) it follows that

$$\begin{cases} \dot{\boldsymbol{A}}(\boldsymbol{\alpha}; \boldsymbol{\beta}) := \lim\limits_{t_n \to 0+} \dfrac{\boldsymbol{A}(\boldsymbol{\alpha}+t_n\boldsymbol{\beta}) - \boldsymbol{A}(\boldsymbol{\alpha})}{t_n} = \boldsymbol{A}'(\boldsymbol{\alpha}; \boldsymbol{\beta}), \\ \dot{\boldsymbol{f}}(\boldsymbol{\alpha}; \boldsymbol{\beta}) := \lim\limits_{t_n \to 0+} \dfrac{\boldsymbol{f}(\boldsymbol{\alpha}+t_n\boldsymbol{\beta}) - \boldsymbol{f}(\boldsymbol{\alpha})}{t_n} = \boldsymbol{f}'(\boldsymbol{\alpha}; \boldsymbol{\beta}) \end{cases} \tag{3.14}$$

and these limits are *independent* of $\{t_n\}$, $t_n \to 0+$. The operation "$\cdot$" can be viewed as a kind of "directional derivative" that coincides with the classical one for smooth functions. Applying "$\cdot$" to the first two equations in the KKT conditions we obtain

$$a_{ij}(\boldsymbol{\alpha})\dot{x}_j(\boldsymbol{\alpha}) = f_i'(\boldsymbol{\alpha}; \boldsymbol{\beta}) - a_{ij}'(\boldsymbol{\alpha}; \boldsymbol{\beta})x_j(\boldsymbol{\alpha}) \quad \forall i \notin \mathcal{I},$$
$$a_{j_ik}(\boldsymbol{\alpha})\dot{x}_k(\boldsymbol{\alpha}) = f_{j_i}'(\boldsymbol{\alpha}; \boldsymbol{\beta}) - a_{j_ik}'(\boldsymbol{\alpha}; \boldsymbol{\beta})x_k(\boldsymbol{\alpha}) + \dot{\lambda}_i \quad \forall j_i \in \mathcal{I},$$

making use of (3.14). The constrained components of $\boldsymbol{x}(\boldsymbol{\alpha})$ determine the following decomposition of $\mathcal{I}$: $\mathcal{I} = \mathcal{I}_+(\boldsymbol{\alpha}) \cup \mathcal{I}_0(\boldsymbol{\alpha}) \cup \mathcal{I}_-(\boldsymbol{\alpha})$, where

$$\begin{aligned} \mathcal{I}_+(\boldsymbol{\alpha}) &= \{j_i \in \mathcal{I} \mid x_{j_i}(\boldsymbol{\alpha}) > -\alpha_i\} && \textit{(set of nonactive constraints)}, \\ \mathcal{I}_0(\boldsymbol{\alpha}) &= \{j_i \in \mathcal{I} \mid x_{j_i}(\boldsymbol{\alpha}) = -\alpha_i,\ \lambda_i(\boldsymbol{\alpha}) = 0\} && \textit{(set of semiactive constraints)}, \\ \mathcal{I}_-(\boldsymbol{\alpha}) &= \{j_i \in \mathcal{I} \mid x_{j_i}(\boldsymbol{\alpha}) = -\alpha_i,\ \lambda_i(\boldsymbol{\alpha}) > 0\} && \textit{(set of strongly active constraints)}. \end{aligned}$$

Next we shall analyze the behavior of the constrained components of $\boldsymbol{x}(\boldsymbol{\alpha})$ and of the vector $\boldsymbol{\lambda}(\boldsymbol{\alpha})$ for small parameter perturbations of the form $\boldsymbol{\alpha} + t\boldsymbol{\beta}$, $t \to 0+$.

Let $j_i \in \mathcal{I}_+(\boldsymbol{\alpha})$. Then owing to the continuity of the mapping $\boldsymbol{x} : \boldsymbol{\alpha} \mapsto \boldsymbol{x}(\boldsymbol{\alpha})$ the index j_i belongs to $\mathcal{I}_+(\boldsymbol{\alpha} + t\boldsymbol{\beta})$ for any $|t| \leq \delta$, where $\delta > 0$ is small enough. From the last condition in the KKT system we have that $\lambda_i(\boldsymbol{\alpha} + t\boldsymbol{\beta}) = 0 = \lambda_i(\boldsymbol{\alpha})$ for any $|t| \leq \delta$, implying $\dot{\lambda}_i(\boldsymbol{\alpha}) = 0$.

If $j_i \in \mathcal{I}_-(\boldsymbol{\alpha})$, then $\lambda_i(\boldsymbol{\alpha}) > 0$ and also $\lambda_i(\boldsymbol{\alpha} + t\boldsymbol{\beta}) > 0$ for $|t| \leq \delta$ using the continuity of the mapping $\boldsymbol{\lambda} : \boldsymbol{\alpha} \mapsto \boldsymbol{\lambda}(\boldsymbol{\alpha})$. The last condition in the KKT system yields $x_{j_i}(\boldsymbol{\alpha} + t\boldsymbol{\beta}) = -\alpha_i - t\beta_i$ so that $\dot{x}_{j_i}(\boldsymbol{\alpha}) = -\beta_i$.

Finally, let $j_i \in \mathcal{I}_0(\boldsymbol{\alpha})$. Then for all $t > 0$,

$$\begin{aligned} &\lambda_i(\boldsymbol{\alpha} + t\boldsymbol{\beta}) \geq \lambda_i(\boldsymbol{\alpha}) = 0, \\ &x_{j_i}(\boldsymbol{\alpha} + t\boldsymbol{\beta}) \geq -\alpha_i - t\beta_i = x_{j_i}(\boldsymbol{\alpha}) - t\beta_i \end{aligned}$$

and consequently $\dot{\lambda}_i(\boldsymbol{\alpha}) \geq 0,\ \dot{x}_{j_i}(\boldsymbol{\alpha}) \geq -\beta_i$.

We now show that

$$\dot{\lambda}_i(\boldsymbol{\alpha})(\dot{x}_{j_i}(\boldsymbol{\alpha}) + \beta_i) = 0 \quad \text{(no sum)} \quad \forall j_i \in \mathcal{I}. \tag{3.15}$$

This is certainly true for $j_i \in \mathcal{I}_+(\boldsymbol{\alpha}) \cup \mathcal{I}_-(\boldsymbol{\alpha})$ since either $\dot{\lambda}_i(\boldsymbol{\alpha}) = 0$ or $\dot{x}_{j_i}(\boldsymbol{\alpha}) = -\beta_i$. Let $j_i \in \mathcal{I}_0(\boldsymbol{\alpha})$. If $\dot{\lambda}_i(\boldsymbol{\alpha}) = 0$, then (3.15) holds true. On the other hand if $\dot{\lambda}_i(\boldsymbol{\alpha}) > 0$, then necessarily $\lambda_i(\boldsymbol{\alpha} + t\boldsymbol{\beta}) > 0$ so that $x_{j_i}(\boldsymbol{\alpha} + t\boldsymbol{\beta}) = -\alpha_i - t\beta_i$ for any $t > 0$ small enough. Hence $\dot{x}_{j_i}(\boldsymbol{\alpha}) = -\beta_i$, and we may conclude that (3.15) holds for any $j_i \in \mathcal{I}$.

With the decomposition of $\mathcal{I}$ into $\mathcal{I}_0(\boldsymbol{\alpha})$, $\mathcal{I}_-(\boldsymbol{\alpha})$, and $\mathcal{I}_+(\boldsymbol{\alpha})$ the following convex set $\mathcal{K}(\boldsymbol{\alpha}, \boldsymbol{\beta})$ will be associated:

$$\mathcal{K}(\boldsymbol{\alpha}, \boldsymbol{\beta}) = \left\{ \boldsymbol{y} \in \mathbb{R}^n \mid y_{j_i} = -\beta_i\ \forall\, j_i \in \mathcal{I}_-(\boldsymbol{\alpha}), \quad y_{j_i} \geq -\beta_i \quad \forall\, j_i \in \mathcal{I}_0(\boldsymbol{\alpha}) \right\}. \tag{3.16}$$

From the previous analysis it follows that $\dot{\boldsymbol{x}}(\{t_n\})$ belongs to $\mathcal{K}(\boldsymbol{\alpha}, \boldsymbol{\beta})$ and, in addition, the pair $(\dot{\boldsymbol{x}}(\{t_n\}), \dot{\boldsymbol{\lambda}}(\{t_n\}))$ satisfies the following KKT type system:

$$\begin{cases} a_{ij}(\boldsymbol{\alpha})\dot{x}_j(\boldsymbol{\alpha}) = f_i'(\boldsymbol{\alpha}) - a_{ij}'(\boldsymbol{\alpha})x_j(\boldsymbol{\alpha}) \quad \forall i \notin \mathcal{I}_0(\boldsymbol{\alpha}) \cup \mathcal{I}_-(\boldsymbol{\alpha}), \\ a_{j_ik}(\boldsymbol{\alpha})\dot{x}_k(\boldsymbol{\alpha}) = f_{j_i}'(\boldsymbol{\alpha}) - a_{j_ik}'(\boldsymbol{\alpha})x_k(\boldsymbol{\alpha}) + \dot{\lambda}_i \quad \forall j_i \in \mathcal{I}_0(\boldsymbol{\alpha}) \cup \mathcal{I}_-(\boldsymbol{\alpha}), \\ \dot{\lambda}_i(\boldsymbol{\alpha}) \geq 0,\ \dot{x}_{j_i}(\boldsymbol{\alpha}) \geq -\beta_i \quad \forall j_i \in \mathcal{I}_0(\boldsymbol{\alpha}),\ \dot{x}_{j_i}(\boldsymbol{\alpha}) = -\beta_i \quad \forall j_i \in \mathcal{I}_-(\boldsymbol{\alpha}), \\ \dot{\lambda}_i(\boldsymbol{\alpha})(\dot{x}_{j_i}(\boldsymbol{\alpha}) + \beta_i) = 0 \quad \forall j_i \in \mathcal{I}_0(\boldsymbol{\alpha}) \cup \mathcal{I}_-(\boldsymbol{\alpha}). \end{cases} \tag{3.17}$$

We are now able to formulate and prove the following theorem.

Theorem 3.2. *Let* (i), (ii), (iv), *and* (v) *be satisfied. Then the solution* $\boldsymbol{x}(\boldsymbol{\alpha})$ *of* $(\mathcal{P}_{in}(\boldsymbol{\alpha}))$ *is directionally differentiable at any* $\boldsymbol{\alpha} \in \mathcal{U}$ *and in any direction* $\boldsymbol{\beta} \in \mathbb{R}^d$. *In addition the directional derivative* $\boldsymbol{x}'(\boldsymbol{\alpha}) := \boldsymbol{x}'(\boldsymbol{\alpha}; \boldsymbol{\beta})$ *can be equivalently characterized as the unique solution of the following minimization problem:*

$$\begin{cases} \text{Find } \boldsymbol{x}'(\boldsymbol{\alpha}) \in \mathcal{K}(\boldsymbol{\alpha}, \boldsymbol{\beta}) \quad \text{such that} \\ \mathcal{G}_{\boldsymbol{\alpha},\boldsymbol{\beta}}(\boldsymbol{x}'(\boldsymbol{\alpha})) \leq \mathcal{G}_{\boldsymbol{\alpha},\boldsymbol{\beta}}(\boldsymbol{y}) \quad \forall \boldsymbol{y} \in \mathcal{K}(\boldsymbol{\alpha}, \boldsymbol{\beta}), \end{cases} \tag{3.18}$$

where $\mathcal{G}_{\boldsymbol{\alpha},\boldsymbol{\beta}}$, $\mathcal{K}(\boldsymbol{\alpha}, \boldsymbol{\beta})$ *are defined by* (3.3), (3.16), *respectively.*

Proof. We have already shown that $\dot{\boldsymbol{x}} := \dot{\boldsymbol{x}}(\{t_n\})$ belongs to $\mathcal{K}(\boldsymbol{\alpha}, \boldsymbol{\beta})$, and in view of (3.17) the vector $\dot{\boldsymbol{x}}$ minimizes $\mathcal{G}_{\alpha,\beta}$ on $\mathcal{K}(\boldsymbol{\alpha}, \boldsymbol{\beta})$. It is seen from the previous analysis that any accumulation point of $\{t^{-1}(\boldsymbol{x}(\boldsymbol{\alpha} + t\boldsymbol{\beta}) - \boldsymbol{x}(\boldsymbol{\alpha}))\}$, $t \to 0+$, possesses this property. But $\mathcal{G}_{\alpha,\beta}$ being strictly convex admits a unique minimizer on $\mathcal{K}(\boldsymbol{\alpha}, \boldsymbol{\beta})$. Therefore the limit of the previous difference quotient for $t \to 0+$ exists and $\dot{\boldsymbol{x}}(\{t_n\}) = \boldsymbol{x}'(\boldsymbol{\alpha}; \boldsymbol{\beta})$. □

Let us comment on the previous results. For the linear state problem ($\mathcal{P}(\boldsymbol{\alpha})$), sensitivity analysis leads to a linear algebraic problem (3.2) or, equivalently, to an *unconstrained* minimization problem for the quadratic function $\mathcal{G}_{\alpha,\beta}$ provided that $\boldsymbol{A}(\boldsymbol{\alpha})$ is symmetric for any $\boldsymbol{\alpha} \in \mathcal{U}$ (see Remark 3.3). In the case of a variational inequality ($\mathcal{P}_{in}(\boldsymbol{\alpha})$), the sensitivity analysis is more involved. This time the directional derivative of $\boldsymbol{x}$ is obtained as the solution of the *constrained* minimization problem. Despite the fact that the control state mapping is directionally differentiable in any direction, it need not be continuously differentiable in $\mathcal{U}$. Indeed, the directional derivative $\boldsymbol{x}'(\boldsymbol{\alpha}; \boldsymbol{\beta})$ being the solution of the quadratic programming problem (3.18) does not satisfy $\boldsymbol{x}'(\boldsymbol{\alpha}; -\boldsymbol{\beta}) = -\boldsymbol{x}'(\boldsymbol{\alpha}, \boldsymbol{\beta})$ in general; i.e., approaching $\boldsymbol{\alpha}$ from the opposite direction, the respective directional derivative may have a jump. At the same time, Theorem 3.2 indicates the source of possible nondifferentiability: it is due to the presence of semiactive constraints or, equivalently, the nonemptiness of $\mathcal{I}_0(\boldsymbol{\alpha})$. If $\mathcal{I}_0(\boldsymbol{\alpha}) = \emptyset$, the respective control state mapping would be continuously differentiable (see Problem 3.4).

Let $J : \mathcal{U} \times \mathbb{R}^n \to \mathbb{R}$ be a function satisfying (iii) and $\mathcal{J} : \mathcal{U} \to \mathbb{R}$ be defined by (3.1) with $\boldsymbol{x}(\boldsymbol{\alpha})$ solving ($\mathcal{P}_{in}(\boldsymbol{\alpha})$). Then $\mathcal{J}$ as a composition of J and $\boldsymbol{x}$ is directionally differentiable at any $\boldsymbol{\alpha} \in \mathcal{U}$ and in any direction $\boldsymbol{\beta} \in \mathbb{R}^d$ but generally not continuously differentiable. For some special choices of J, however, it may happen that $\mathcal{J}$ is *once* continuously differentiable. Take, for example,

$$J(\boldsymbol{\alpha}, \boldsymbol{y}) = \frac{1}{2}\boldsymbol{y}^{\mathrm{T}}\boldsymbol{A}(\boldsymbol{\alpha})\boldsymbol{y} - \boldsymbol{y}^{\mathrm{T}}\boldsymbol{f}(\boldsymbol{\alpha}).$$

Then

$$\mathcal{J}(\boldsymbol{\alpha}) = \frac{1}{2}(\boldsymbol{x}(\boldsymbol{\alpha}))^{\mathrm{T}}\boldsymbol{A}(\boldsymbol{\alpha})\boldsymbol{x}(\boldsymbol{\alpha}) - (\boldsymbol{x}(\boldsymbol{\alpha}))^{\mathrm{T}}\boldsymbol{f}(\boldsymbol{\alpha})$$

is a C^1-function. To see that, we first compute the directional derivative $\mathcal{J}'(\boldsymbol{\alpha}; \boldsymbol{\beta})$:

$$\begin{aligned}\mathcal{J}'(\boldsymbol{\alpha}; \boldsymbol{\beta}) &= (\boldsymbol{x}'(\boldsymbol{\alpha}))^{\mathrm{T}}\boldsymbol{A}(\boldsymbol{\alpha})\boldsymbol{x}(\boldsymbol{\alpha}) \\ &\quad + \frac{1}{2}(\boldsymbol{x}(\boldsymbol{\alpha}))^{\mathrm{T}}\boldsymbol{A}'(\boldsymbol{\alpha})\boldsymbol{x}(\boldsymbol{\alpha}) - (\boldsymbol{x}'(\boldsymbol{\alpha}))^{\mathrm{T}}\boldsymbol{f}(\boldsymbol{\alpha}) - (\boldsymbol{x}(\boldsymbol{\alpha}))^{\mathrm{T}}\boldsymbol{f}'(\boldsymbol{\alpha}),\end{aligned}$$

making use of (iv). From (KKT($\boldsymbol{\alpha}$)) and the definition of $\mathcal{K}(\boldsymbol{\alpha}, \boldsymbol{\beta})$ we see that

$$(\boldsymbol{x}'(\boldsymbol{\alpha}))^{\mathrm{T}}(\boldsymbol{A}(\boldsymbol{\alpha})\boldsymbol{x}(\boldsymbol{\alpha}) - \boldsymbol{f}(\boldsymbol{\alpha})) = -\beta_i\lambda_i(\boldsymbol{\alpha})$$

and consequently

$$\mathcal{J}'(\boldsymbol{\alpha}; \boldsymbol{\beta}) = \frac{1}{2}(\boldsymbol{x}(\boldsymbol{\alpha}))^{\mathrm{T}}\boldsymbol{A}'(\boldsymbol{\alpha})\boldsymbol{x}(\boldsymbol{\alpha}) - (\boldsymbol{x}(\boldsymbol{\alpha}))^{\mathrm{T}}\boldsymbol{f}'(\boldsymbol{\alpha}) - \beta_i\lambda_i(\boldsymbol{\alpha}). \tag{3.19}$$

From (ii) and the continuity of $\boldsymbol{\lambda} : \boldsymbol{\alpha} \mapsto \boldsymbol{\lambda}(\boldsymbol{\alpha})$ it follows that the mapping $\boldsymbol{\alpha} \mapsto \mathcal{J}'(\boldsymbol{\alpha}, \cdot)$ is also continuous in $\mathcal{U}$ so that $\mathcal{J} \in C^1(\mathcal{U})$.

Directional derivatives of solutions $\boldsymbol{x}(\boldsymbol{\alpha})$ to algebraic inequalities, which are a finite element discretization of the scalar variational inequalities studied in Subsection 2.5.4, can be computed directly from Theorem 3.2 by setting

$$\mathcal{K}(\boldsymbol{\alpha}, \boldsymbol{\beta}) = \{\boldsymbol{y} \in \mathbb{R}^n \mid y_{j_i} = 0 \, \forall j_i \in \mathcal{I}_-(\boldsymbol{\alpha}),\ y_{j_i} \geq 0 \, \forall j_i \in \mathcal{I}_0(\boldsymbol{\alpha})\},$$

where

$$\mathcal{I}_-(\boldsymbol{\alpha}) = \{j_i \in \mathcal{I} \mid x_{j_i}(\boldsymbol{\alpha}) = 0,\ \lambda_i(\boldsymbol{\alpha}) > 0\},$$
$$\mathcal{I}_0(\boldsymbol{\alpha}) = \{j_i \in \mathcal{I} \mid x_{j_i}(\boldsymbol{\alpha}) = 0,\ \lambda_i(\boldsymbol{\alpha}) = 0\}.$$

Observe that $\mathcal{K}(\boldsymbol{\alpha}, \boldsymbol{\beta})$ does not depend on $\boldsymbol{\beta}$; information on $\boldsymbol{\beta}$ is encoded in the functional $\mathcal{G}_{\boldsymbol{\alpha},\boldsymbol{\beta}}$. Sensitivity analysis in contact problems (see Subsection 2.5.5) is slightly more complicated since the closed, convex set $\mathcal{K}(\boldsymbol{\alpha})$ is of the following type:

$$\mathcal{K}(\boldsymbol{\alpha}) = \{\boldsymbol{x} \in \mathbb{R}^n \mid x_{j_i} \geq \varphi_i(\boldsymbol{\alpha}) \, \forall j_i \in \mathcal{I}\}, \tag{3.20}$$

where $\varphi_i : \mathcal{U} \to \mathbb{R}$, $i \in \mathcal{I}$, are continuously differentiable functions in $\mathcal{U}$ (see (2.183) with $\varphi_i(\boldsymbol{\alpha}) = -s_{\varkappa}(b_i, \boldsymbol{\alpha})$). If its proof is adapted, Theorem 3.2 extends to this case, too. This is left as an exercise.

3.2 Sensitivity analysis in thickness optimization

This section deals with sensitivity analysis in sizing optimization. We shall show how to compute the derivatives of stiffness matrices and of right-hand sides with respect to discrete sizing variables. Knowledge of these derivatives is indispensable to obtain sensitivities of solutions and cost functionals.

To illustrate how to proceed let us consider thickness optimization of the elastic beam studied in Section 2.1. Instead of the compliance of the beam, we take the following cost functional (see Problem 2.4):

$$J(e, y) = \frac{1}{2}\int_0^{\ell} (y - y_d)^2 \, dx + \frac{1}{2}\int_0^{\ell} e^2 \, dx, \tag{3.21}$$

where $y_d \in C([0, \ell])$ is given such that $y_d(0) = y_d(\ell) = 0$. The set U_h^{ad} and the space V_h used for the discretization of (2.2) are defined by (2.15) and (2.16), respectively. The discrete state problem leads to the following system of $2(d-1)$ algebraic equations (see (2.18)):

$$\boldsymbol{K}(\boldsymbol{e})\boldsymbol{q}(\boldsymbol{e}) = \boldsymbol{f}(\boldsymbol{e}), \tag{3.22}$$

where $\boldsymbol{e} = (e_0, \ldots, e_d) \in \mathbb{R}^{d+1}$ is the discrete sizing variable. (To make the presentation more general, we allow here the force vector to depend on $\boldsymbol{e}$, too. One could consider the loads caused by the weight of the beam, for example.) Recall that e_i, $i = 0, \ldots, d$, are the nodal values of $e_h \in U_h^{ad}$. Using a numerical integration formula for the evaluation of J we obtain its algebraic representation in the form

$$\widetilde{\mathcal{J}}(\boldsymbol{e}) := \mathcal{J}(\boldsymbol{e}, \boldsymbol{q}(\boldsymbol{e})) = \frac{1}{2}\sum_{i=1}^{d-1} \omega_i (q_{2i-1}(\boldsymbol{e}) - y_d(a_i))^2 + \frac{1}{2}\sum_{i=0}^{d} \omega_i e_i^2, \tag{3.23}$$

where $\{\omega_i\}_{i=0}^d$ are the weights of the selected numerical integration scheme and $\boldsymbol{q}(\boldsymbol{e})$ solves (3.22) (the components of $\boldsymbol{q}(\boldsymbol{e})$ are arranged as in (2.19)).

Let $\boldsymbol{e} \in \mathcal{U}$ and $\overline{\boldsymbol{e}} \in \mathbb{R}^{d+1}$. Then arguing as in Section 3.1 we obtain the following expressions for the directional derivatives of $\boldsymbol{q}$ and $\widetilde{\mathcal{J}}$:

$$\boldsymbol{q}'(\boldsymbol{e};\overline{\boldsymbol{e}}) = \boldsymbol{K}(\boldsymbol{e})^{-1}\left(\boldsymbol{f}'(\boldsymbol{e};\overline{\boldsymbol{e}}) - \boldsymbol{K}'(\boldsymbol{e};\overline{\boldsymbol{e}})\boldsymbol{q}(\boldsymbol{e})\right), \tag{3.24}$$

$$\widetilde{\mathcal{J}}'(\boldsymbol{e};\overline{\boldsymbol{e}}) = \left(\nabla_{\boldsymbol{e}}\mathcal{J}(\boldsymbol{e},\boldsymbol{q}(\boldsymbol{e}))\right)^{\mathrm{T}}\overline{\boldsymbol{e}} + \boldsymbol{p}(\boldsymbol{e})^{\mathrm{T}}\left(\boldsymbol{f}'(\boldsymbol{e};\overline{\boldsymbol{e}}) - \boldsymbol{K}'(\boldsymbol{e};\overline{\boldsymbol{e}})\boldsymbol{q}(\boldsymbol{e})\right), \tag{3.25}$$

where $\boldsymbol{p}(\boldsymbol{e}) \in \mathbb{R}^{2(d-1)}$ is the adjoint state:

$$\boldsymbol{K}(\boldsymbol{e})\boldsymbol{p}(\boldsymbol{e}) = \nabla_{\boldsymbol{q}}\mathcal{J}(\boldsymbol{e},\boldsymbol{q}(\boldsymbol{e})), \tag{3.26}$$

making use of the symmetry of $\boldsymbol{K}(\boldsymbol{e})$. Choosing $\overline{\boldsymbol{e}}$ as the kth canonical basis vector in $\mathbb{R}^{d+1}$ we obtain the partial derivatives $\partial\boldsymbol{q}/\partial e_k$, $\partial\widetilde{\mathcal{J}}/\partial e_k$ from (3.24) and (3.25). The partial gradients of $\mathcal{J}$ needed in (3.25) and (3.26) are easy to compute:

$$\frac{\partial\mathcal{J}(\boldsymbol{e},\boldsymbol{q})}{\partial e_k} = \omega_k e_k, \quad k = 0,\ldots,d, \tag{3.27}$$

and

$$\frac{\partial\mathcal{J}(\boldsymbol{e},\boldsymbol{q})}{\partial q_{2k-1}} = \omega_k(q_{2k-1} - y_d(a_k)), \tag{3.28}$$

$$\frac{\partial\mathcal{J}(\boldsymbol{e},\boldsymbol{q})}{\partial q_{2k}} = 0, \tag{3.29}$$

$k = 1,\ldots,d-1$. To determine

$$\frac{\partial\boldsymbol{K}(\boldsymbol{e})}{\partial e_k} = \left(\frac{\partial k_{ij}(\boldsymbol{e})}{\partial e_k}\right)_{i,j=1}^{n}, \quad n = 2(d-1), \quad k = 0,\ldots,d,$$

we use (see (2.20))

$$k_{ij}(\boldsymbol{e}) = \int_I \beta e_h^3 \varphi_i'' \varphi_j''\,dx, \quad i,j = 1,\ldots,n, \tag{3.30}$$

where $e_h \in U_h^{ad}$ and $\{\varphi_j\}_{j=1}^n$ is the basis of V_h. Since U_h^{ad} is realized by piecewise linear functions, any $e_h \in U_h^{ad}$ can be expressed in the form

$$e_h(x) = \sum_{i=0}^{d} e_i\psi_i(x), \quad \boldsymbol{e} = (e_0,\ldots,e_d) \in \mathcal{U},$$

where $\{\psi_i\}_{i=0}^d$ is the Courant basis associated with the equidistant partition Δ_h of $[0,\ell]$; i.e.,

$$\psi_k(x) = \begin{cases} h^{-1}(x - a_{k-1}), & x \in [a_{k-1}, a_k], \\ 1 - h^{-1}(x - a_k), & x \in [a_k, a_{k+1}], \\ 0 & \text{otherwise}, \end{cases}$$

$k = 1, \dots, d-1$, with the appropriate modifications for $k = 0, d$. From this and (3.30),

$$\frac{\partial k_{ij}(\boldsymbol{e})}{\partial e_k} = \int_{a_{k-1}}^{a_k} 3\beta h^{-1}(x - a_{k-1}) e_h^2 \varphi_i'' \varphi_j'' \, dx + \int_{a_k}^{a_{k+1}} 3\beta(1 - h^{-1}(x - a_k)) e_h^2 \varphi_i'' \varphi_j'' \, dx, \tag{3.31}$$

$i, j = 1, \dots, n$ and $k = 1, \dots, d-1$, again with the modifications for $k = 0, d$.

If f does not depend on the thickness, then

$$\frac{\partial \boldsymbol{f}}{\partial e_k} = 0, \quad k = 0, \dots, d.$$

Remark 3.7.

(i) Taking the compliance $J(y) = \int_0^\ell f\, y \, dx$ as the cost functional and using the same integration formulas for the evaluation of both J and the right side of (2.2), we obtain

$$\mathcal{J}(\boldsymbol{q}) = \boldsymbol{f}^{\mathrm{T}} \boldsymbol{q},$$

so that $\boldsymbol{q}(\boldsymbol{e}) = \boldsymbol{p}(\boldsymbol{e})$, as follows from (3.26), and no adjoint state is needed.

(ii) In multicriteria optimization or in problems involving a number of state constraints, sensitivities for several functions $\mathcal{J}_i$, $i = 1, \dots, M$, are needed. If the number M is bigger than the number of design variables d, the adjoint equation technique is not any more advantageous since M adjoint equations have to be solved. In this case we compute the derivatives from (3.24) and substitute them directly into

$$\frac{\partial \widetilde{\mathcal{J}}_i(\boldsymbol{e})}{\partial e_k} = \frac{\partial \mathcal{J}_i(\boldsymbol{e}, \boldsymbol{q}(\boldsymbol{e}))}{\partial e_k} + (\nabla_{\boldsymbol{q}} \mathcal{J}_i(\boldsymbol{e}, \boldsymbol{q}(\boldsymbol{e})))^{\mathrm{T}} \frac{\partial \boldsymbol{q}(\boldsymbol{e})}{\partial e_k}, \quad k = 0, \dots, d, \ i = 1, \dots, M,$$

where $\widetilde{\mathcal{J}}_i(\boldsymbol{e}) = \mathcal{J}_i(\boldsymbol{e}, \boldsymbol{q}(\boldsymbol{e}))$.

3.3 Sensitivity analysis in shape optimization

Sensitivity analysis in shape optimization deals with computations of derivatives of solutions to state problems as well as control and cost functionals with respect to shape variations. But first, basic questions such as how to describe changes in the geometry and how to differentiate functions with varying domain of their definition have to be clarified. We present the approach based on the *material derivative* idea of continuum mechanics. This presentation will be formal, meaning that it is correct provided that all data we need are sufficiently smooth. For a rigorous mathematical treatment of this topic going beyond the scope of this textbook we refer to [SZ92].

3.3.1 The material derivative approach in the continuous setting

A bounded domain $\Omega \subset \mathbb{R}^n$ with Lipschitz boundary $\partial\Omega$ is thought to be a collection of material particles changing their position in time. A space occupied by them at time t will determine a new configuration Ω_t. The change in the geometry of Ω will be given by an evolutionary process deforming the initial configuration Ω. To formalize this idea mathematically, we introduce a one-parameter family of mappings $\{F_t\}$, $t \in [0, t_0]$, $F_t : \Omega \to \mathbb{R}^n$, defining the motion of each material particle $x \in \Omega$ and such that $F_0 =$ identity (id). At time t the particle $x \in \Omega$ will occupy a new position x_t:

$$x_t = F_t(x), \quad x \in \Omega, \quad t \in [0, t_0]. \tag{3.32}$$

The new configuration of Ω at time t is given by

$$\Omega_t = F_t(\Omega); \tag{3.33}$$

i.e., Ω_t is the image of Ω with respect to F_t. The set $Q \subset \mathbb{R}^{n+1}$, where $Q = \cup_{t\in[0,t_0]}\{t\} \times \Omega_t$, determines the evolution of Ω in space and time $t \in [0, t_0]$. To keep the shape of all Ω_t "reasonable," each F_t, $t \in [0, t_0]$, has to be a one-to-one transformation of Ω onto Ω_t such that

$$F_t(\operatorname{int} \Omega) = \operatorname{int} \Omega_t, \quad F_t(\partial\Omega) = \partial\Omega_t. \tag{3.34}$$

In the rest of this textbook we shall consider a special class of mappings F_t being a perturbation of the identity, namely

$$F_t = \mathrm{id} + t\mathcal{V}, \quad t > 0, \tag{3.35}$$

where $\mathcal{V} \in (H^{1,\infty}(\Omega))^n$ is the so-called velocity field. It can be shown that for t_0 small enough F_t of the form (3.35) is a one-to-one mapping of Ω onto Ω_t satisfying (3.34) and preserving the Lipschitz continuity of $\partial\Omega_t$.

Remark 3.8. The form of the velocity field $\mathcal{V}$ realizing shape variations depends on how shapes of admissible domains are parametrized. Consider, for example, the family of domains shown in Figure 2.2, whose moving parts $\Gamma(\alpha)$ of the boundaries are parametrized by functions $\alpha \in \widetilde{U}^{ad}$, $\widetilde{U}^{ad}$ as in (2.96). Then it is natural to take $\mathcal{V}$ in the form

$$\mathcal{V} = (\mathcal{V}_1, 0), \quad \mathcal{V}_1(x) = \frac{x_1}{\alpha(x_2)}\psi(x_2), \quad x = (x_1, x_2) \in \Omega(\alpha), \tag{3.36}$$

where $\alpha \in \widetilde{U}^{ad}$ and $\psi : [0, 1] \to \mathbb{R}$ is a Lipschitz continuous function. It is readily seen that for this choice of $\mathcal{V}$ we have

$$\Omega_t(\alpha) = \left\{(x_1, x_2) \in \mathbb{R}^2 \mid 0 < x_1 < \alpha(x_2) + t\psi(x_2), \quad x_2 \in]0, 1[\right\}.$$

In computations we use shapes described by functions belonging to a finite dimensional space $\mathcal{D}_\varkappa$, $\varkappa > 0$, spanned by functions $\xi_1, \ldots, \xi_{d(\varkappa)}$, $d(\varkappa) = \dim \mathcal{D}_\varkappa$. Then the coefficients of the linear combination of $\{\xi_j\}_{j=1}^{d(\varkappa)}$ form a vector of the discrete design variables.

Let DF_t be the Jacobian of F_t and denote $I_t = \det DF_t$. Important formulas concerning their differentiation are listed in the next lemma.

LEMMA 3.2.

(i) $DF_t|_{t=0} = \mathrm{id}$, (ii) $\left.\dfrac{d}{dt}F_t\right|_{t=0} = \mathcal{V}$,

(iii) $\left.\dfrac{d}{dt}DF_t\right|_{t=0} = D\mathcal{V} = \left(\dfrac{\partial \mathcal{V}_i}{\partial x_j}\right)_{i,j=1}^{n}$, (iv) $\left.\dfrac{d}{dt}(DF_t)^{\mathrm{T}}\right|_{t=0} = D\mathcal{V}^{\mathrm{T}}$,

(v) $\left.\dfrac{d}{dt}(DF_t^{-1})\right|_{t=0} = -D\mathcal{V}$, (vi) $\left.\dfrac{d}{dt}I_t\right|_{t=0} = \operatorname{div}\mathcal{V}$.

Proof. The proof is left as an easy exercise. □

Let us consider state problem $(\mathcal{P})$ in a domain $\Omega \subset \mathbb{R}^n$. Let $(\mathcal{P}_t)$, $t \in]0, t_0]$, be a family of problems related to $(\mathcal{P})$ but solved in $\Omega_t = F_t(\Omega)$ with F_t given by (3.35). Solutions of $(\mathcal{P}_t)$ will be denoted by $u_t : \Omega_t \to \mathbb{R}^m$. The function u_t can be viewed as the restriction of another function $u : Q \to \mathbb{R}^m$ to $\{t\} \times \Omega_t$:

$$u_t(x_t) = u(t, x + t\mathcal{V}(x)), \quad x \in \Omega,\ t \in [0, t_0]. \tag{3.37}$$

Suppose that u is smooth enough in a δ-neighborhood Q_δ of Q. Then the classical chain rule of differentiation applies:

$$\dot{u}(0, x) := \left.\frac{d}{dt}u(t, x + t\mathcal{V}(x))\right|_{t=0} = \frac{\partial}{\partial t}u(0, x) + \nabla_x u(0, x) \cdot \mathcal{V}(x). \tag{3.38}$$

CONVENTION: *$(\mathcal{P}_t)$ for $t = 0$ coincides with $(\mathcal{P})$, and $u_0(x_0) = u(0, x) := u(x)$, $x \in \Omega$, solves $(\mathcal{P})$. The symbol u now has two meanings: it designates the function u defined by* (3.37) *and designates u's restriction to the time level $t = 0$.*

Using this convention, (3.38) can be written in the following abbreviated form:

$$\dot{u} = \frac{\partial u}{\partial t} + \nabla_x u \cdot \mathcal{V} \quad \text{in } \Omega. \tag{3.39}$$

The total derivative $\dot{u}$ is termed the *pointwise material derivative* of u. It characterizes the behavior of u at $x \in \Omega$ in the velocity direction $\mathcal{V}(x)$. To be precise we should write $\dot{u}(x, \mathcal{V}(x))$. The partial derivative $\partial u/\partial t$ will be denoted by u' in what follows and it will be called the *pointwise shape derivative* of u. Under certain smoothness assumptions the shape and spatial derivatives commute:

$$\left(\frac{\partial u}{\partial x_i}\right)' = \frac{\partial}{\partial x_i}(u'). \tag{3.40}$$

The material derivative concept can be extended to less regular functions belonging to Sobolev spaces. For example, let $u_t \circ F_t \in H^k(\Omega)$ for any $t \in [0, t_0]$. Then the material derivative $\dot{u}$ can be defined as an element of $H^k(\Omega)$ that is the limit of the finite difference quotient in the norm of $H^k(\Omega)$:

$$\left\| \frac{u_t \circ F_t - u}{t} - \dot{u} \right\|_{H^k(\Omega)} \to 0, \quad t \to 0+.$$

With $\dot{u}$ in hand one also can define the shape derivative u' for less regular functions by

$$u' := \dot{u} - \nabla u \cdot \mathcal{V} \quad \text{in } \Omega, \tag{3.41}$$

which is inspired by (3.39).

Next we shall explain how to compute the shape and material derivatives of u. It is clear that knowledge of only one of them is needed to determine the other one in view of (3.41) (the term $\nabla u \cdot \mathcal{V}$ results from u, a solution to $(\mathcal{P})$). Relations satisfied by u' and $\dot{u}$ will be obtained by differentiating $(\mathcal{P}_t)$ with respect to t at $t = 0$. Since state problems are defined in a weak form, i.e., a form involving integrals, we first show how to compute the derivative of integrals in which both the integrands and the domains of integration depend on the parameter $t \geq 0$.

Let

$$J_t = \int_{\Omega_t} f(t, x_t)\, dx_t, \quad C_t = \int_{\partial\Omega_t} f(t, x_t)\, ds_t,$$

where $\Omega_t = F_t(\Omega)$ and $\partial\Omega_t = F_t(\partial\Omega)$, and let $f : Q_\delta \to \mathbb{R}$ be a sufficiently smooth function, where Q_δ is a δ-neighborhood of Q. Denote

$$\dot{J} := \dot{J}(\Omega; \mathcal{V}) = \frac{d}{dt} J_t \bigg|_{t=0+}, \qquad \dot{C} := \dot{C}(\partial\Omega; \mathcal{V}) = \frac{d}{dt} C_t \bigg|_{t=0+}.$$

These derivatives can be viewed as directional type derivatives characterizing the behavior of J and C when Ω "moves" in the direction $\mathcal{V}$ defining F_t. They can be computed as follows.

Lemma 3.3.

$$\dot{J} = \int_\Omega \dot{f}\, dx + \int_\Omega f \operatorname{div} \mathcal{V}\, dx, \tag{3.42}$$

$$\dot{J} = \int_\Omega f'\, dx + \int_{\partial\Omega} f \mathcal{V} \cdot \nu\, ds, \tag{3.43}$$

$$\dot{C} = \int_{\partial\Omega} f'\, ds + \int_{\partial\Omega} \left(\frac{\partial f}{\partial \nu} + f H \right) \mathcal{V} \cdot \nu\, ds, \tag{3.44}$$

where $\dot{f} := \dot{f}(0, x)$ and $f' := f'(0, x)$ are the material and shape derivatives, respectively, of f at $t = 0$ and H denotes the mean curvature of $\partial\Omega$.

Proof. For the proof of (3.44), which is quite technical, we refer to [HCK86]. Expression (3.43) follows from (3.41), (3.42), and Green's formula. Thus it remains to prove (3.42). Using the theorem of substitution for integrals and applying the classical result of differentiation of integrals with respect to parameters, we obtain

$$\frac{d}{dt} J_t \bigg|_{t=0+} = \frac{d}{dt} \left(\int_\Omega f(t, x + t\mathcal{V}(x)) I_t\, dx \right) \bigg|_{t=0+} = \int_\Omega \frac{d}{dt} [f(t, x + t\mathcal{V}(x)) I_t] \bigg|_{t=0+} dx.$$

From this and (vi) of Lemma 3.2 we arrive at (3.42). $\square$

We shall illustrate how to get sensitivities in particular boundary value problems.

EXAMPLE 3.1. (Nonhomogeneous Dirichlet boundary value problem.) Let $f \in L^2_{loc}(\mathbb{R}^n)$, $g \in H^1_{loc}(\mathbb{R}^n)$ be given and consider

$$\begin{cases} -\Delta u_t = f & \text{in } \Omega_t = F_t(\Omega), \\ u_t = g & \text{on } \partial\Omega_t = F_t(\partial\Omega) \end{cases} \tag{$\mathcal{P}'_t$}$$

or, in the weak form:

$$\begin{cases} \text{Find } u_t \in H^1(\Omega_t),\ u_t = g \text{ on } \partial\Omega_t \text{ such that} \\ \displaystyle\int_{\Omega_t} \nabla u_t \cdot \nabla v_t \, dx_t = \int_{\Omega_t} f v_t \, dx_t \quad \forall v_t \in H^1_0(\Omega_t). \end{cases} \tag{$\mathcal{P}_t$}$$

Let $v \in H^1_0(\Omega)$ be arbitrary and take v_t in the form $v_t = v \circ F_t^{-1} \in H^1_0(\Omega_t)$. Since v_t is constant along each streamline $x + t\mathcal{V}(x)$, $x \in \Omega$, we have $\dot{v} = 0$ in Ω so that

$$v' = -\nabla v \cdot \mathcal{V} \quad \text{in } \Omega. \tag{3.45}$$

From this, (3.40), and (3.43) it follows that

$$\begin{aligned} \frac{d}{dt}\left(\int_{\Omega_t} \nabla u_t \cdot \nabla v_t \, dx_t\right)\bigg|_{t=0+} &= \int_\Omega \nabla u' \cdot \nabla v \, dx \\ &- \int_\Omega \nabla u \cdot \nabla(\nabla v \cdot \mathcal{V}) \, dx + \int_{\partial\Omega} (\nabla u \cdot \nabla v)(\mathcal{V} \cdot \nu) \, ds. \end{aligned} \tag{3.46}$$

Applying Green's formula to the second integral on the right of (3.46), we obtain

$$\begin{aligned} \int_\Omega \nabla u \cdot \nabla(\nabla v \cdot \mathcal{V}) \, dx &= -\int_\Omega \Delta u \nabla v \cdot \mathcal{V} \, dx + \int_{\partial\Omega} \frac{\partial u}{\partial \nu} \nabla v \cdot \mathcal{V} \, ds \\ &= \int_\Omega \Delta u v' \, dx + \int_{\partial\Omega} \frac{\partial u}{\partial \nu} \nabla v \cdot \mathcal{V} \, ds, \end{aligned} \tag{3.47}$$

making use of (3.45) again.

We now arrange the curvilinear integrals in (3.46) and (3.47). It holds that

$$(\nabla u \cdot \nabla v)(\mathcal{V} \cdot \nu) = \left(\frac{\partial u}{\partial \nu}\frac{\partial v}{\partial \nu} + \frac{\partial u}{\partial s}\frac{\partial v}{\partial s}\right)(\mathcal{V} \cdot \nu) = \frac{\partial u}{\partial \nu}\frac{\partial v}{\partial \nu}(\mathcal{V} \cdot \nu),$$

$$\frac{\partial u}{\partial \nu}(\nabla v \cdot \mathcal{V}) = \frac{\partial u}{\partial \nu}\left[\frac{\partial v}{\partial \nu}(\mathcal{V} \cdot \nu) + \frac{\partial v}{\partial s}(\mathcal{V} \cdot s)\right] = \frac{\partial u}{\partial \nu}\frac{\partial v}{\partial \nu}(\mathcal{V} \cdot \nu),$$

since $v \in H^1_0(\Omega)$, yielding $\partial v/\partial s = 0$ on $\partial\Omega$, where $\partial/\partial s$ stands for the derivative along $\partial\Omega$. From this, (3.46), and (3.47) we obtain the following expression for the derivative of the first integral in $(\mathcal{P}_t)$:

$$\frac{d}{dt}\left(\int_{\Omega_t} \nabla u_t \cdot \nabla v_t \, dx_t\right)\bigg|_{t=0+} = \int_\Omega \nabla u' \cdot \nabla v \, dx - \int_\Omega \Delta u v' \, dx. \tag{3.48}$$

Differentiation of the second integral in $(\mathcal{P}_t)$ is easy. Since f does not depend on t, we have $f' = 0$ in Ω so that

$$\frac{d}{dt}\left(\int_{\Omega_t} f v_t \, dx_t\right)\bigg|_{t=0+} = \int_{\Omega} f v' \, dx, \tag{3.49}$$

as follows from (3.43) using also that $v = 0$ on $\partial\Omega$. From (3.48), (3.49), and the fact that $-\Delta u = f$ in Ω, we finally obtain

$$\int_{\Omega} \nabla u' \cdot \nabla v \, dx = 0 \quad \forall v \in H_0^1(\Omega); \tag{3.50}$$

i.e., u' is harmonic: $\Delta u' = 0$ in Ω. To specify the boundary condition satisfied by u' we use (3.41):

$$\begin{aligned} u' &= \dot{u} - \nabla u \cdot \mathcal{V} = \dot{g} - \nabla u \cdot \mathcal{V} = \nabla g \cdot \mathcal{V} - \nabla u \cdot \mathcal{V} \\ &= \frac{\partial}{\partial \nu}(g - u)(\mathcal{V} \cdot \nu) \quad \text{on } \partial\Omega \end{aligned}$$

since $g' = 0$ in Ω and $g - u \in H_0^1(\Omega)$. In summary, we proved that the shape derivative u' satisfies the nonhomogeneous Dirichlet boundary value problem

$$\begin{cases} -\Delta u' = 0 & \text{in } \Omega, \\ u' = \dfrac{\partial}{\partial \nu}(g - u)(\mathcal{V} \cdot \nu) & \text{on } \partial\Omega. \end{cases} \tag{3.51}$$

Example 3.2. (Nonhomogeneous Neumann boundary value problem.) Let $f \in L^2_{loc}(\mathbb{R}^n)$ and $g : \mathbb{R}^n \to \mathbb{R}$ be such that $g_{|\partial\Omega_t} \in L^2(\partial\Omega_t)$ $\forall t \in [0, t_0]$. Consider the problem

$$\begin{cases} -\Delta u_t + u_t = f & \text{in } \Omega_t = F_t(\Omega), \\ \dfrac{\partial u_t}{\partial \nu_t} = g & \text{on } \partial\Omega_t = F_t(\partial\Omega) \end{cases} \tag{$\mathcal{P}_t'$}$$

or, in the weak form:

$$\begin{cases} \text{Find } u_t \in H^1(\Omega_t) \quad \text{such that} \\ \displaystyle\int_{\Omega_t} (\nabla u_t \cdot \nabla v_t + u_t v_t) \, dx_t = \int_{\Omega_t} f v_t \, dx_t + \int_{\partial\Omega_t} g v_t \, ds_t \quad \forall v_t \in H^1(\Omega_t). \end{cases} \tag{$\mathcal{P}_t$}$$

Since any function $v_t \in H^1(\Omega_t)$ can be obtained as the restriction of an appropriate function $v \in H^1(\mathbb{R}^n)$ to Ω_t, one can use $v_{|\Omega_t}$ as a test function in $(\mathcal{P}_t)$. Differentiating both sides of $(\mathcal{P}_t)$ with respect to t at $t = 0+$, we obtain the following equation satisfied by the shape derivative of u:

$$\begin{aligned} \int_{\Omega} (\nabla u' \cdot \nabla v + u' v) \, dx + \int_{\partial\Omega} (\nabla u \cdot \nabla v + u v)(\mathcal{V} \cdot \nu) \, ds &= \int_{\partial\Omega} f v (\mathcal{V} \cdot \nu) \, ds \\ &+ \int_{\partial\Omega} \left(\frac{\partial}{\partial \nu}(g v) + (g v H)\right)(\mathcal{V} \cdot \nu) \, ds \quad \forall v \in H^1(\Omega), \end{aligned} \tag{3.52}$$

making use of (3.43), (3.44), and the fact that v, f, and g do not depend on t, implying $v' = f' = g' = 0$ in $\mathbb{R}^n$. $\square$

We now show how to differentiate functionals depending on solutions u_t to state problems $(\mathcal{P}_t)$. To this end let us consider two sufficiently smooth functions $L : (t, y) \mapsto L(t, y) \in \mathbb{R}$, $t \in [0, t_0]$, $y \in \mathbb{R}^m$, and $f : Q_\delta \to \mathbb{R}^m$ (recall that Q_δ is a δ-neighborhood of Q and $Q = \cup_{t\in[0,t_0]}\{t\} \times \Omega_t$) and denote

$$E_t = \int_{\Omega_t} L(t, f_t)\, dx_t, \quad \Omega_t = F_t(\Omega), \quad f_t = f_{|\{t\}\times\Omega_t}.$$

Arguing as in the proof of (3.42) and (3.43), one can easily show that

$$\begin{aligned} \dot{E} := \dot{E}(\Omega; \mathcal{V}) &= \left.\frac{d}{dt} E_t\right|_{t=0+} \\ &= \int_\Omega \left[\frac{\partial}{\partial t} L(0, f) + \nabla_y L(0, f) \cdot \dot{f}\right] dx + \int_\Omega L(0, f) \operatorname{div} \mathcal{V}\, dx \end{aligned} \tag{3.53}$$

and

$$\dot{E} = \int_\Omega \left[\frac{\partial}{\partial t} L(0, f) + \nabla_y L(0, f) \cdot f'\right] dx + \int_{\partial\Omega} L(0, f)(\mathcal{V} \cdot \nu)\, ds, \tag{3.54}$$

where $f := f(0, x)$ and $\dot{f}$ and f' stand for the material and shape derivatives, respectively, of f at $t = 0$. If L does not depend on t explicitly, then

$$\dot{E} = \int_\Omega \nabla_y L(f) \cdot \dot{f}\, dx + \int_\Omega L(f) \operatorname{div} \mathcal{V}\, dx, \tag{3.55}$$

$$\dot{E} = \int_\Omega \nabla_y L(f) \cdot f'\, dx + \int_{\partial\Omega} L(f)(\mathcal{V} \cdot \nu)\, ds. \tag{3.56}$$

Remark 3.9. From (3.53) and (3.54) we see that the *same* derivative $\dot{E}$ can be equivalently expressed in *two different ways*.

Example 3.3. Let $L : \mathbb{R} \to \mathbb{R}$ be a sufficiently smooth function and $u_t \in H_0^1(\Omega_t)$ be the solution of the homogeneous Dirichlet boundary value problem from Example 3.1 ($g = 0$). Denote

$$E_t = \int_{\Omega_t} L(u_t)\, dx_t.$$

Then (3.56) yields

$$\dot{E} = \int_\Omega L_{,y}(u) u'\, dx + \int_{\partial\Omega} L(u)(\mathcal{V} \cdot \nu)\, ds, \qquad L_{,y} := \frac{d}{dy} L. \tag{3.57}$$

To get rid of u' from (3.57) we use the adjoint state technique again. Let p be the solution to the following adjoint state problem:

$$\begin{cases} -\Delta p = L_{,y}(u) & \text{in } \Omega, \\ \quad\;\; p = 0 & \text{on } \partial\Omega. \end{cases} \tag{3.58}$$

Then Green's formula and (3.58) yield

$$0 = \int_\Omega \nabla p \cdot \nabla u' \, dx = -\int_\Omega \Delta p u' \, dx + \int_{\partial\Omega} \frac{\partial p}{\partial \nu} u' \, ds$$
$$= \int_\Omega L_{,y}(u) u' \, dx + \int_{\partial\Omega} \frac{\partial p}{\partial \nu} u' \, ds. \tag{3.59}$$

The first integral in (3.59) vanishes since $p \in H_0^1(\Omega)$ and $u' \in H^1(\Omega)$ satisfies (3.50). From this, (3.57), (3.59), and $(3.51)_2$ with $g = 0$ we obtain the final expression for $\dot{E}$:

$$\dot{E} = \int_{\partial\Omega} \frac{\partial p}{\partial \nu} \frac{\partial u}{\partial \nu} (\mathcal{V} \cdot \nu) \, ds + \int_{\partial\Omega} L(u)(\mathcal{V} \cdot \nu) \, ds. \tag{3.60}$$

Take, for example, the compliance functional

$$E_t = \int_{\Omega_t} f u_t \, dx_t,$$

where f is the right-hand side in $(\mathcal{P}_t)$. Then from (3.58) we see that $p = u$ in Ω and, since $L(u) := fu = 0$ on $\partial\Omega$, expression (3.60) becomes

$$\dot{E} = \int_{\partial\Omega} \left(\frac{\partial u}{\partial \nu}\right)^2 (\mathcal{V} \cdot \nu) \, ds. \qquad \square \tag{3.61}$$

The same sensitivity results for functionals can be obtained by using a duality technique. To illustrate how one can proceed, consider Example 3.3. We have

$$E_t = \int_{\Omega_t} L(u_t) \, dx_t = \inf_{v_t \in H_0^1(\Omega_t)} \sup_{q_t \in H_0^1(\Omega_t)} \mathcal{L}_t(v_t, q_t),$$

where the Lagrangian $\mathcal{L}_t : H_0^1(\Omega_t) \times H_0^1(\Omega_t) \to \mathbb{R}$ is defined by

$$\mathcal{L}_t(v_t, q_t) = L(v_t) + \int_{\Omega_t} (f q_t - \nabla q_t \cdot \nabla v_t) \, dx_t.$$

Let $(\overline{v}_t, \overline{q}_t) \in H_0^1(\Omega_t) \times H_0^1(\Omega_t)$ be such that $E_t = \mathcal{L}_t(\overline{v}_t, \overline{q}_t)$. Then necessarily $\overline{v}_t = u_t$ solves $(\mathcal{P}_t)$ and

$$\delta_{v_t} \mathcal{L}_t(\overline{v}_t, \overline{q}_t) = 0 \iff \int_{\Omega_t} \nabla \overline{q}_t \cdot \nabla v_t \, dx_t = \int_{\Omega_t} L_{,y}(\overline{v}_t) v_t \, dx_t \quad \forall v_t \in H_0^1(\Omega_t), \tag{3.62}$$

where δ_{v_t} stands for the partial variation of $\mathcal{L}_t$ with respect to v_t. From (3.62) we recover the adjoint state (3.58) at time t so that $(\overline{v}_t, \overline{q}_t) = (u_t, p_t)$. Hence

$$E_t = \int_{\Omega_t} L(u_t) \, dx_t + \int_{\Omega_t} (f p_t - \nabla u_t \cdot \nabla p_t) \, dx_t.$$

From this, (3.43), and (3.56),

$$\dot{E} = \int_\Omega L_{,y}(u)u'\,dx + \int_{\partial\Omega} L(u)(\mathcal{V}\cdot\nu)\,ds + \int_\Omega fp'\,dx$$
$$- \int_\Omega \nabla u'\cdot\nabla p\,dx - \int_\Omega \nabla u\cdot\nabla p'\,dx - \int_{\partial\Omega} (\nabla u\cdot\nabla p)(\mathcal{V}\cdot\nu)\,ds, \quad (3.63)$$

using also that $f' = 0$ in Ω and $p = 0$ on $\partial\Omega$. Green's formula, (3.58), and the definition of $(\mathcal{P})$ yield

$$\int_\Omega \nabla u'\cdot\nabla p\,dx = -\int_\Omega u'\Delta p\,dx + \int_{\partial\Omega} u'\frac{\partial p}{\partial\nu}\,ds = \int_\Omega L_{,y}(u)u'\,dx + \int_{\partial\Omega} u'\frac{\partial p}{\partial\nu}\,ds$$

and

$$\int_\Omega \nabla u\cdot\nabla p'\,dx = -\int_\Omega \Delta u p'\,dx + \int_{\partial\Omega} \frac{\partial u}{\partial\nu}p'\,ds = \int_\Omega fp'\,dx + \int_{\partial\Omega} \frac{\partial u}{\partial\nu}p'\,ds.$$

Inserting these expressions into (3.63) we obtain

$$\dot{E} = \int_{\partial\Omega} L(u)(\mathcal{V}\cdot\nu)\,ds - \int_{\partial\Omega} u'\frac{\partial p}{\partial\nu}\,ds - \int_{\partial\Omega} p'\frac{\partial u}{\partial\nu}\,ds - \int_{\partial\Omega} \frac{\partial u}{\partial\nu}\frac{\partial p}{\partial\nu}(\mathcal{V}\cdot\nu)\,ds.$$

Finally using $(3.51)_2$ for both u' and p' we find (3.60).

From the previous examples we see that only the normal component of $\mathcal{V}$ and the normal derivatives of states and adjoint states on the boundaries appear in integrands defining $\dot{E}$. The fact that sensitivities depend only on boundary data is not accidental. This is a general result that can be rigorously justified (see [SZ92]). In computations, however, the boundary integral form of sensitivities turns out not to be convenient. Indeed, classical variational formulations of state problems together with low order finite elements give a poor approximation of the normal fluxes. To overcome this difficulty we may use the alternative formula (3.53), which involves only volume integrals (see Problems 3.11 and 3.12).

On the basis of the previous results one can formulate necessary optimality conditions satisfied by optimal solutions. Their interpretation may reveal some important properties that are usually hidden in the original setting.

Consider, for example, a shape optimization problem with the homogeneous Dirichlet boundary value state problem

$$\begin{cases} -\Delta u(\Omega) = f & \text{in } \Omega \in \mathcal{O}, \\ u(\Omega) = 0 & \text{on } \partial\Omega \end{cases} \quad (\mathcal{P}(\Omega))$$

with the compliance $J(\Omega) = \int_\Omega fu(\Omega)\,dx$ as the cost and

$$\mathcal{O} = \left\{\Omega \subset \mathbb{R}^n \mid \int_\Omega dx = c\right\},$$

where c is a positive constant. To release the constant volume constraint, we introduce the Lagrangian $\mathcal{L} : \mathcal{O}\times\mathbb{R}\to\mathbb{R}$ as follows:

$$\mathcal{L}(\Omega,\lambda) = J(\Omega) + \lambda\left(\int_\Omega dx - c\right).$$

Suppose that $\Omega^* \in \mathcal{O}$ is an optimal shape and denote $\Omega_t^* = F_t(\Omega^*)$. Then there exists $\overline{\lambda} \in \mathbb{R}$ such that

$$\dot{\mathcal{L}}(\Omega^*, \overline{\lambda}; \mathcal{V}) := \frac{d}{dt}\mathcal{L}(\Omega_t^*, \overline{\lambda})\bigg|_{t=0+} = 0 \tag{3.64}$$

holds for any regular velocity field $\mathcal{V}$. From the definition of $\mathcal{L}$, (3.43), and (3.61) we see that (3.64) is equivalent to

$$\int_{\partial\Omega^*} \left[\left(\frac{\partial u(\Omega^*)}{\partial \nu} \right)^2 + \overline{\lambda} \right] (\mathcal{V} \cdot \nu)\, ds = 0 \quad \forall \mathcal{V},$$

implying that the absolute value of the normal flux $\partial u(\Omega^*)/\partial \nu$ across the boundary of the optimal domain Ω^* is *constant*.

The next example interprets the optimality conditions in the contact shape optimization problem studied in Subsection 2.5.5, where we claimed that shape optimization with respect to the cost functional J given by (2.175) results in an appropriate distribution of contact stresses along the optimal contact part. In what follows we shall clarify the meaning of this statement.

Let $\Omega(\alpha) \in \widetilde{\mathcal{O}}$ be given. In view of the special parametrization of shapes of admissible domains (see Figure 2.6), the velocity field $\mathcal{V} \in (H^{1,\infty}(\Omega(\alpha)))^2$ will be chosen as follows:

$$\mathcal{V} = (0, \mathcal{V}_2), \quad \mathcal{V}_2 = 0 \quad \text{on } \Gamma_u(\alpha) \cup \Gamma_P(\alpha), \tag{3.65}$$

meaning that only the contact part $\Gamma_C(\alpha)$ varies. It can be shown (see [HN96, p. 153]) that

$$\begin{aligned} \dot{E} := \dot{E}(\Omega(\alpha); \mathcal{V}) = \frac{1}{2}\int_{\Gamma_C(\alpha)} \tau_{ij}(u)\varepsilon_{ij}(u)\mathcal{V}_2\nu_2\, ds - \int_{\Gamma_C(\alpha)} f_i u_i \mathcal{V}_2 \nu_2\, ds \\ - \int_{\Gamma_C(\alpha)} \tau_{2i}(u)\nu_i \mathcal{V}_2\, ds - \int_{\Gamma_C(\alpha)} \tau_{2i}(u)\nu_i \frac{\partial u_2}{\partial x_2}\mathcal{V}_2\, ds, \end{aligned} \tag{3.66}$$

where $u := u(\alpha) \in K(\alpha)$ solves $(\mathcal{P}(\alpha))$. We now make several simplifications in (3.66). The right-hand side of (3.66) contains one dominating term, namely the third one, while the remaining ones represent lower order terms. This claim can be justified by using one of the axioms of linear elasticity, which says that the deformation gradients are small compared to unity. The same holds for the displacement field u since $\Gamma_u(\alpha) \neq \emptyset$. Neglecting the lower order terms in (3.66) we have

$$\dot{E} \approx -\int_{\Gamma_C(\alpha)} \tau_{2i}(u)\nu_i \mathcal{V}_2\, ds. \tag{3.67}$$

The second simplification results from the fact that the contact surfaces can be identified with the shape of a rigid support. Thus $\nu \approx (0, -1)$ in our particular case, implying

$$\dot{E} \approx \int_{\Gamma_C(\alpha)} \tau_{22}(u)\mathcal{V}_2\, ds. \tag{3.68}$$

Suppose that domains belonging to $\mathcal{O}$ are subject only to the constant volume constraint and *no other* constraints are present. As in the previous example, the volume constraint will be released by using the Lagrange multiplier technique. Let $\Omega(\alpha^*) \in \mathcal{O}$ be an optimal shape. Then there exists $\bar{\lambda} \in \mathbb{R}$ such that

$$\dot{E}(\Omega(\alpha^*); \mathcal{V}) + \bar{\lambda}\frac{d}{dt}\left(\int_{\Omega_t(\alpha^*)} dx - c\right)\bigg|_{t=0+} = 0 \tag{3.69}$$

for any $\mathcal{V}$ of the form (3.65). Further,

$$\frac{d}{dt}\left(\int_{\Omega_t(\alpha^*)} dx\right)\bigg|_{t=0+} = \int_{\Gamma_C(\alpha^*)} \mathcal{V}_2 \nu_2 \, ds \approx -\int_{\Gamma_C(\alpha^*)} \mathcal{V}_2 \, ds$$

in consideration of $\nu \approx (0, -1)$. Thus, from (3.68) and (3.69),

$$\int_{\Gamma_C(\alpha^*)} (\tau_{22}(u(\alpha^*)) - \bar{\lambda})\mathcal{V}_2 \, ds \approx 0$$

for any $\mathcal{V}_2$ satisfying (3.65). Therefore $\tau_{22}(u(\alpha^*)) \approx \bar{\lambda} =$ const. on $\Gamma_C(\alpha^*)$. Since $\tau_{22}(u(\alpha^*))$ represents the contact (normal) stress on $\Gamma_C(\alpha^*)$ we may conclude that *the contact stress along the optimal contact part is almost constant.* This is certainly a very important property from a practical point of view. Shape optimization with the cost functional J given by (2.175) avoids stress concentrations between bodies in contact.

Remark 3.10. It is worth noticing that from (3.66) it follows that the cost functional J, which is equal to the value of the total potential energy in the equilibrium state, is *once continuously differentiable*. Compare with (3.19) where the same property has been proved for solutions of $(\mathcal{P}_{in}(\boldsymbol{\alpha}))$.

3.3.2 Isoparametric approach for discrete problems

This subsection deals with sensitivity analysis of discretized optimal shape design problems. We describe how to differentiate stiffness and mass matrices and force vectors with respect to discrete design variables. Our approach will be based on the isoparametric technique, enabling us to compute these derivatives by using elementary matrix operations. We start with a scalar elliptic problem in 2D. The reader will find that after minor modifications the formulas are also valid in the three-dimensional case.

Consider the following state problem:

$$\begin{cases} -\Delta u(\alpha) + u(\alpha) = f & \text{in } \Omega(\alpha) \subset \mathbb{R}^2, \\ \dfrac{\partial u(\alpha)}{\partial \nu} = 0 & \text{on } \partial\Omega(\alpha). \end{cases} \tag{$\mathcal{P}(\alpha)$}$$

Using a finite element discretization of $(\mathcal{P}(\alpha))$ and an appropriate parametrization of $\Omega(\alpha)$ by a finite number of discrete design variables $\alpha_1, \dots, \alpha_d$ forming the vector $\boldsymbol{\alpha}$, we obtain the following system of linear algebraic equations:

$$\boldsymbol{K}(\boldsymbol{\alpha})\,\boldsymbol{q}(\boldsymbol{\alpha}) + \boldsymbol{M}(\boldsymbol{\alpha})\,\boldsymbol{q}(\boldsymbol{\alpha}) = \boldsymbol{f}(\boldsymbol{\alpha}),$$

where $\boldsymbol{K}(\boldsymbol{\alpha})$ and $\boldsymbol{M}(\boldsymbol{\alpha})$ are the stiffness and mass matrices, respectively, and $\boldsymbol{f}(\boldsymbol{\alpha})$ is the force vector.

Our aim is to find sensitivities of $\boldsymbol{K}(\boldsymbol{\alpha})$, $\boldsymbol{M}(\boldsymbol{\alpha})$, and $\boldsymbol{f}(\boldsymbol{\alpha})$ with respect to the design parameter α_k, $k = 1, \ldots, d$, i.e., to find $\partial \boldsymbol{K}(\boldsymbol{\alpha})/\partial\alpha_k$, $\partial \boldsymbol{M}(\boldsymbol{\alpha})/\partial\alpha_k$, and $\partial \boldsymbol{f}(\boldsymbol{\alpha})/\partial\alpha_k$. In what follows we denote $(\cdot)' := \partial(\cdot)/\partial\alpha_k$. For the sake of simplicity of notation the argument $\boldsymbol{\alpha}$ will be omitted.

The global vector $\boldsymbol{q}$ of nodal values, arranged relative to the global numbering of nodes, is related to the element (or local) nodal value vector $\boldsymbol{q}^e$ by

$$\boldsymbol{q}^e = \boldsymbol{P}^e \boldsymbol{q}, \tag{3.70}$$

where $\boldsymbol{P}^e$ are Boolean matrices whose elements are only ones and zeros. These matrices are used only for notational convenience and are not explicitly formed in actual computations.

The global stiffness and mass matrices and the force vector can be assembled with the aid of the local ones as follows:

$$\boldsymbol{K} = \sum_e (\boldsymbol{P}^e)^{\mathrm{T}} \boldsymbol{K}^e \boldsymbol{P}^e, \quad \boldsymbol{M} = \sum_e (\boldsymbol{P}^e)^{\mathrm{T}} \boldsymbol{M}^e \boldsymbol{P}^e, \quad \boldsymbol{f} = \sum_e (\boldsymbol{P}^e)^{\mathrm{T}} \boldsymbol{f}^e, \tag{3.71}$$

where the summation is carried out over all elements of a given partition $\mathcal{T}$ of $\overline{\Omega}$. The elements of $\boldsymbol{K}^e$, $\boldsymbol{M}^e$, and $\boldsymbol{f}^e$ are given by

$$k_{ij}^e = \int_{T_e} \nabla\varphi_i \cdot \nabla\varphi_j \, dx, \quad m_{ij}^e = \int_{T_e} \varphi_i \varphi_j \, dx, \quad f_i^e = \int_{T_e} f \varphi_i \, dx, \tag{3.72}$$

respectively. Here φ_i, $i = 1, \ldots, p$, are the shape functions associated with the element $T_e \in \mathcal{T}$. In practical computations the integrals in (3.72) are usually evaluated by using a numerical integration formula. Therefore, the integrals are transformed to a fixed "parent element" $\widehat{T}$. Let $F_e : \widehat{T} \to T_e$ be a one-to-one mapping of $\widehat{T}$ onto T_e. Further, let

$$\boldsymbol{N}^e = \begin{pmatrix} \varphi_1 \\ \vdots \\ \varphi_p \end{pmatrix} \quad \text{and} \quad \boldsymbol{G}^e = \begin{pmatrix} \partial\varphi_1/\partial x_1 & \cdots & \partial\varphi_p/\partial x_1 \\ \partial\varphi_1/\partial x_2 & \cdots & \partial\varphi_p/\partial x_2 \end{pmatrix} \tag{3.73}$$

be matrices whose elements are the shape functions and their derivatives associated with the element T_e. Then the local matrices and vectors can be written in a compact form as follows:

$$\boldsymbol{K}^e = \int_{\widehat{T}} \boldsymbol{G}^{\mathrm{T}} \boldsymbol{G} |\boldsymbol{J}| \, d\xi, \quad \boldsymbol{M}^e = \int_{\widehat{T}} \boldsymbol{N}\boldsymbol{N}^{\mathrm{T}} |\boldsymbol{J}| \, d\xi, \quad \boldsymbol{f}^e = \int_{\widehat{T}} f \boldsymbol{N} |\boldsymbol{J}| \, d\xi, \tag{3.74}$$

where $|\boldsymbol{J}|$ denotes the determinant of the (transposed) Jacobian $\boldsymbol{J}$ of the mapping $F_e : \widehat{T} \to T_e$. To simplify our notation, the superscript e by the matrices behind the sign of the integrals here and in what follows will be omitted.

Remark 3.11. To be more precise we should write

$$\boldsymbol{K}^e = \int_{\widehat{T}} (\boldsymbol{G}^{\mathrm{T}} \boldsymbol{G})(F_e(\xi)) |\boldsymbol{J}(\xi)| \, d\xi$$

and similarly for the other integrals. However, the abbreviated form of (3.74) is commonly accepted. For this reason we shall also omit letters denoting the arguments of functions.

The derivatives of the global matrices $\boldsymbol{K}$ and $\boldsymbol{M}$ and the vector $\boldsymbol{f}$ now can be computed element by element using (3.71) and (3.74). Since the domain of integration in (3.74) is fixed, we obtain

$$(\boldsymbol{K}^e)' = \int_{\widehat{T}} \Big((\boldsymbol{G}')^{\mathrm{T}} \boldsymbol{G} |\boldsymbol{J}| + \boldsymbol{G}^{\mathrm{T}} \boldsymbol{G}' |\boldsymbol{J}| + \boldsymbol{G}^{\mathrm{T}} \boldsymbol{G} |\boldsymbol{J}|' \Big)\, d\xi, \tag{3.75}$$

$$(\boldsymbol{M}^e)' = \int_{\widehat{T}} \Big(\boldsymbol{N}' \boldsymbol{N}^{\mathrm{T}} |\boldsymbol{J}| + \boldsymbol{N} (\boldsymbol{N}')^{\mathrm{T}} |\boldsymbol{J}| + \boldsymbol{N} \boldsymbol{N}^{\mathrm{T}} |\boldsymbol{J}|' \Big)\, d\xi, \tag{3.76}$$

$$(\boldsymbol{f}^e)' = \int_{\widehat{T}} \left(\left(\frac{\partial f}{\partial x_1} x_1' + \frac{\partial f}{\partial x_2} x_2' \right) \boldsymbol{N} |\boldsymbol{J}| + f \boldsymbol{N}' |\boldsymbol{J}| + f \boldsymbol{N} |\boldsymbol{J}|' \right) d\xi. \tag{3.77}$$

Thus it suffices to find expressions for $\boldsymbol{N}'$, $\boldsymbol{G}'$, x', and $|\boldsymbol{J}|'$. A naive way to do this would be to express the integrands as explicit functions of the design variable $\boldsymbol{\alpha}$ and to perform the differentiation. This, however, will result in very specific expressions of daunting complexity.

Instead we use the *chain rule of differentiation*. Let (X_{i1}, X_{i2}), $i = 1, \ldots, p$, be the nodal coordinates of $T_e = F_e(\widehat{T})$. The integrands in (3.75)–(3.77) are completely defined by the nodal points of T_e. Thus,

$$\boldsymbol{N}' = \sum_{i,j} \frac{\partial \boldsymbol{N}}{\partial X_{ij}} X'_{ij}, \qquad |\boldsymbol{J}|' = \sum_{i,j} \frac{\partial |\boldsymbol{J}|}{\partial X_{ij}} X'_{ij},$$

$$\boldsymbol{G}' = \sum_{i,j} \frac{\partial \boldsymbol{G}}{\partial X_{ij}} X'_{ij}, \qquad x' = \sum_{i,j} \frac{\partial F_e}{\partial X_{ij}} X'_{ij}.$$

In practical computations, elements are implemented by using the *isoparametric technique*. Let

$$\widehat{\boldsymbol{N}} = \begin{pmatrix} \widehat{\varphi}_1 \\ \vdots \\ \widehat{\varphi}_p \end{pmatrix} \quad \text{and} \quad \widehat{\boldsymbol{G}} = \begin{pmatrix} \partial \widehat{\varphi}_1/\partial \xi_1 & \cdots & \partial \widehat{\varphi}_p/\partial \xi_1 \\ \partial \widehat{\varphi}_1/\partial \xi_2 & \cdots & \partial \widehat{\varphi}_p/\partial \xi_2 \end{pmatrix}$$

be the matrices made of the shape functions and their derivatives for the standard p-noded *Lagrangian* finite element of a chosen parent element $\widehat{T}$. The one-to-one mapping F^e between $\widehat{T}$ and T_e is defined by means of the shape functions on $\widehat{T}$ and the nodal coordinates of T_e as follows:

$$(F_e)_j(\xi) = \sum_{i=1}^{p} \widehat{\varphi}_i(\xi) X_{ij}, \quad \xi \in \widehat{T}, \quad j = 1, 2. \tag{3.78}$$

For a general element T_e the shape functions are given by

$$\varphi_i(x) = \widehat{\varphi}_i(F_e^{-1}(x)), \quad x \in T_e, \tag{3.79}$$

so that

$$\nabla_x \varphi_i(x) = \boldsymbol{J}^{-1} \nabla_\xi \widehat{\varphi}_i(F_e^{-1}(x)). \tag{3.80}$$

From this we see that the matrices $\boldsymbol{G}$ and $\widehat{\boldsymbol{G}}$ are related by

$$\boldsymbol{G} = \boldsymbol{J}^{-1}\widehat{\boldsymbol{G}}, \tag{3.81}$$

and the (transposed) Jacobian of the mapping F_e is given by

$$\boldsymbol{J} = \widehat{\boldsymbol{G}}\boldsymbol{X}, \tag{3.82}$$

where

$$\boldsymbol{X} = \begin{pmatrix} X_{11} & X_{12} \\ X_{21} & X_{22} \\ \vdots & \vdots \\ X_{p1} & X_{p2} \end{pmatrix}$$

is the matrix containing the nodal coordinates of T_e. For curved elements, the shape functions of a general element are usually represented by rational functions.

Before giving simple formulas for $\boldsymbol{N}'$, $\boldsymbol{G}'$, x', and $|\boldsymbol{J}|'$, we need the following classical result.

Lemma 3.4. *(Jacobi's formula for the derivative of a determinant.) Let $\boldsymbol{A}$ be a nonsingular $m \times m$ matrix function. Then*

$$|\boldsymbol{A}|' = |\boldsymbol{A}| \operatorname{tr}(\boldsymbol{A}^{-1}\boldsymbol{A}'). \tag{3.83}$$

Proof. Let $\boldsymbol{A}^* = (a^*_{ij})^m_{i,j=1}$ be the adjugate of $\boldsymbol{A}$, i.e., a matrix with the elements

$$a^*_{ij} = (-1)^{i-j} \det\left(\boldsymbol{A} \text{ without its } j\text{th row and } i\text{th column}\right).$$

With the aid of $\boldsymbol{A}^*$ one can represent the determinant and the inverse of $\boldsymbol{A}$ as follows:

$$|\boldsymbol{A}| = \sum_{k=1}^{m} a_{ik}a^*_{ki} \text{ for any } i = 1, \dots, m, \quad \boldsymbol{A}^{-1} = \frac{1}{|\boldsymbol{A}|}\boldsymbol{A}^*.$$

From this we get the desired result. Indeed,

$$|\boldsymbol{A}|' = \sum_{i=1}^{m}\sum_{j=1}^{m} \frac{\partial|\boldsymbol{A}|}{\partial a_{ij}} a'_{ij} = \sum_{i=1}^{m}\sum_{j=1}^{m} a^*_{ji}a'_{ij} = \operatorname{tr}(\boldsymbol{A}^*\boldsymbol{A}') = |\boldsymbol{A}| \operatorname{tr}(\boldsymbol{A}^{-1}\boldsymbol{A}'). \quad \square$$

Theorem 3.3. *For isoparametric finite elements,*

$$\boldsymbol{G}' = -\boldsymbol{G}\boldsymbol{X}'\boldsymbol{G}, \tag{3.84}$$

$$\boldsymbol{N}' = \boldsymbol{0}, \tag{3.85}$$

$$x' = (\boldsymbol{X}')^{\mathrm{T}}\widehat{\boldsymbol{N}}, \tag{3.86}$$

$$|\boldsymbol{J}|' = |\boldsymbol{J}| \operatorname{tr}(\boldsymbol{G}\boldsymbol{X}'). \tag{3.87}$$

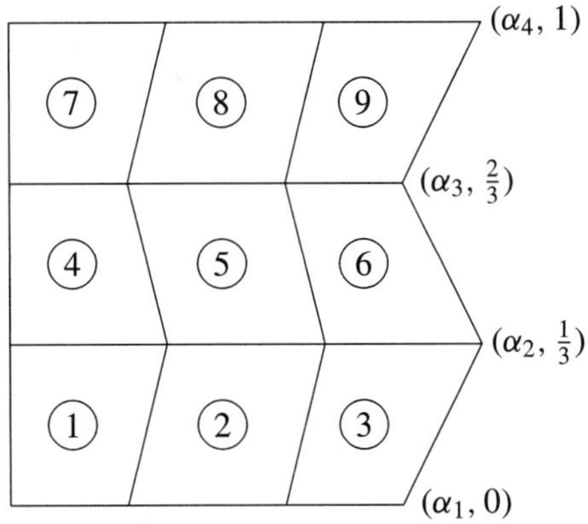

Figure 3.1. *Partition into quadrilateral elements.*

Proof. Since $\widehat{\boldsymbol{G}}$ does not depend on $\boldsymbol{\alpha}$ we obtain, from (3.81),

$$(\boldsymbol{J}\boldsymbol{G})' = \boldsymbol{J}'\boldsymbol{G} + \boldsymbol{J}\boldsymbol{G}' = \widehat{\boldsymbol{G}}' = \boldsymbol{0},$$

so that

$$\boldsymbol{G}' = -\boldsymbol{J}^{-1}\boldsymbol{J}'\boldsymbol{G} = -\boldsymbol{J}^{-1}\widehat{\boldsymbol{G}}\boldsymbol{X}'\boldsymbol{G} = -\boldsymbol{G}\boldsymbol{X}'\boldsymbol{G},$$

making use of (3.81) and (3.82). Let $\xi \in \widehat{T}$. From (3.79) it follows that $\varphi_i(x(\xi)) = \widehat{\varphi}_i(\xi)$; i.e., $\boldsymbol{N}$ is independent of the nodal coordinates of T_e, and therefore of design, too. This yields (3.85).

Equation (3.86) now immediately follows from the relation $x = \boldsymbol{X}^{\mathrm{T}}\widehat{\boldsymbol{N}}$. Finally, applying Lemma 3.4 to $|\boldsymbol{J}|$ and using (3.81) and (3.82), we get the derivative of $|\boldsymbol{J}|$:

$$|\boldsymbol{J}|' = |\boldsymbol{J}| \operatorname{tr}(\boldsymbol{J}^{-1}\boldsymbol{J}') = |\boldsymbol{J}| \operatorname{tr}(\boldsymbol{J}^{-1}\widehat{\boldsymbol{G}}\boldsymbol{X}') = |\boldsymbol{J}| \operatorname{tr}(\boldsymbol{G}\boldsymbol{X}'). \qquad \square$$

Example 3.4. Let us consider the very simple situation shown in Figure 3.1. We have four design variables $\alpha_1, \ldots, \alpha_4$ and nine quadrilateral elements (each horizontal line is divided into three segments of the same length). The nodal coordinate matrix corresponding to the four-noded element T_5 is given by

$$\boldsymbol{X}^{(5)} = \begin{pmatrix} \alpha_2/3 & 1/3 \\ 2\alpha_2/3 & 1/3 \\ 2\alpha_3/3 & 2/3 \\ \alpha_3/3 & 2/3 \end{pmatrix}.$$

The derivatives of the matrix $\boldsymbol{X}^{(5)}$ with respect to the design variables are readily obtained:

$$\frac{\partial \boldsymbol{X}^{(5)}}{\partial \alpha_1} = \frac{\partial \boldsymbol{X}^{(5)}}{\partial \alpha_4} = \boldsymbol{0}, \quad \frac{\partial \boldsymbol{X}^{(5)}}{\partial \alpha_2} = \begin{pmatrix} 1/3 & 0 \\ 2/3 & 0 \\ 0 & 0 \\ 0 & 0 \end{pmatrix}, \quad \frac{\partial \boldsymbol{X}^{(5)}}{\partial \alpha_3} = \begin{pmatrix} 0 & 0 \\ 0 & 0 \\ 2/3 & 0 \\ 1/3 & 0 \end{pmatrix}.$$

The benefit of the chain rule technique is obvious: all we need are the derivatives of $\boldsymbol{X}^e$. The calculation of $(\boldsymbol{X}^e)'$ can be done completely independently. On the other hand, the parametrization of the shape is not explicitly visible from formulas (3.84)–(3.87). *This is important in practical implementation: if one changes the parametrization of shapes, only the "mesh module" that calculates $\boldsymbol{X}$ and $\boldsymbol{X}'$ needs modifications. On the other hand, the same mesh module can be used without any modification for different state problems.*

Consider now a plane linear elasticity problem:

$$\partial \tau_{ij}(u)/\partial x_j + f_i = 0 \quad \text{in } \Omega \subset \mathbb{R}^2, \quad i = 1, 2; \tag{3.88}$$

$$u_i = 0 \quad \text{on } \partial\Omega, \quad i = 1, 2; \tag{3.89}$$

$$\tau_{ij}(u) = c_{ijkl}\varepsilon_{kl}(u), \quad i, j, k, l = 1, 2; \tag{3.90}$$

$$\varepsilon_{ij}(u) = \frac{1}{2}(\partial u_i/\partial x_j + \partial u_j/\partial x_i), \quad i, j = 1, 2. \tag{3.91}$$

We assume, for simplicity, that the elasticity coefficients c_{ijkl} and the body forces $f = (f_1, f_2)$ are *constant.*

Applying the isoparametric element technique in the same way as in the scalar case, we get similar expressions for the local stiffness matrix $\boldsymbol{K}^e$ and the local force vector $\boldsymbol{f}^e$. The matrix $\boldsymbol{K}^e$ can be written in the following compact form:

$$\boldsymbol{K}^e = \int_{\widehat{T}} \boldsymbol{B}^{\mathrm{T}}\boldsymbol{D}\boldsymbol{B}|\boldsymbol{J}|\,d\xi,$$

where

$$\boldsymbol{B} = \begin{pmatrix} \partial\varphi_1/\partial x_1 & 0 & \partial\varphi_2/\partial x_1 & 0 & \cdots & \partial\varphi_p/\partial x_1 & 0 \\ 0 & \partial\varphi_1/\partial x_2 & 0 & \partial\varphi_2/\partial x_2 & \cdots & 0 & \partial\varphi_p/\partial x_2 \\ \partial\varphi_1/\partial x_2 & \partial\varphi_1/\partial x_1 & \partial\varphi_2/\partial x_2 & \partial\varphi_2/\partial\varphi_1 & \cdots & \partial\varphi_p/\partial x_2 & \partial\varphi_p/\partial x_1 \end{pmatrix} \tag{3.92}$$

and φ_i, $i = 1, \dots, p$, are the shape functions of T_e given by (3.79). The symmetric matrix $\boldsymbol{D}$ contains the elasticity coefficients of the linear Hooke's law (3.90) expressed in matrix form as follows:

$$\begin{pmatrix} \tau_{11} \\ \tau_{22} \\ \tau_{12} \end{pmatrix} = \begin{pmatrix} d_{11} & d_{12} & d_{13} \\ d_{12} & d_{22} & d_{23} \\ d_{13} & d_{23} & d_{33} \end{pmatrix} \begin{pmatrix} \varepsilon_{11} \\ \varepsilon_{22} \\ 2\varepsilon_{12} \end{pmatrix}.$$

The local force vector corresponding to the body force $f = (f_1, f_2)^{\mathrm{T}}$ is given by

$$\boldsymbol{f}^e = \int_{\widehat{T}} \boldsymbol{\Phi}^{\mathrm{T}} f|\boldsymbol{J}|\,d\xi,$$

where

$$\boldsymbol{\Phi} = \begin{pmatrix} \varphi_1 & 0 & \varphi_2 & 0 & \cdots & \varphi_p & 0 \\ 0 & \varphi_1 & 0 & \varphi_2 & \cdots & 0 & \varphi_p \end{pmatrix}.$$

Taking into account that $\boldsymbol{D}' = f' = \boldsymbol{\Phi}' = \boldsymbol{0}$, we obtain the following formulas for the derivatives of the local stiffness matrix and the force vector:

$$(\boldsymbol{K}^e)' = \int_{\widehat{T}} \Big((\boldsymbol{B}')^{\mathrm{T}}\boldsymbol{D}\boldsymbol{B}|\boldsymbol{J}| + \boldsymbol{B}^{\mathrm{T}}\boldsymbol{D}\boldsymbol{B}'|\boldsymbol{J}| + \boldsymbol{B}^{\mathrm{T}}\boldsymbol{D}\boldsymbol{B}|\boldsymbol{J}|' \Big)\,d\xi,$$

$$(\boldsymbol{f}^e)' = \int_{\widehat{T}} \boldsymbol{\Phi}^{\mathrm{T}} f|\boldsymbol{J}|'\,d\xi.$$

The nonzero elements of $\boldsymbol{B}$ can be assembled from those of the matrix $\boldsymbol{G}$. Similarly, $\boldsymbol{B}'$ can be obtained from $\boldsymbol{G}'$ using (3.84).

Remark 3.12. Suppose now that boundary tractions $P = (P_1, P_2)$ act on a part $\Gamma_P(\alpha)$ and that the respective boundary conditions are of the form

$$\left.\begin{aligned} \tau_{ij}(u)\nu_i\nu_j &= P_\nu := P \cdot \nu \\ \tau_{ij}(u)\nu_i t_j &= P_t := P \cdot t \end{aligned}\right\} \quad \text{on } \Gamma_P(\alpha). \tag{3.93}$$

Let $T_e' \subset \partial T_e$ be a side of T_e placed on $\Gamma_P(\alpha)$, $T_e' = F_e(\widehat{T}')$, where $\widehat{T}' = \{(x_1, x_2) \mid x_1 \in (-1, 1),\ x_2 = 0\}$. Then the local contribution corresponding to (3.93) can be written in the form (see [HO77])

$$\boldsymbol{p}^e = \int_{-1}^{1} \boldsymbol{\Phi}^{\mathrm{T}}(P_t J_{11} - P_\nu J_{12},\ P_\nu J_{11} + P_t J_{12})^{\mathrm{T}}\, d\widehat{s}, \tag{3.94}$$

where $\boldsymbol{\Phi}$ is as above and J_{ij} denotes the elements of $\boldsymbol{J}$. Assuming P to be constant on $\Gamma_P(\alpha)$, we easily obtain from (3.94)

$$(\boldsymbol{p}^e)' = \int_{-1}^{1} \boldsymbol{\Phi}^{\mathrm{T}}(P_t J_{11}' - P_\nu J_{12}',\ P_\nu J_{11}' + P_t J_{12}')^{\mathrm{T}}\, d\widehat{s}. \tag{3.95}$$

Problems

Problem 3.1. Perform sensitivity analysis for a nonlinear state equation from Remark 3.4.

Problem 3.2. Suppose that the matrix function $\boldsymbol{A} : \mathcal{U} \to \mathbb{R}^{n\times n}$ satisfies (iv) and (v') of Remark 3.5. Prove that there exists a continuous function $m := m(\boldsymbol{\alpha}) > 0$ for any $\boldsymbol{\alpha} \in \mathcal{U}$ such that

$$\boldsymbol{y}^{\mathrm{T}}\boldsymbol{A}(\boldsymbol{\alpha})\boldsymbol{y} \ge m(\boldsymbol{\alpha})\|\boldsymbol{y}\|^2 \quad \forall \boldsymbol{y} \in \mathbb{R}^n,\ \forall \boldsymbol{\alpha} \in \mathcal{U}.$$

Problem 3.3. Prove Lemma 3.1.

Problem 3.4. Prove that the mapping $\boldsymbol{x} : \boldsymbol{\alpha} \mapsto \boldsymbol{x}(\boldsymbol{\alpha}) \in \mathcal{K}(\boldsymbol{\alpha})$, where $\boldsymbol{x}(\boldsymbol{\alpha})$ solves $(\mathcal{P}_{in}(\boldsymbol{\alpha}))$ from Section 3.1, is continuously differentiable provided that the set of semiactive constraints $\mathcal{I}_0(\boldsymbol{\alpha})$ is empty.

Problem 3.5. Let $\boldsymbol{A} \in C^1(\mathcal{U} \times \mathbb{R}^n; \mathbb{R}^{n\times n})$ be a generally nonlinear matrix function once continuously differentiable with respect to both variables, $\boldsymbol{f} \in C^1(\mathcal{U}, \mathbb{R}^n)$, and consider the following variational inequality:

$$\left\{\begin{aligned} &\text{Find } \boldsymbol{x}(\boldsymbol{\alpha}) \in \mathcal{K}(\boldsymbol{\alpha}) \quad \text{such that} \\ &(\boldsymbol{y} - \boldsymbol{x}(\boldsymbol{\alpha}))^{\mathrm{T}}\boldsymbol{A}(\boldsymbol{\alpha}, \boldsymbol{x}(\boldsymbol{\alpha}))\,\boldsymbol{x}(\boldsymbol{\alpha}) \ge (\boldsymbol{y} - \boldsymbol{x}(\boldsymbol{\alpha}))^{\mathrm{T}}\boldsymbol{f}(\boldsymbol{\alpha}) \quad \forall \boldsymbol{y} \in \mathcal{K}(\boldsymbol{\alpha}), \end{aligned}\right. \tag{3.96}$$

where $\mathcal{K}(\boldsymbol{\alpha})$ is as in (3.11). Suppose that $\boldsymbol{A}$ is strictly monotone uniformly with respect to $\mathcal{U}$; i.e., $\exists m = \text{const.} > 0$:

$$(\boldsymbol{y} - \boldsymbol{x})^{\mathrm{T}}(\boldsymbol{A}(\boldsymbol{\alpha}, \boldsymbol{y})\, \boldsymbol{y} - \boldsymbol{A}(\boldsymbol{\alpha}, \boldsymbol{x})\, \boldsymbol{x}) \geq m\|\boldsymbol{y} - \boldsymbol{x}\|^2 \quad \forall \boldsymbol{y}, \boldsymbol{x} \in \mathbb{R}^n \quad \forall \boldsymbol{\alpha} \in \mathcal{U}.$$

Then (3.96) has a unique solution $\boldsymbol{x}(\boldsymbol{\alpha})$ for any $\boldsymbol{\alpha} \in \mathcal{U}$. Prove that the mapping $\boldsymbol{x} : \boldsymbol{\alpha} \mapsto \boldsymbol{x}(\boldsymbol{\alpha})$ is Lipschitz continuous in $\mathcal{U}$ and directionally differentiable at any $\boldsymbol{\alpha} \in \mathcal{U}$ and in any direction $\boldsymbol{\beta}$. Prove the counterpart of Theorem 3.2.

Problem 3.6. Let $\boldsymbol{x}(\boldsymbol{\alpha})$ be a solution of $(\mathcal{P}_{in}(\boldsymbol{\alpha}))$ with $\mathcal{K}(\boldsymbol{\alpha})$ defined by (3.20). Show that Theorem 3.2 holds true with $\mathcal{K}(\boldsymbol{\alpha}, \boldsymbol{\beta})$ given by

$$\mathcal{K}(\boldsymbol{\alpha}, \boldsymbol{\beta}) = \{\boldsymbol{y} \in \mathbb{R}^n \mid y_{j_i} = \varphi_i'(\boldsymbol{\alpha}; \boldsymbol{\beta})\ \forall j_i \in \mathcal{I}_-(\boldsymbol{\alpha}),\ y_{j_i} \geq \varphi_i'(\boldsymbol{\alpha}; \boldsymbol{\beta})\ \forall j_i \in \mathcal{I}_0(\boldsymbol{\alpha})\},$$

where $\varphi_i'(\boldsymbol{\alpha}, \boldsymbol{\beta})$ is the directional derivative of φ_i at $\boldsymbol{\alpha}$ and in the direction $\boldsymbol{\beta}$ and

$$\mathcal{I}_-(\boldsymbol{\alpha}) = \{j_i \in \mathcal{I} \mid x_{j_i}(\boldsymbol{\alpha}) = \varphi_i(\boldsymbol{\alpha}),\ \lambda_i(\boldsymbol{\alpha}) > 0\},$$
$$\mathcal{I}_0(\boldsymbol{\alpha}) = \{j_i \in \mathcal{I} \mid x_{j_i}(\boldsymbol{\alpha}) = \varphi_i(\boldsymbol{\alpha}),\ \lambda_i(\boldsymbol{\alpha}) = 0\}.$$

Problem 3.7. Consider the same thickness optimization problem as in Section 3.2 but with the set U_h^{ad} defined by (2.32). Calculate the derivatives of the respective stiffness matrix $\boldsymbol{K}(\boldsymbol{e})$.

Problem 3.8. Consider the thickness optimization problem for an elastic unilaterally supported beam and its discretization studied in Subsection 2.5.1. Compute the directional derivatives of $\boldsymbol{q} : \boldsymbol{e} \mapsto \boldsymbol{q}(\boldsymbol{e})$, where $\boldsymbol{q}(\boldsymbol{e})$ solves the algebraic inequality resulting from the finite element approximation of $(\mathcal{P}(\boldsymbol{e}))$.

Problem 3.9. Prove Lemma 3.2.

Problem 3.10. Denote $A_t = (DF_t^{-1})^{\mathrm{T}} DF_t^{-1} I_t$, where $F_t : \mathbb{R}^n \to \mathbb{R}^n$ is given by (3.35). Compute $\mathcal{A} := \frac{d}{dt} A_t\big|_{t=0}$ by using Lemma 3.2.

Problem 3.11. Let $L : [0, t_0] \times R^m \to \mathbb{R}$ and $f : Q_\delta \to \mathbb{R}^m$ be two sufficiently smooth functions, where Q_δ is a δ-neighborhood of Q and $Q = \cup_{t \in [0,t_0]}\{t\} \times \Omega_t$, $\Omega_t = F_t(\Omega)$, with F_t given by (3.35). Compute

$$\dot{E}(\partial\Omega; \mathcal{V}) := \frac{d}{dt}\left(\int_{\partial\Omega_t} L(t, f_t)\, ds_t\right)\bigg|_{t=0+}, \quad f_t = f_{|\{t\}\times\partial\Omega_t}.$$

Problem 3.12. Consider the homogeneous Dirichlet boundary value problem

$$\begin{cases} -\Delta u(\Omega) = f & \text{in } \Omega, \\ u(\Omega) = 0 & \text{on } \partial\Omega, \end{cases} \qquad (\mathcal{P}(\Omega))$$

where $f \in L^2_{loc}(\mathbb{R}^n)$. Prove that the material derivative $\dot{u}$ satisfies

$$\begin{cases} -\Delta \dot{u} = \operatorname{div}(f\mathcal{V} + \mathcal{A}\nabla u) & \text{in } \Omega, \\ \dot{u} = 0 & \text{on } \partial\Omega, \end{cases}$$

where $\mathcal{A}$ is the same as in Problem 3.10.

Problem 3.13. Consider the following optimization problems:

$$\text{(i)} \ \min_{\Omega \subset \mathbb{R}^n} \left\{ J(\Omega, u(\Omega)) = \int_\Omega (u(\Omega) - u_{ad})^2 \, dx \right\}, \quad u_{ad} \in L^2_{loc}(\mathbb{R}^n),$$

$$\text{(ii)} \ \min_{\Omega \subset \mathbb{R}^n} \left\{ J(\Omega, u(\Omega)) = \int_\Omega |\nabla u(\Omega) - z_{ad}|^2 \, dx \right\}, \quad z_{ad} \in (L^2_{loc}(\mathbb{R}^n))^2,$$

where $u(\Omega)$ solves

$$\begin{cases} -\Delta u(\Omega) = f & \text{in } \Omega, \quad f \in L^2_{loc}(\mathbb{R}^n), \\ u(\Omega) = g & \text{on } \partial\Omega, \quad g \in H^1_{loc}(\mathbb{R}^n). \end{cases} \qquad (\mathcal{P}(\Omega))$$

Express $\dot{J}(\Omega; \mathcal{V})$ by using both the shape and material derivatives of u.

Problem 3.14. Consider the following optimal shape design problem:

$$\min_{\Omega \subset \mathbb{R}^n} \left\{ J(\partial\Omega, u(\Omega)) = \int_{\partial\Omega} (u(\Omega) - u_{ad})^2 \, ds \right\},$$

where $u(\Omega)$ solves

$$\begin{cases} -\Delta u(\Omega) + u(\Omega) = f & \text{in } \Omega, \\ \dfrac{\partial u(\Omega)}{\partial \nu_\Omega} = g & \text{on } \partial\Omega, \end{cases} \qquad (\mathcal{P}(\Omega))$$

$f \in L^2_{loc}(\mathbb{R}^n)$, u_{ad}, $g : \mathbb{R}^n \to \mathbb{R}$ are such that $u_{ad|\partial\Omega}$ and $g_{|\partial\Omega}$ belong to $L^2(\partial\Omega)$ for any admissible Ω. Compute $\dot{J}(\partial\Omega; \mathcal{V})$ by using both the adjoint state and duality techniques.

Chapter 4

Numerical Minimization Methods

Unlike authors of many other books on structural and shape optimization, we discuss basic nonlinear programming algorithms only very briefly. Nonlinear programming is central to operations research and a vast literature is available. The reader not familiar with the subject should consult [DS96], [Fle87], [GMW81], and [BSS93], for example.

We will focus on methods that we will use for the numerical realization of the "upper" optimization level in examples presented in Chapters 7 and 8. We do not consider methods, such as preconditioned conjugate gradients, intended for solving very large and sparse quadratic (or almost quadratic) programming problems with very simple constraints (or without constraints) arising from the discretization of state problems.

As we have seen previously, the algebraic form of all discrete sizing and optimal shape design problems leads to a minimization problem of the following type:

$$\min_{x \in \mathcal{U}} f(\boldsymbol{x}), \tag{$\mathbb{P}$}$$

where $f : \mathcal{U} \to \mathbb{R}$ is a *continuous function* and $\mathcal{U} \subset \mathbb{R}^n$ is a *nonempty* set representing constraints. In this chapter we briefly discuss typical gradient type and global optimization methods based on function evaluations, which will be used for the realization of ($\mathbb{P}$). In the second part of this chapter we shall also mention methods of multiobjective optimization.

4.1 Gradient methods for unconstrained optimization

We start with gradient type methods for unconstrained optimization when $\mathcal{U} = \mathbb{R}^n$. Suppose that f is *once continuously differentiable* in $\mathbb{R}^n$. Then the necessary condition for $\boldsymbol{x}^*$ to solve ($\mathbb{P}$) is to be a *stationary point* of f; i.e., $\boldsymbol{x}^*$ satisfies the system of n generally nonlinear equations

$$\nabla f(\boldsymbol{x}^*) = \boldsymbol{0}. \tag{4.1}$$

Therefore any minimizer of f in $\mathbb{R}^n$ is a stationary point at the same time. The opposite is true for convex functions. Gradient-based methods realize a sequence $\{\boldsymbol{x}_k\}$, $\boldsymbol{x}_k \in \mathbb{R}^n$,

$k = 0, 1, \ldots,$ starting from an initial point $\boldsymbol{x}_0$ following the formula

$$\boldsymbol{x}_{k+1} = \boldsymbol{x}_k + \rho_k \boldsymbol{d}_k, \tag{4.2}$$

where $\rho_k > 0$ is a given number and $\boldsymbol{d}_k \in \mathbb{R}^n$ are chosen in such a way that $f(\boldsymbol{x}_{k+1}) \le f(\boldsymbol{x}_k)\ \forall\, k = 0, 1, \ldots.$ From the Taylor expansion of f around $\boldsymbol{x}_k$ it follows that a decrease in f is locally achieved when $\boldsymbol{d}_k$ is the so-called descent direction of f at $\boldsymbol{x}_k$:

$$\boldsymbol{d}_k{}^{\mathrm{T}} \nabla f(\boldsymbol{x}_k) < 0. \tag{4.3}$$

The simplest and oldest method is steepest descent, in which $\boldsymbol{d}_k = -\nabla f(\boldsymbol{x}_k)\ \forall k = 0, 1, \ldots$ and ρ_k is defined from the condition

$$f(\boldsymbol{x}_k + \rho_k \boldsymbol{d}_k) = \min_{\rho \ge 0} f(\boldsymbol{x}_k + \rho \boldsymbol{d}_k). \tag{4.4}$$

The one-dimensional minimization of f in (4.4) is called a *line search*. The steepest descent method corresponds to a linear approximation of f. Due to its simplicity it is not surprising that the method is not too efficient. It can be shown that if $f \in C^2(\mathbb{R}^n)$ is *convex*, then convergence is *linear*; i.e.,

$$\limsup_{k \to \infty} \|\boldsymbol{x}_k - \boldsymbol{x}^*\| \le \left(\frac{\varkappa - 1}{\varkappa + 1} \right)^k \|\boldsymbol{x}_0 - \boldsymbol{x}^*\|, \tag{4.5}$$

where $\varkappa > 0$ is the spectral condition number of the Hessian of f at the minimum point $\boldsymbol{x}^*$. A better result is obtained for *conjugate gradient methods*. In this case the descent direction $\boldsymbol{d}_k$ at $\boldsymbol{x}_k$ is computed by using the gradient of f at $\boldsymbol{x}_k$ and the previous descent direction $\boldsymbol{d}_{k-1}$:

$$\begin{aligned} \boldsymbol{d}_0 &= -\boldsymbol{g}_0, \\ \boldsymbol{d}_k &= -\boldsymbol{g}_k + \alpha_k \boldsymbol{d}_{k-1}, \quad k \ge 1, \end{aligned}$$

where $\boldsymbol{g}_k = \nabla f(\boldsymbol{x}_k)$ and $\alpha_k = \|\boldsymbol{g}_k\|^2 / \|\boldsymbol{g}_{k-1}\|^2$. It is known that under appropriate assumptions on f and with an appropriate choice of ρ_k the ratio on the right-hand side of (4.5) can be replaced by a much more favorable one, namely $(\sqrt{\varkappa} - 1)/(\sqrt{\varkappa} + 1)$, where $\varkappa$ has the same meaning as in (4.5). Let us mention that this method with exact line searches (4.4), when applied to a quadratic function, terminates at its stationary point $\boldsymbol{x}_k$ after $k \le n$ iterations.

The speed of convergence can be improved by using methods based on quadratic approximations of f. As a typical representative of such a class of methods we briefly discuss the sequential quadratic programming (SQP) method. Before that, however, Newton's and quasi-Newton methods are presented to help the reader understand the SQP algorithm. The reasons for selecting the SQP method are the following:

- It is widely accepted that the SQP method is very efficient and reliable for *general* nonlinear programming problems provided that accurate derivatives of objective and constraint functions are available.
- Well-tested and reliable implementations of the SQP method are available either free of charge or at nominal cost for academic research.*

*More information on nonlinear programming software packages (properties, commercial status, contact information, etc.) can be found at www-fp.mcs.anl.gov/otc/guide/softwareguide/ or www.ici.ro/camo. URLs were current as of January 2003.

Comparison of several gradient-based methods in structural optimization problems has been done in [SZZ94].

4.1.1 Newton's method

Let us consider an *unconstrained* nonlinear optimization problem ($\mathbb{P}$) with $f : \mathbb{R}^n \to \mathbb{R}$ that is at least *twice continuously differentiable* in $\mathbb{R}^n$. Let $\boldsymbol{x}_0 \in \mathbb{R}^n$ be an initial guess for the solution of ($\mathbb{P}$). Then f can be approximated near $\boldsymbol{x}_0$ by its second order Taylor expansion

$$f(\boldsymbol{x}_0 + \boldsymbol{d}) \approx f(\boldsymbol{x}_0) + \boldsymbol{g}(\boldsymbol{x}_0)^{\mathrm{T}}\boldsymbol{d} + \frac{1}{2}\boldsymbol{d}^{\mathrm{T}}\boldsymbol{H}(\boldsymbol{x}_0)\boldsymbol{d}, \quad \boldsymbol{d} \in \mathbb{R}^n, \tag{4.6}$$

where $\boldsymbol{g}(\boldsymbol{x}) = \nabla f(\boldsymbol{x})$ and $\boldsymbol{H}(\boldsymbol{x}) = \left\{\partial^2 f(\boldsymbol{x})/\partial x_i \partial x_j\right\}_{i,j=1}^n$ are the gradient and the Hessian matrix of f at $\boldsymbol{x} \in \mathbb{R}^n$, respectively. If $\boldsymbol{H}(\boldsymbol{x}_0)$ is *positive definite*, the (unique) minimum of the quadratic approximation on the right of (4.6) with respect to $\boldsymbol{d}$ is obtained as the solution of the linear system of equations $\boldsymbol{H}(\boldsymbol{x}_0)\boldsymbol{d} = -\boldsymbol{g}(\boldsymbol{x}_0)$. A new approximation for the minimum in ($\mathbb{P}$) is obtained by setting $\boldsymbol{x}_1 = \boldsymbol{x}_0 + \boldsymbol{d}$.

Suppose that $\boldsymbol{H}(\boldsymbol{x})$ is *regular* for all $\boldsymbol{x} \in \mathbb{R}^n$. Repeating the previous approach we obtain the following algorithm.

Algorithm 4.1. *(Basic Newton's method for unconstrained optimization.)*

1. Choose an initial guess $\boldsymbol{x}_0 \in \mathbb{R}^n$ and a tolerance parameter $\varepsilon > 0$ for the stopping criterion. Set $k := 0$.
2. Calculate the gradient $\boldsymbol{g}(\boldsymbol{x}_k)$ and the Hessian matrix $\boldsymbol{H}(\boldsymbol{x}_k)$ of f at $\boldsymbol{x}_k$.
3. Calculate the search direction $\boldsymbol{d}_k$ by solving the linear system of equations
$$\boldsymbol{H}(\boldsymbol{x}_k)\,\boldsymbol{d}_k = -\boldsymbol{g}(\boldsymbol{x}_k).$$
4. Update the approximated minimum by
$$\boldsymbol{x}_{k+1} := \boldsymbol{x}_k + \boldsymbol{d}_k. \tag{4.7}$$
5. If $\|\boldsymbol{g}(\boldsymbol{x}_{k+1})\| \le \varepsilon$, then stop. Otherwise set $k := k + 1$ and go to step 2.

Thus Newton's method corresponds to the choice $\boldsymbol{d}_k = -[\boldsymbol{H}(\boldsymbol{x}_k)]^{-1}\boldsymbol{g}(\boldsymbol{x}_k)$ and $\rho_k = 1$ for all $k \in \mathbb{N}$ in (4.2) (notice that $\boldsymbol{d}_k$ is a descent direction only if $\boldsymbol{H}(\boldsymbol{x}_k)$ is positive definite).

It can be shown (see, e.g., Problem 4.1 and [GMSW89, p. 11]) that if $\boldsymbol{x}_0$ is sufficiently close to $\boldsymbol{x}^*$ and f satisfies certain (quite restrictive) assumptions, then the sequence $\{\boldsymbol{x}_k\}$ generated by Algorithm 4.1 converges *quadratically* to $\boldsymbol{x}^*$; i.e.,

$$\lim_{k\to\infty} \boldsymbol{x}_k = \boldsymbol{x}^*, \qquad \limsup_{k\to\infty} \frac{\|\boldsymbol{x}_{k+1} - \boldsymbol{x}^*\|}{\|\boldsymbol{x}_k - \boldsymbol{x}^*\|^2} = \text{const.} > 0.$$

4.1.2 Quasi-Newton methods

The calculation of the Hessian matrix requires the evaluation of $n(n+1)/2$ partial derivatives. Hand coding of the second derivatives is very error prone and tedious. Therefore their computation is recommended only if automatic differentiation techniques described in Chapter 5 are available. A finite difference approximation of the Hessian is usually very expensive. If the gradient $\boldsymbol{g} := \nabla f$ is available analytically, a simple finite difference approximation of the Hessian $\boldsymbol{H}(\boldsymbol{x})$ requires n additional evaluations of $\boldsymbol{g}(\boldsymbol{x})$. Another serious drawback of the classical Newton method is that the Hessian $\boldsymbol{H}(\boldsymbol{x}_k)$ may not be positive definite; i.e., $\boldsymbol{d}_k = -[\boldsymbol{H}(\boldsymbol{x}_k)]^{-1}\boldsymbol{g}(\boldsymbol{x}_k)$ is not generally a descent direction.

To overcome these disadvantages the so-called quasi-Newton methods have been developed. Instead of the Hessian they use its approximation by a *symmetric, positive definite matrix.* Let f be three times continuously differentiable in $\mathbb{R}^n$. Using the Taylor expansion for the gradient $\boldsymbol{g}(\boldsymbol{x}+\boldsymbol{d}) = \boldsymbol{g}(\boldsymbol{x}) + \boldsymbol{H}(\boldsymbol{x})\boldsymbol{d} + O(\|\boldsymbol{d}\|^2)$ one can obtain an approximation of the second order derivative of f in the direction $\boldsymbol{d}$ without explicitly forming approximations of the individual elements of the Hessian:

$$f''(\boldsymbol{x};\boldsymbol{d},\boldsymbol{d}) = \boldsymbol{d}^{\mathrm{T}}\boldsymbol{H}(\boldsymbol{x})\boldsymbol{d} \approx \boldsymbol{d}^{\mathrm{T}}(\boldsymbol{g}(\boldsymbol{x}+\boldsymbol{d}) - \boldsymbol{g}(\boldsymbol{x})).$$

Therefore we replace the Hessians in Newton's method with a sequence $\{\boldsymbol{B}_k\}$ of symmetric matrices satisfying the so-called quasi-Newton condition

$$\boldsymbol{B}_{k+1}(\boldsymbol{x}_{k+1} - \boldsymbol{x}_k) = \boldsymbol{g}(\boldsymbol{x}_{k+1}) - \boldsymbol{g}(\boldsymbol{x}_k). \tag{4.8}$$

If $n = 1$, then the numbers B_k are unique and the method is just the secant method for the single nonlinear equation $f'(x) = 0$.

If $n > 1$, then relation (4.8) does not determine $\boldsymbol{B}_k$ in a unique way. The identity matrix may be chosen as the initial guess $\boldsymbol{B}_0$. After that a new Hessian approximation at $\boldsymbol{x}_1$ is formed by updating $\boldsymbol{B}_0$, taking into account the additional curvature information obtained during the step $\boldsymbol{x}_1 = \boldsymbol{x}_0 + \boldsymbol{d}$. It is believed that the most efficient update is the *Broyden–Fletcher–Goldfarb–Shanno* (BFGS) rank-two update given by

$$\boldsymbol{B}_{k+1} = \boldsymbol{B}_k - \frac{\boldsymbol{B}_k\boldsymbol{s}_k(\boldsymbol{B}_k\boldsymbol{s}_k)^{\mathrm{T}}}{\boldsymbol{s}_k^{\mathrm{T}}\boldsymbol{B}_k\boldsymbol{s}_k} + \frac{\boldsymbol{y}_k(\boldsymbol{y}_k)^{\mathrm{T}}}{(\boldsymbol{y}_k)^{\mathrm{T}}\boldsymbol{s}_k}. \tag{4.9}$$

Here we write in shorthand $\boldsymbol{s}_k = \boldsymbol{x}_{k+1} - \boldsymbol{x}_k$ and $\boldsymbol{y}_k = \boldsymbol{g}(\boldsymbol{x}_{k+1}) - \boldsymbol{g}(\boldsymbol{x}_k)$. If the natural step length is used, i.e., $\boldsymbol{x}_{k+1} = \boldsymbol{x}_k + \boldsymbol{d}_k$, then $\boldsymbol{s}_k = \boldsymbol{d}_k$ in (4.9).

It can be proved that if $\boldsymbol{B}_k$ is positive definite, then $\boldsymbol{B}_{k+1}$ is also positive definite provided that

$$(\boldsymbol{y}_k)^{\mathrm{T}}\boldsymbol{d}_k > 0.$$

Replacing the Hessian in Newton's method with the BFGS approximation, we obtain the following algorithm.

ALGORITHM 4.2. *(Quasi-Newton method with BFGS update.)*

1. Choose an initial guess $\boldsymbol{x}_0 \in \mathbb{R}^n$ and a tolerance parameter $\varepsilon > 0$ for the stopping criterion. Set $\boldsymbol{B}_0 := \boldsymbol{I}$ and $k := 0$.
2. Calculate the search direction $\boldsymbol{d}_k$ by solving the linear system of equations
$$\boldsymbol{B}_k\boldsymbol{d}_k = -\boldsymbol{g}(\boldsymbol{x}_k).$$
3. Update the approximated minimum by
$$\boldsymbol{x}_{k+1} := \boldsymbol{x}_k + \boldsymbol{d}_k. \tag{4.10}$$

4. If $\|g(x_{k+1})\| \le \varepsilon$, then stop.
5. Update the Hessian approximation by

$$B_{k+1} := B_k - \frac{B_k s_k (B_k s_k)^{\mathrm{T}}}{s_k^{\mathrm{T}} B_k s_k} + \frac{y_k (y_k)^{\mathrm{T}}}{(y_k)^{\mathrm{T}} s_k},$$

where $s_k = x_{k+1} - x_k$, $g_k = \nabla f(x_k)$, and $y_k = g_{k+1} - g_k$.

Remark 4.1. The solution of the linear system in step 2 requires $O(n^3)$ arithmetic operations. It would be possible to formulate the BFGS quasi-Newton method yielding an approximation to the *inverse* Hessian matrix. In that case, step 2 would be replaced by a matrix-vector product requiring only $O(n^2)$ arithmetic operations. The preferred way is to update the *Cholesky factors* $L_k D_k L_k^{\mathrm{T}}$ of B_k instead. The factors L_{k+1} and D_{k+1} of the updated matrix B_{k+1} can be formed using $O(n^2)$ operations. If the Cholesky factors are available, then the linear system in step 2 can be solved using $O(n^2)$ operations. This approach is more numerically stable and in addition allows one to estimate the condition number of the Hessian matrix very easily. The latter is very useful in practice.

4.1.3 Ensuring convergence

From elementary numerical analysis it is well known that Newton's method for solving a nonlinear equation converges to a single root only if an initial guess is sufficiently close to the root. Unfortunately the same holds true in the case of Algorithms 4.1 and 4.2, even if f is convex. One cannot, of course, tolerate this in practice. The methods should be *globally convergent*, meaning that the sequence $\{x_k\}$ generated by (4.2) is such that $\liminf_{k\to\infty} \|g(x_k)\| = 0$ for any choice of $x_0 \in \mathbb{R}^n$. One way to achieve this goal is to add a *line search technique* and to ensure that d_k, $k = 0, 1, \dots,$ are *uniform descent directions:*

$$\exists \varepsilon_0 > 0: \quad -d_k^{\mathrm{T}} g(x_k) \ge \varepsilon_0 \|d_k\| \, \|g(x_k)\| \quad \forall k \in \mathbb{N}, \tag{4.11}$$

meaning that the angle between d_k and $g(x_k)$ is uniformly bounded away from orthogonality.

Assume that the search direction d_k satisfies (4.11). Instead of (4.7) or (4.10), the new approximation for solving ($\mathbb{P}$) is calculated as

$$x_{k+1} = x_k + \rho_k d_k, \tag{4.12}$$

where ρ_k is defined by (4.4).

In practice the exact minimization to determine ρ_k is not usually used. Convergence can be guaranteed if a sufficient decrease in f is obtained. Moreover, $\rho_k = 1$ is used whenever possible so that the good convergence rates of Newton's method or the quasi-Newton method near the optimum can be exploited. A sufficient decrease in f can be attained if ρ_k satisfies the following *Wolfe conditions:*

$$f(x_{k+1}) - f(x_k) \le \beta \rho_k d_k^{\mathrm{T}} g(x_k), \tag{4.13}$$

$$d_k^{\mathrm{T}} g(x_{k+1}) \ge \gamma d_k^{\mathrm{T}} g(x_k), \tag{4.14}$$

where $0 < \beta < \frac{1}{2}$ and $\beta < \gamma < 1$. It can be shown that under (4.11), (4.13), and (4.14) the descent direction method (4.2) is globally convergent provided f is bounded from below and the gradient of f is Lipschitz continuous in $\mathbb{R}^n$ (see [Fle87]).

Unfortunately, the *line search variant* of Newton's method is globally convergent only if the minimized function is *uniformly convex* in $\mathbb{R}^n$; i.e.,

$$\exists \alpha = \text{const.} > 0 : \quad \boldsymbol{y}^{\mathrm{T}} \boldsymbol{H}(\boldsymbol{x}) \boldsymbol{y} \geq \alpha \|\boldsymbol{y}\|^2, \quad \boldsymbol{x}, \boldsymbol{y} \in \mathbb{R}^n.$$

To ensure global convergence for nonconvex functions, the so-called trust region concept has to be used. Let us describe it in brief. Denote by

$$Q_k(\boldsymbol{s}) = \frac{1}{2} \boldsymbol{s}^{\mathrm{T}} \boldsymbol{H}(\boldsymbol{x}_k) \boldsymbol{s} + \boldsymbol{g}(\boldsymbol{x}_k)^{\mathrm{T}} \boldsymbol{s}, \quad \boldsymbol{s} \in \mathbb{R}^n,$$

the quadratic approximation of the difference $f(\boldsymbol{x}_k + \boldsymbol{s}) - f(\boldsymbol{x}_k)$ in a vicinity of $\boldsymbol{x}_k$ and let

$$\omega_k(\boldsymbol{s}) = \frac{f(\boldsymbol{x}_k + \boldsymbol{s}) - f(\boldsymbol{x}_k)}{Q_k(\boldsymbol{s})}$$

be the ratio of the real and expected decreases in f at $\boldsymbol{x}_k$. We realize the sequence $\{\boldsymbol{x}_k\}$ by

$$\boldsymbol{x}_{k+1} = \boldsymbol{x}_k + \boldsymbol{s}_k, \quad \boldsymbol{x}_0 \in \mathbb{R}^n \text{ given}, \tag{4.15}$$

with $\boldsymbol{s}_k \in B_k = \{\boldsymbol{s} \in \mathbb{R}^n \mid \|\boldsymbol{s}\| \leq \Delta_k\}$ being the solution of

$$Q_k(\boldsymbol{s}_k) = \min_{\boldsymbol{s} \in B_k} Q_k(\boldsymbol{s}).$$

The sequence $\{\Delta_k\}$ defining the radius of B_k is defined as follows:

$$\begin{aligned} &\text{If } \omega_k(\boldsymbol{s}_k) < \underline{\rho} \text{ then } \underline{\beta}\|\boldsymbol{s}_k\| \leq \Delta_{k+1} \leq \overline{\beta}\|\boldsymbol{s}_k\|; \\ &\text{otherwise } \Delta_k \leq \Delta_{k+1} \leq \overline{\gamma}\Delta_k, \end{aligned}$$

where $0 < \underline{\beta} \leq \overline{\beta} < 1 < \overline{\gamma}$ and $0 < \underline{\rho} < 1$. It can be shown that, if f is bounded from below, if it has a bounded Hessian in $\mathbb{R}^n$, and if the level sets

$$L_{\overline{f}} = \{\boldsymbol{x} \in \mathbb{R}^n \mid f(\boldsymbol{x}) \leq \overline{f}\}$$

are compact for any $\overline{f} \in \mathbb{R}$, then for any $\boldsymbol{x}_0 \in \mathbb{R}^n$ there is an accumulation point $\boldsymbol{x}^*$ of $\{\boldsymbol{x}_k\}$ defined by (4.15) such that $\boldsymbol{g}(\boldsymbol{x}^*) = \boldsymbol{0}$ and $\boldsymbol{H}(\boldsymbol{x}^*)$ is positive semidefinite. If, in addition, $\boldsymbol{H}(\boldsymbol{x}^*)$ is positive definite, the whole sequence $\{\boldsymbol{x}_k\}$ tends to $\boldsymbol{x}^*$, a local minimizer of f.

4.2 Methods for constrained optimization

The numerical realization of sizing or shape optimization problems usually leads to a nonlinear programming problem with *constraints*. Let us consider problem ($\mathbb{P}$) with *equality constraints* defining $\mathcal{U}$:

$$\mathcal{U} = \{\boldsymbol{x} \in \mathbb{R}^n \mid \boldsymbol{c}(\boldsymbol{x}) = (c_1(\boldsymbol{x}), \ldots, c_m(\boldsymbol{x}))^{\mathrm{T}} = \boldsymbol{0}\}, \tag{4.16}$$

where $f : \mathbb{R}^n \to \mathbb{R}$ and $\boldsymbol{c} : \mathbb{R}^n \to \mathbb{R}^m$ are *twice continuously differentiable functions* in $\mathbb{R}^n$. In what follows we denote by $\boldsymbol{A}(\boldsymbol{x}) \in \mathbb{R}^{m \times n}$ the Jacobian of the vector-valued function $\boldsymbol{c}$ at $\boldsymbol{x} \in \mathbb{R}^n$.

A nonlinear programming problem with $\mathcal{U}$ given by inequality constraints $c_i(\boldsymbol{x}) \le 0$ for $i = 1, \ldots, m$ can be converted to a problem with equality constraints by using an *active set strategy*: if it is known a priori which constraints are active at the optimum (i.e., $c_i(\boldsymbol{x}^*) = 0,\ i \in \mathcal{I}$), the rest of the constraints can be removed from the problem and a new one with only equality constraints is obtained. In practice the set $\mathcal{I}$ is not known a priori but it can be treated as an additional unknown of the problem and adjusted iteratively during the numerical optimization process. Next we shall consider problem ($\mathbb{P}$) with the equality constraints (4.16).

The basic principle invoked in solving a constrained optimization problem is that of replacing the original problem with a sequence of simpler *subproblems* that are related in a known way to the original problem ($\mathbb{P}$).

The classical *penalty method* is an example of the above-mentioned principle. The original constrained problem ($\mathbb{P}$) is replaced with a sequence of unconstrained ones:

$$\min_{\boldsymbol{x}\in\mathbb{R}^n} \left\{ f_k(\boldsymbol{x}) := f(\boldsymbol{x}) + \frac{1}{\varepsilon_k}\|\boldsymbol{c}(\boldsymbol{x})\|^2 \right\}, \qquad (\mathbb{P}_{\varepsilon_k})$$

where $\varepsilon_k \to 0+$ as $k \to \infty$. Then each convergent subsequence of $\{\boldsymbol{x}_k^*\}$ of solutions to ($\mathbb{P}_{\varepsilon_k}$) tends to a solution of ($\mathbb{P}$) (see Problem 4.2). Each problem ($\mathbb{P}_{\varepsilon_k}$) can now be solved by using Newton's method or the quasi-Newton method, for example. Unfortunately, when ε_k is very small, the Hessian $\boldsymbol{H}_k$ of f_k becomes *ill conditioned* at the solution $\boldsymbol{x}^*$ of ($\mathbb{P}$) since

$$\boldsymbol{H}_k(\boldsymbol{x}^*) = \boldsymbol{H}(\boldsymbol{x}^*) + \frac{2}{\varepsilon_k}\boldsymbol{A}(\boldsymbol{x}^*)^{\mathrm{T}}\boldsymbol{A}(\boldsymbol{x}^*)$$

is dominated by the singular matrix $\boldsymbol{A}(\boldsymbol{x}^*)^{\mathrm{T}}\boldsymbol{A}(\boldsymbol{x}^*)$. To avoid ill-conditioning of ($\mathbb{P}_{\varepsilon_k}$) caused by small ε_k the so-called augmented Lagrangian methods have been developed (see [Fle87]).

The most efficient and numerically stable methods are based on seeking a point satisfying the sufficient optimality conditions for constrained optimization that are mentioned below. Before doing that let us recall the following basic notion: a point $\boldsymbol{x} \in \mathcal{U}$ is said to be *regular* if the gradient vectors of the constraint functions $c_i,\ i = 1, \ldots, m$, are *linearly independent* at $\boldsymbol{x}$; i.e., the Jacobian matrix $\boldsymbol{A}(\boldsymbol{x})$ has full rank equal to m.

Let $\mathcal{L} : \mathbb{R}^n \times \mathbb{R}^m \to \mathbb{R}$ be the Lagrangian corresponding to ($\mathbb{P}$):

$$\mathcal{L}(\boldsymbol{x}, \boldsymbol{\lambda}) = f(\boldsymbol{x}) - \boldsymbol{\lambda}^{\mathrm{T}}\boldsymbol{c}(\boldsymbol{x}). \qquad (4.17)$$

Theorem 4.1. *(Karush–Kuhn–Tucker (KKT) first order necessary optimality condition.) Let the objective and constraint functions of problem* ($\mathbb{P}$) *be continuously differentiable at a regular point* $\boldsymbol{x}^* \in \mathcal{U}$. *A necessary condition for* $\boldsymbol{x}^*$ *to be a local minimizer in* ($\mathbb{P}$) *is that there exists a vector* $\boldsymbol{\lambda}^* \in \mathbb{R}^m$ *such that*

$$\nabla f(\boldsymbol{x}^*) - \boldsymbol{A}(\boldsymbol{x}^*)^{\mathrm{T}}\boldsymbol{\lambda}^* = \boldsymbol{0}. \qquad (4.18)$$

Since $\boldsymbol{x}^*$ is regular, then $\boldsymbol{\lambda}^*$ (called the vector of Lagrange multipliers) is *unique*.

Theorem 4.2. *(KKT second order sufficient optimality conditions.) Let the objective function and constraint functions of problem* ($\mathbb{P}$) *be twice continuously differentiable at a* *regular point* $\boldsymbol{x}^* \in \mathcal{U}$. *A sufficient condition for* $\boldsymbol{x}^*$ *to be a local minimizer in* $(\mathbb{P})$ *is that there exists a vector* $\boldsymbol{\lambda}^* \in \mathbb{R}^m$ *such that*

(i) $\nabla f(\boldsymbol{x}^*) - \boldsymbol{A}(\boldsymbol{x}^*)^{\mathrm{T}}\boldsymbol{\lambda}^* = \boldsymbol{0}$;

(ii) $\boldsymbol{d}^{\mathrm{T}}\nabla_x^2\mathcal{L}(\boldsymbol{x}^*, \boldsymbol{\lambda}^*)\boldsymbol{d} > 0$ *for any* $\boldsymbol{d} \neq \boldsymbol{0}$ *satisfying* $\boldsymbol{A}(\boldsymbol{x}^*)\boldsymbol{d} = \boldsymbol{0}$, *where* $\nabla_x^2\mathcal{L}$ *stands for the partial Hessian of* $\mathcal{L}$ *with respect to* $\boldsymbol{x}$.

Remark 4.2. Denote by $\boldsymbol{Z}(\boldsymbol{x}^*)$ an $n \times (n-m)$ matrix whose columns form a basis of the null space $\mathcal{N}(\boldsymbol{x}^*)$ of $\boldsymbol{A}(\boldsymbol{x}^*)$. Then condition (ii) in Theorem 4.2 can be equivalently expressed as follows:

(ii') *The reduced Hessian* $\boldsymbol{Z}(\boldsymbol{x}^*)^{\mathrm{T}}\nabla_x^2\mathcal{L}(\boldsymbol{x}^*, \boldsymbol{\lambda}^*)\boldsymbol{Z}(\boldsymbol{x}^*)$ *is positive definite.*

4.2.1 Sequential quadratic programming methods

Instead of quadratic approximations of f, we now use quadratic approximations of the Lagrangian $\mathcal{L}$. Since $(\boldsymbol{x}^*, \boldsymbol{\lambda}^*)$ is a saddle point of $\mathcal{L}$ we have to add some constraints into the quadratic programming subproblem to ensure its solvability. Condition (ii) in Theorem 4.2 suggests that a set of linear constraints might be appropriate.

Let $(\boldsymbol{x}_k, \boldsymbol{\lambda}_k) \in \mathbb{R}^n \times \mathbb{R}^m$ be a current approximation of $(\boldsymbol{x}^*, \boldsymbol{\lambda}^*)$. A point $\boldsymbol{x}_k + \boldsymbol{d}_k$ is intended to be a better approximation for $\boldsymbol{x}^*$. Therefore $\boldsymbol{x}_k + \boldsymbol{d}_k$ should also be feasible; i.e., $\boldsymbol{c}(\boldsymbol{x}_k + \boldsymbol{d}_k) \approx \boldsymbol{0}$. Making use of the Taylor expansion of $\boldsymbol{c}$ at $\boldsymbol{x}_k$ we get

$$\boldsymbol{c}(\boldsymbol{x}_k + \boldsymbol{d}_k) = \boldsymbol{c}(\boldsymbol{x}_k) + \boldsymbol{A}(\boldsymbol{x}_k)\boldsymbol{d}_k + O(\|\boldsymbol{d}_k\|^2) = \boldsymbol{0}. \tag{4.19}$$

Ignoring the higher order terms in (4.19), the search direction $\boldsymbol{d}_k$ should satisfy

$$\boldsymbol{A}(\boldsymbol{x}_k)\boldsymbol{d}_k = -\boldsymbol{c}(\boldsymbol{x}_k).$$

We are now able to formulate a quadratic programming subproblem with equality constraints whose solution determines $\boldsymbol{d}_k$:

$$\begin{aligned} &\min_{\boldsymbol{d}\in\mathbb{R}^n} \frac{1}{2}\boldsymbol{d}^{\mathrm{T}}\boldsymbol{B}_k\boldsymbol{d} + \boldsymbol{g}(\boldsymbol{x}_k)^{\mathrm{T}}\boldsymbol{d} \\ &\text{subject to} \\ &\boldsymbol{A}(\boldsymbol{x}_k)\boldsymbol{d} = -\boldsymbol{c}(\boldsymbol{x}_k), \end{aligned} \tag{$\mathbb{P}_k$}$$

where $\boldsymbol{g}(\boldsymbol{x}_k) := \nabla f(\boldsymbol{x}_k)$ and the matrix $\boldsymbol{B}_k$ is the exact Hessian (or its approximation) with respect to $\boldsymbol{x}$ of $\mathcal{L}$ at $(\boldsymbol{x}_k, \boldsymbol{\lambda}_k)$.

Let $\boldsymbol{Z}_k \in \mathbb{R}^{n\times(n-m)}$ and $\boldsymbol{Y}_k \in \mathbb{R}^{n\times m}$ denote matrices whose columns form the basis of the null space $\mathcal{N}(\boldsymbol{x}_k)$ of $\boldsymbol{A}(\boldsymbol{x}_k)$ and the range space $\mathcal{R}(\boldsymbol{x}_k)$ of $\boldsymbol{A}(x^k)^{\mathrm{T}}$, respectively. If the reduced Hessian $(\boldsymbol{Z}_k)^{\mathrm{T}}\boldsymbol{B}_k\boldsymbol{Z}_k$ is positive definite, subproblem $(\mathbb{P}_k)$ has a unique solution $\boldsymbol{d}_k$. Indeed, the vector $\boldsymbol{d}_k$ can be decomposed as follows:

$$\boldsymbol{d}_k = \boldsymbol{Z}_k\boldsymbol{d}_Z + \boldsymbol{Y}_k\boldsymbol{d}_Y, \quad \boldsymbol{d}_Z \in \mathbb{R}^{n-m}, \quad \boldsymbol{d}_Y \in \mathbb{R}^m,$$

making use of the fact that $\mathbb{R}^n = \mathcal{N}(\boldsymbol{x}_k) \oplus \mathcal{R}(\boldsymbol{x}_k)$. The range space component of $\boldsymbol{d}_Y$ is given by the unique solution of the linear system

$$\boldsymbol{A}(\boldsymbol{x}_k)\boldsymbol{Y}_k\boldsymbol{d}_Y = -\boldsymbol{c}(\boldsymbol{x}_k),$$

while the null space component $\boldsymbol{d}_Z$ solves

$$(\boldsymbol{Z}_k)^{\mathrm{T}}\boldsymbol{B}_k\boldsymbol{Z}_k\boldsymbol{d}_Z = -(\boldsymbol{Z}_k)^{\mathrm{T}}\boldsymbol{g}(\boldsymbol{x}_k) - (\boldsymbol{Z}_k)^{\mathrm{T}}\boldsymbol{B}_k\boldsymbol{Y}^k\boldsymbol{d}_Y. \tag{4.20}$$

The Lagrange multiplier $\boldsymbol{\mu}_k \in \mathbb{R}^m$ corresponding to the set of linear constraints in $(\mathbb{P}_k)$ satisfies the compatible overdetermined system

$$\boldsymbol{A}(\boldsymbol{x}_k)^{\mathrm{T}}\boldsymbol{\mu}_k = \boldsymbol{B}_k\boldsymbol{d}_k + \boldsymbol{g}(\boldsymbol{x}_k). \tag{4.21}$$

The reason for using $\boldsymbol{g}(\boldsymbol{x}_k)$ instead of $\nabla_{\boldsymbol{x}}\mathcal{L}(\boldsymbol{x}_k, \boldsymbol{\lambda}_k)$ in $(\mathbb{P}_k)$ is the following: First, the solution $\boldsymbol{d}_k$ of $(\mathbb{P}_k)$ is the same if $\boldsymbol{g}(\boldsymbol{x}_k)$ is replaced by $\nabla_{\boldsymbol{x}}\mathcal{L}(\boldsymbol{x}_k, \boldsymbol{\lambda}_k)$ in (4.20) because

$$(\boldsymbol{Z}_k)^{\mathrm{T}}\nabla_{\boldsymbol{x}}\mathcal{L}(\boldsymbol{x}_k, \boldsymbol{\lambda}_k) = (\boldsymbol{Z}_k)^{\mathrm{T}}\boldsymbol{g}(\boldsymbol{x}_k) - (\boldsymbol{Z}_k)^{\mathrm{T}}\boldsymbol{A}(\boldsymbol{x}_k)^{\mathrm{T}}\boldsymbol{\lambda}_k = (\boldsymbol{Z}_k)^{\mathrm{T}}\boldsymbol{g}(\boldsymbol{x}_k),$$

employing the fact that $\boldsymbol{A}(\boldsymbol{x}_k)\boldsymbol{Z}_k = \boldsymbol{0}$. Second, when $\boldsymbol{d}_k \to \boldsymbol{0}$ then (4.21) leads to the optimality condition for problem $(\mathbb{P})$:

$$\boldsymbol{A}(\boldsymbol{x}^*)^{\mathrm{T}}\boldsymbol{\mu}^* = \boldsymbol{g}(\boldsymbol{x}^*).$$

Therefore the vector of the Lagrange multipliers in $(\mathbb{P}_k)$ can be taken as an approximation of the Lagrange multipliers in the original problem $(\mathbb{P})$.

Algorithm 4.3. *(Simple SQP algorithm.)*

1. Choose an initial guess $(\boldsymbol{x}_0, \boldsymbol{\lambda}_0) \in \mathbb{R}^n \times \mathbb{R}^m$ and a tolerance parameter $\varepsilon > 0$ for the stopping criterion. Set $k := 0$.
2. If $\|\boldsymbol{g}(\boldsymbol{x}_k) - \boldsymbol{A}(\boldsymbol{x}_k)^{\mathrm{T}}\boldsymbol{\lambda}_k\| \leq \varepsilon$ and $\|\boldsymbol{c}(\boldsymbol{x})\| \leq \varepsilon$, then stop.
3. Let $(\boldsymbol{d}_k, \boldsymbol{\mu}_k)$ be the solution and the vector of Lagrange multipliers, respectively, of the quadratic programming subproblem

$$\begin{aligned} &\min \frac{1}{2}\boldsymbol{d}^{\mathrm{T}}\boldsymbol{B}_k\boldsymbol{d} + \boldsymbol{g}(\boldsymbol{x}_k)^{\mathrm{T}}\boldsymbol{d} \\ &\text{subject to} \\ &\boldsymbol{A}(\boldsymbol{x}_k)\boldsymbol{d} = -\boldsymbol{c}(\boldsymbol{x}_k). \end{aligned} \tag{4.22}$$

4. Set

$$\boldsymbol{x}_{k+1} := \boldsymbol{x}_k + \boldsymbol{d}_k, \qquad \boldsymbol{\lambda}_{k+1} := \boldsymbol{\mu}_k.$$

5. Calculate a new Hessian $\boldsymbol{B}_{k+1}$ (exact or its BFGS approximation).
6. Set $k := k + 1$ and go to step 2.

It can be shown (see [GMW81, p. 239]) that, if $\boldsymbol{B}_k$ is the exact Hessian (with respect to $\boldsymbol{x}$) of $\mathcal{L}$ and the initial guess $(\boldsymbol{x}_0, \boldsymbol{\lambda}_0)$ is sufficiently close to $(\boldsymbol{x}^*, \boldsymbol{\lambda}^*)$, the sequences $\{\boldsymbol{x}_k\}$, $\{\boldsymbol{\lambda}_k\}$ defined by Algorithm 4.3 converge *quadratically* to $\boldsymbol{x}^*$ and $\boldsymbol{\lambda}^*$, respectively, satisfying the first order KKT condition (4.18).

As in Newton's method, the convergence properties of Algorithm 4.3 will be improved if a line search is added to step 4. In contrast to unconstrained optimization the decrease in

f itself is not a measure of progress in constrained nonlinear programming. Usually step 4 is replaced by the following step:

4′. Set

$$\begin{pmatrix} \boldsymbol{x}_{k+1} \\ \boldsymbol{\lambda}_{k+1} \end{pmatrix} = \begin{pmatrix} \boldsymbol{x}_k \\ \boldsymbol{\lambda}_k \end{pmatrix} + \rho_k \begin{pmatrix} \boldsymbol{d}_k \\ \boldsymbol{\mu}_k - \boldsymbol{\lambda}_k \end{pmatrix}, \tag{4.23}$$

where $\rho_k > 0$ is chosen in such a way that a "sufficient decrease" in the *augmented Lagrangian merit function*

$$\Phi_r(\boldsymbol{x}, \boldsymbol{\lambda}) := \mathcal{L}(\boldsymbol{x}, \boldsymbol{\lambda}) + \frac{r}{2}\|\boldsymbol{c}(\boldsymbol{x})\|^2 \tag{4.24}$$

is achieved. Here $r > 0$ is a penalty parameter. It can be shown that under additional assumptions the sequence $\{(\boldsymbol{x}_k, \boldsymbol{\lambda}_k)\}$ defined by (4.23) converges to $(\boldsymbol{x}^*, \boldsymbol{\lambda}^*)$ satisfying the KKT first order necessary optimality conditions. Furthermore, there exists finite r_0 such that $\boldsymbol{x}^*$ is the unconstrained minimizer of $\Phi_r(\boldsymbol{x}, \boldsymbol{\lambda}^*)$ $\forall r > r_0$. For details we refer to [Sch82].

4.2.2 Available sequential quadratic programming software

There are several SQP implementations available. Most of the implementations assume that the size of the optimization problem is moderate because the matrices are stored as dense ones. The following subroutines are widely used and easily accessible:

- NLPQL, by K. Schittkowski, University of Bayreuth, Germany. NLPQL solves the following nonlinear programming problem:

$$\min_{\boldsymbol{x}\in\mathbb{R}^n} f(\boldsymbol{x})$$

subject to

$$\begin{aligned} c_i(\boldsymbol{x}) &= 0, \quad i = 1, \dots, m_0; \\ c_i(\boldsymbol{x}) &\geq 0, \quad i = m_0 + 1, \dots, m; \\ x_i^{low} &\leq x_i \leq x_i^{upp}, \quad i = 1, \dots, n. \end{aligned}$$

  For details on the availability of NLPQL contact the author at klaus.schittkowski@uni-bayreuth.de. NLPQL is also included in the commercial IMSL subroutine library [IMS94].

- DONLP2, by P. Spellucci, Technical University of Darmstad, Germany. DONLP2 solves the following nonlinear programming problem:

$$\min_{\boldsymbol{x}\in\mathbb{R}^n} f(\boldsymbol{x})$$

subject to

$$\begin{aligned} c_i(\boldsymbol{x}) &= 0, \quad i = 1, \dots, m_0; \\ c_i(\boldsymbol{x}) &\geq 0, \quad i = m_0 + 1, \dots, m. \end{aligned}$$

  The code is available online free of charge for research purposes from ftp://plato.la.asu.edu/pub/donlp2/.

- E04UCF, included in the commercial NAG subroutine library [NAG97]. E04UCF solves the following nonlinear programming problem:

$$\min_{x \in \mathbb{R}^n} f(x)$$

subject to

$$x_i^{low} \le x_i \le x_i^{upp}, \quad i = 1, \dots, n;$$

$$\ell_i^A \le \sum_{j=1}^n a_{ij} x_j \le u_i^A, \quad i = 1, \dots, m_A;$$

$$\ell_i^c \le c_i(x) \le u_i^c, \quad i = 1, \dots, m_c.$$

In E04UCF the linear constraints, defined by a matrix $A = (a_{ij}) \in \mathbb{R}^{m_A \times n}$, and general nonlinear constraints are described separately. All constraints have both lower and upper bounds. If some bound is not present one can use $\pm M$ instead, where $M > 0$ is a "large" number. Equality constraints are obtained if the upper and lower bounds coincide.

NLPQL allows the use of *reverse communication*; i.e., the values of objective and constraint functions and their derivatives can be evaluated in the (sub)program from which the optimizer is called. This is useful especially in sizing and shape optimization (SSO) since the calculation of function values of f, c_i and of their derivatives is difficult to separate into different subroutines.

4.3 On optimization methods using function values only

In this section we shall present two types of optimization methods that do not use any gradient information and have a potential to reveal *global minima* of functions with several local minima. One of the main reasons for their popularity is that they are very simple to implement. A structural analyst with some experience in finite element analysis but no knowledge of nonlinear programming is able to implement these methods in a few hours. Since they do not use gradients, one can use them for minimization of nondifferentiable functions. However, they need considerable fine tuning to perform well. Moreover, they should not be used when the problem is "too simple." In this case gradient-based methods with a finite difference approximation of derivatives together with some global strategy (several initial guesses, for example) are usually much more efficient than methods based on function evaluations only.

Solutions of ($\mathbb{P}$), i.e., global minimizers of f with respect to $\mathcal{U}$, representing solutions of discretized problems may hardly be detected by gradient methods. This happens, for example, if the objective function f is multimodal; i.e., many local minima exist in the feasible region. If a global minimizer x^* of f is needed, global optimization methods represent an appropriate tool to find it. Due to the limited accuracy of floating-point representations of real numbers by computers, the global optimization problem is considered as solved if an element of the level set

$$L_{f(x^*)+\varepsilon} = \{x \in \mathcal{U} \mid f(x) \le f(x^*) + \varepsilon\}, \quad \varepsilon > 0,$$

has been found by using an appropriate algorithm. Unfortunately, the theoretical results indicate that, in general, the global optimization problem is NP-complete; i.e., there is no efficient (i.e., polynomial time) approximation algorithm capable of solving an arbitrary global optimization problem. For details see, e.g., [Bäc96, pp. 35–62].

In many applications the feasible region $\mathcal{U}$ is usually defined by simple box constraints; i.e., $\mathcal{U} = \prod_{i=1}^{n}[x_i^{low}, x_i^{upp}]$, $x_i^{low}, x_i^{upp} \in \mathbb{R}$, $i = 1, 2, \ldots, n$. Specialized stochastic algorithms have been developed especially for such problems in the past decades. Some of them imitate the evolutionary process in organic populations [BS93, Gol89, Mic92]; some of them are based on the physical process of annealing [MRR$^+$53, Ing93]. Empirical results confirm that objective functions appearing in industrial applications are such that stochastic algorithms yield a good approximation of global minima.

4.3.1 Modified controlled random search algorithm

Random search belongs to the class of stochastic algorithms of global optimization. Random search algorithms are reviewed in [TŽ89]. The controlled random search (CRS) algorithm was proposed by Price [Pri76] in 1976. The CRS algorithm is based on ideas of the Nelder–Mead simplex method [NM64]. It starts with a population $\mathcal{P}$ of N points ($N \gg n$) taken at random in the feasible region $\mathcal{U}$. A new trial point $\boldsymbol{x}$ is generated from a simplex S in $\mathbb{R}^n$, whose $n+1$ vertices belong to the population $\mathcal{P}$ in $\mathcal{U}$, using the relation

$$\boldsymbol{x} = \boldsymbol{g} - Y(\boldsymbol{z} - \boldsymbol{g}),$$

where $\boldsymbol{z}$ is one vertex of S, $\boldsymbol{g}$ the center of gravity of the face defined by the remaining vertices of S, and Y a multiplicative factor. The point $\boldsymbol{x}$ may be considered as resulting from the reflection of $\boldsymbol{z}$ with respect to $\boldsymbol{g}$. Let $\boldsymbol{x}_{max}$ be the point with the largest objective function value among N points currently stored. If $f(\boldsymbol{x}) < f(\boldsymbol{x}_{max})$, then $\boldsymbol{x}_{max} := \boldsymbol{x}$; i.e., the worst point in $\mathcal{P}$ is replaced by the new trial point. The process continues until a stopping condition is fulfilled.

The *modified* CRS (MCRS) algorithm is described in [KT95]. The principal modification of the CRS algorithm consists of randomizing the factor Y when searching for a new trial point. Consider the procedure *Reflection* formally written as follows.

procedure Reflection(**in:** $\mathcal{P}$, **out:** $\boldsymbol{x}$)
repeat
    set S := set of $(n+1)$ points selected from $\mathcal{P}$ at random;
    set $\boldsymbol{x} := \mathbf{g} - Y(\mathbf{z} - \mathbf{g})$;
until $\boldsymbol{x} \in \mathcal{U}$;

The multiplication factor Y is a random variable. Then the MCRS algorithm can be written very simply.

ALGORITHM 4.4. *(MCRS.)*

$\mathcal{P}$:= population of N points in $\mathcal{U}$ generated at random;
repeat
    Reflection($\mathcal{P}, \boldsymbol{x}$);
    if $f(\boldsymbol{x}) < f(\boldsymbol{x}_{max})$ **then** $\boldsymbol{x}_{max} := \boldsymbol{x}$;
until stopping criterion is true;

Several distributions of Y have been tested. It was found that good results are obtained with Y distributed uniformly in $[0, \alpha[$ with α ranging from 4 to 8 [KT95]. The MCRS algorithm has been successfully applied to parameter estimation in nonlinear regression models [KTK00] and in shape optimization [HJKT00] (see also Chapter 6).

No particular stopping criterion is defined. However, in most optimization problems the stopping criterion is expressed as

$$f(\boldsymbol{x}_{max}) - f(\boldsymbol{x}_{min}) \leq \varepsilon, \tag{4.25}$$

where $\boldsymbol{x}_{min}$ is the point with the smallest value of the objective function among all the N points of $\mathcal{P}$ held in the memory and $\varepsilon > 0$ is an input parameter. The MCRS algorithm is very easy to implement. It has three input tuning parameters:

- the number N of points in $\mathcal{P}$,
- the value of α defining the range of Y,
- the value of ε for the stopping condition.

A correct setup of the tuning parameters depends on the nature of the optimization problem to be solved. The higher the values of N and α the more thorough the search for global minima is. Empirical observations indicate that the values $\alpha = 8$ and $N = \max(5n, n^2)$ are acceptable in many applications.

Some other modifications of the CRS algorithms can be found in [ATV97] and in several references therein.

4.3.2 Genetic algorithms

Genetic algorithms (GAs) are stochastic methods that can be used to solve problems approximately in search, optimization, and machine learning. GAs can be presented using the concept of natural evolution: a randomly initialized population of individuals (approximate solutions to a mathematical problem) evolves, mimicking Darwin's principle of survival of the fittest. In a GA a new generation of individuals is produced using the simulated genetic operations *crossover* and *mutation*. The probability of survival of the generated individuals depends on their fitness: the best ones survive with a high probability, the worst die rapidly. For more detailed treatment of the topic see [Hol75], [Gol89], for example.

In what follows we shall apply the general idea of GAs to the problem of finding an approximate solution to a shape optimization problem, for example. Let us consider a population of admissible shapes. Every shape is uniquely defined by a set of floating-point numbers (design parameters), which can be represented as a single finite bit string ("digital chromosome"). We assume that it is possible to evaluate the fitness of each shape, where the fitness is a real number indicating how "good" a shape is compared with the others in the population. To simulate the process of breeding a new generation of shapes from the current one, the following steps are used:

- *Reproduction* according to fitness: the more fit the shape is, the more likely it is to be chosen as a parent.
- *Recombination*: the bit strings of the parents are *paired, crossed,* and *mutated* to produce offspring bit strings. Every new bit string uniquely determines a new shape.
- *Replacement*: the new population of shapes replaces the old one.

Let us now consider a discrete shape optimization problem reduced to the solution of the nonlinear programming problem ($\mathbb{P}$) with $\mathcal{U}$ given by simple box constraints. In the traditional GA the chromosome string is an array of bits (zeros and ones) and the genetic operations are performed bitwise. Instead of this we use the so-called real coding. The real-coded GA processes a population $\mathcal{P}$ of vectors $\{\boldsymbol{x}^{(i)}\}$, $\boldsymbol{x}^{(i)} \in \mathbb{R}^n$, $i = 1, \ldots, N$, where N is the population size. The genetic operations to vectors are now performed componentwise using standard arithmetic operations. Next we shall give a specific meaning to the general concepts mentioned above in our specific context.

Genetic type algorithms are used for the maximization of a fitness function F characterizing the state of a population. Therefore if one can use GAs in SSO formulated as minimization problems for a cost functional f, one has to set $F = -f$, using the well-known fact that $\max_{\mathcal{U}}(-f) = -\min_{\mathcal{U}} f$. For multiobjective optimization problems, which will be discussed later, a completely different choice is useful. Therefore we define the fitness evaluation phase as the following procedure:

procedure Evaluate_fitness(**in:** $N, \mathcal{P}, f$, **out:** $\mathcal{F}$)
set $\mathcal{F} \equiv \{F_1, \ldots, F_N\} := \{-f(\boldsymbol{x}^{(1)}), \ldots, -f(\boldsymbol{x}^{(N)})\}$;

To select individuals to become a parent in breeding we can use, for example, the so-called tournament selection rule. For each "tournament" a fixed number n_T of individuals is selected randomly from $\mathcal{P}$. The individual with the highest fitness value wins the tournament, i.e., is selected to be a parent. If there are several such individuals, then the first one to enter the tournament wins. This selection rule is conveniently written as the following procedure:

procedure Choose_parent(**in:** $\mathcal{P}, \mathcal{F}, N, n_T$, **out:** $\widetilde{\boldsymbol{x}}$)
select $k \in \{1, \ldots, N\}$ at random;
do $i = 2, \ldots, n_T$
    select $m \in \{1, \ldots, N\}$ at random;
    if $F_m > F_k$ **then** set $k := m$;
end do
set $\widetilde{\boldsymbol{x}} := \boldsymbol{x}^{(k)}$;

parent $\boldsymbol{x}^a$	5.12	4.13	3.12	1.78	4.33
parent $\boldsymbol{x}^b$	0.83	2.53	6.11	7.25	4.44
offspring $\boldsymbol{y}^a$	5.12	4.13	6.11	7.25	4.44
offspring $\boldsymbol{y}^b$	0.83	2.53	3.12	1.78	4.33

Figure 4.1. *An example of one-site crossover.*

In the breeding process the offspring can be implemented by using, e.g., the following *one-site crossover.* With a given probability $P_{\rm co}$ the components of $\boldsymbol{x}^a$, $\boldsymbol{x}^b$ are crossed in the following way:

$$\begin{aligned} y_k^a &= x_k^a,\ y_k^b = x_k^b, \quad k = 1, \dots, m \\ y_k^a &= x_k^b,\ y_k^b = x_k^a, \quad k = m+1, \dots, n, \end{aligned}$$

where $1 \le m \le n$ are again randomly chosen. Otherwise, the values of the parents $\boldsymbol{x}^a$, $\boldsymbol{x}^b$ are copied to $\boldsymbol{y}^a$, $\boldsymbol{y}^b$. Figure 4.1 shows an example of this type of one-site crossover. The crossover operation can be written as the following procedure:

procedure Crossover(**in:** P_{co}, $\boldsymbol{x}^a$, $\boldsymbol{x}^b$, **out:** $\boldsymbol{y}^a$, $\boldsymbol{y}^b$)
select $r \in [0, 1]$ at random;
if $r \le P_{co}$ **then**
  select $m \in \{1, \dots, n\}$ at random;
  $\boldsymbol{y}^a := (x_1^a, \dots, x_m^a, x_{m+1}^b, \dots, x_m^b)$
  $\boldsymbol{y}^b := (x_1^b, \dots, x_m^b, x_{m+1}^a, \dots, x_n^a)$
else
  $\boldsymbol{y}^a := \boldsymbol{x}^a$;
  $\boldsymbol{y}^b := \boldsymbol{x}^b$;
end if;

The *mutation* can be understood as a way to escape from local minima. Mating two "locally" good parents may produce a better offspring but only locally. A random variation of the properties of the offspring may produce a design that would be impossible to get using just crossover operations to selected parents. The mutation of a single floating-point string $\boldsymbol{x}$ can be done, e.g., in the following way, which uses a special distribution promoting small mutations [MNP+97]. The components of the individual under consideration are gone through one by one and mutated with a given probability P_m. We denote the component x_i after the mutation by x_i^m. It is computed in three steps:

procedure Mutate(**in:** N, $\boldsymbol{x}^{low}$, $\boldsymbol{x}^{upp}$, p, P_m, **in/out:** $\mathcal{P}$)
do $k = 1, \dots, N$
  set $\widetilde{\boldsymbol{x}} := \boldsymbol{x}^{(k)}$;
  do $i = 1, \dots, n$
    select $r \in [0, 1]$ at random;
    if $r \le P_m$ **then**
      set $t := (x_i - x_i^{low})/(x_i^{upp} - x_i^{low})$,

$$\text{set } t_m := \begin{cases} t - t\left(\dfrac{t-r}{t}\right)^p, & r \le t, \\[2ex] t + (1-t)\left(\dfrac{r-t}{1-t}\right)^p, & r > t; \end{cases}$$

      set $\widetilde{x}_i := (1 - t_m)x_i^{low} + t_m x_i^{upp}$;
    end if
  end do
  replace $\boldsymbol{x}^{(k)}$ by $\widetilde{\boldsymbol{x}}$ in $\mathcal{P}$;
end do;

The input parameter $p \geq 1$ ("mutation exponent") defines the distribution of mutations and $\boldsymbol{x}^{low}$ and $\boldsymbol{x}^{upp} \in \mathbb{R}^n$ are the vectors defining the lower and upper bounds for $\mathcal{U}$, respectively. If $p = 1$, then the mutation is uniform. The probability of small mutations grows if the value of p grows.

Because we have already defined all of the necessary genetic operations, we can present the following simple GA for solving the nonlinear programming problem $(\mathbb{P})$ with box constraints.

ALGORITHM 4.5. *(Simple GA.)*

procedure GA(**in:** $N, n_T, P_{co}, P_m, p, \boldsymbol{x}^{low}, \boldsymbol{x}^{upp}, f$, **out:** $\mathcal{P}$)
$\mathcal{P} :=$ population of N points $\boldsymbol{x}^{low} \leq \boldsymbol{x}^{(i)} \leq \boldsymbol{x}^{upp}, i = 1, \ldots, N$, generated at random;
Evaluate_fitness($N, \mathcal{P}, f, \mathcal{F}$);
repeat
  set $\mathcal{P}^{new} := \emptyset$;
  do $k = 1, \ldots, N/2$
    Choose_parent($\mathcal{P}, \mathcal{F}, N, n_T, \boldsymbol{x}^a$);
    Choose_parent($\mathcal{P}, \mathcal{F}, N, n_T, \boldsymbol{x}^b$);
    Crossover($P_{co}, \boldsymbol{x}^a, \boldsymbol{x}^b, \boldsymbol{y}^a, \boldsymbol{y}^b$);
    set $\mathcal{P}^{new} := \mathcal{P}^{new} \cup \{\boldsymbol{y}^a, \boldsymbol{y}^b\}$;
  end do
  Mutate($N, \boldsymbol{x}^{low}, \boldsymbol{x}^{upp}, p, P_m, \mathcal{P}^{new}$);
  set $\mathcal{P} := \mathcal{P}^{new}$;
  Evaluate_fitness($N, \mathcal{P}, f, \mathcal{F}$);
until stopping criterion is true;

There is no natural stopping criterion for Algorithm 4.5. Usually the iteration process is terminated after a given number of generations has been produced or there is no improvement in the fitness of the best individual in the population. As a result the algorithm gives the final population, i.e., a set of approximate solutions to $(\mathbb{P})$. Unlike in the optimization methods discussed before, the best individual in the offspring population may be worse than the best one in the parent population. This fact further complicates the choice of a stopping criterion. A remedy to this is to add an *elitism mechanism* to Algorithm 4.5; i.e., a few best parents are always copied into the new population. In this way the best individuals are not lost in the reproduction and a decrease in the value of the objective function corresponding to the best individual in each generation becomes monotonic.

In practical shape optimization applications GAs require quite a lot of cost function evaluations. However, the inherent parallelism in GAs makes them attractive. In GAs, the objective functions can be evaluated at each generation independently. In a parallel implementation based on the master-slave prototype, the master process computes the genetic operations and the slave processes compute the object function values. Therefore the amount of communication between the master and the slave processes is rather small and a network of low cost workstations can be used to execute the different processes.

4.4 On multiobjective optimization methods

In classical optimal shape design problems only one objective function based on one scientific discipline is minimized. Real-life problems are rarely so simple. Often there exist

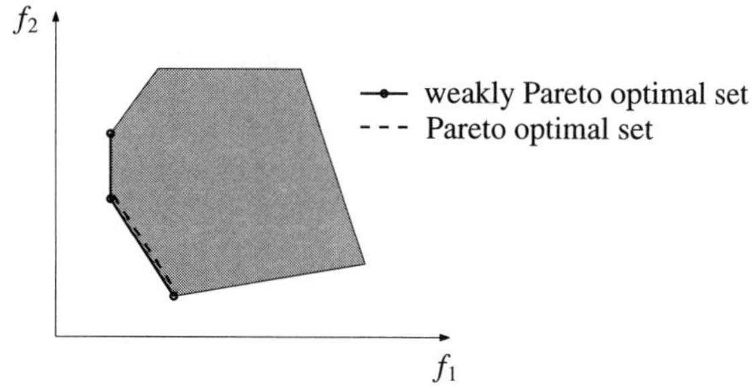

Figure 4.2. *Pareto and weakly Pareto optimal set.*

several (conflicting) objective functions that should be simultaneously optimized. For example, in aerospace engineering one wants to design the shape of an airfoil to minimize its drag and at the same time maximize its lift. Such multiobjective optimization problems (also known in the literature as *multicriteria* or *vector* optimization problems) require tools different from the standard optimization techniques for single (or scalar) objective optimization; see, for example, [Mie99].

4.4.1 Setting of the problem

Consider a multiobjective optimization problem of the form

$$\begin{aligned} &\min \{f_1(\boldsymbol{x}), f_2(\boldsymbol{x}), \dots, f_k(\boldsymbol{x})\} \\ &\text{subject to } \boldsymbol{x} \in \mathcal{U}, \end{aligned} \tag{4.26}$$

where $f_i : \mathbb{R}^n \to \mathbb{R}$, $i = 1, \dots, k$, are objective functions and $\emptyset \neq \mathcal{U} \subset \mathbb{R}^n$ is a set of feasible design variables. We do not specify constraint functions defining $\mathcal{U}$. Let Z be the image of $\mathcal{U}$ under the mapping $\boldsymbol{f} = (f_1, \dots, f_k) : \mathcal{U} \to \mathbb{R}^k$. The elements z of Z are called *criterion vectors*.

The components of an *ideal criterion vector* $\boldsymbol{z}^* = (z_1^*, \dots, z_k^*) \in \mathbb{R}^k$ are optima of the individual objective functions; i.e., $z_i^* = \min_{\boldsymbol{x} \in \mathcal{U}} f_i(\boldsymbol{x})$, $i = 1, \dots, k$. Unfortunately, $\boldsymbol{z}^*$ is usually infeasible; i.e., it is not possible to minimize all the objective functions at the same point from $\mathcal{U}$ because of the conflicting character of f_i, $i = 1, \dots, k$. Instead we introduce the concept of (weak) Pareto optimality.

Definition 4.1. *A design variable $\boldsymbol{x}^* \in \mathcal{U}$ and the corresponding criterion vector $\boldsymbol{f}(\boldsymbol{x}^*)$ are Pareto optimal if there does not exist another design variable $\boldsymbol{x} \in \mathcal{U}$ such that $f_i(\boldsymbol{x}) \leq f_i(\boldsymbol{x}^*)$ $\forall i = 1, \dots, k$ and $f_j(\boldsymbol{x}) < f_j(\boldsymbol{x}^*)$ for at least one objective function f_j. Weak Pareto optimality is defined by employing k strict inequalities.*

We call the set of all Pareto optimal solutions the *Pareto optimal set*. This set can be nonconvex and nonconnected. The geometrical interpretation of Pareto optimal and weakly Pareto optimal points in $f_1 f_2$ space is shown in Figure 4.2.

There are usually many Pareto optimal solutions. From a mathematical point of view, every design in the Pareto optimal set would make an equally good solution of the multiobjective optimization problem. However, in practical design problems one has to choose just one design as the solution. Selecting one design from the Pareto optimal set requires extra information that is not contained in the objective functions. The selection is done by a decision maker. A *decision maker* is a person who is supposed to have some extra information on the optimization problem not contained in the mathematical formulation, enabling the selection of a particular solution.

4.4.2 Solving multiobjective optimization problems by scalarization

In general, multiobjective optimization problems are solved by *scalarization*: converting the problem into one or several scalar optimization problems with a real-valued objective function depending possibly on some parameters. Thus standard methods of constrained nonlinear optimization can be used.

One has to keep in mind that descent methods for scalar optimization may find only a local minimizer. In addition, it is important to scale the problem properly; i.e., the values of the cost functions have to be of the same order of magnitude. The following two methods are examples of scalarizing functions. For more methods and further information see [Mie99].

Method of global criterion

In the method of global criterion, the decision maker is not involved. The distance between some reference point in $\mathbb{R}^k$ and the set Z is minimized. A natural choice for the reference point is the ideal criterion vector z^*. Thus the original vector optimization problem is replaced by the following standard scalar nonlinear programming problem:

$$\min_{x \in \mathcal{U}} \|f(x) - z^*\|_p, \tag{4.27}$$

where $\|\cdot\|_p$ denotes the norm in $\mathbb{R}^k$ defined as follows: $\|x\|_p = (\sum_{i=1}^k |x_i|^p)^{1/p}$ if $1 \le p < \infty$ and $\|x\|_\infty = \max_{i=1,\dots,k} |x_i|$, $x = (x_1, \dots, x_k)$. It can be shown that any solution of (4.27) is Pareto optimal if $1 \le p < \infty$. If $p = \infty$, the solution of (4.27) is weakly Pareto optimal (see [Mie99] and also Problem 4.3 for $1 \le p < \infty$).

COROLLARY 4.1. *If $\mathcal{U}$ is a nonempty compact subset and all f_i, $i = 1, \dots, k$, are continuous in $\mathcal{U}$, the problem* (4.27) *has a solution. Consequently, the Pareto optimal set for problem* (4.26) *is nonempty.*

Let $p = \infty$, $f(x) \ge \mathbf{0}$, and $z^* = \mathbf{0}$. Then (4.27) reduces to the following min-max problem:

$$\min_{x \in \mathcal{U}} \max_{1 \le i \le k} \{f_i(x)\}, \tag{4.28}$$

which can be transformed into a smooth one as follows:

$$\begin{gathered} \min_{x \in \mathcal{U},\ x_{n+1} \in \mathbb{R}} x_{n+1}, \\ f_i(x) - x_{n+1} \le 0, \quad i = 1, \dots, k. \end{gathered} \tag{4.29}$$

A weak point in the previous approach is that it requires the knowledge of an ideal objective vector z^*. In practical optimization problems it can be quite difficult to find global minima of individual objective functions. To overcome this practical difficulty one may use the so-called achievement scalarizing function approach (for further details we refer to [Mie99] and references therein).

Weighting method

The idea of the weighting method is to associate with each objective function f_i a weighting factor $w_i \geq 0$ and to minimize the weighted sum of the objective functions. We further suppose that the weights are normalized: $\sum_{i=1}^{k} w_i = 1$. The multiobjective optimization problem is then replaced by the following scalar nonlinear programming problem:

$$\min \sum_{i=1}^{k} w_i f_i(\boldsymbol{x}) \qquad (4.30)$$
$$\text{subject to } \boldsymbol{x} \in \mathcal{U}.$$

If $w_i > 0 \; \forall i = 1, \ldots, k$, then any solution $\boldsymbol{x}^*$ of (4.30) is Pareto optimal (see Problem 4.4). Here either the decision maker specifies the weights or the weights are varied and the decision maker must select the best of the solutions obtained.

4.4.3 On interactive methods for multiobjective optimization

In *interactive methods* the decision maker works with an interactive computer program. The basic idea is that the computer program tries to determine the preference structure of the decision maker iteratively in an interactive way. After each iteration the program presents a new (weakly) Pareto optimal solution and the decision maker is asked to provide some information. After a finite number of steps the method should give a solution that the decision maker is satisfied with. For a survey on interactive multiobjective methods we refer to [Mie99].

We will briefly discuss the NIMBUS method (Nondifferentiable Interactive Multiobjective BUndle-based optimization System) of Miettinen and Mäkelä [MM95], [MM00]. In this method the decision maker examines the values of the objective functions calculated at a current point $\boldsymbol{x}^c$ and divides the set $\mathcal{I} = \{1, \ldots, k\}$ into up to five subsets denoted by $I^<$, $I^\leq$, $I^=$, $I^>$, I^0 as follows:

1. $i \in I^< \iff$ the value of f_i should be decreased;
2. $i \in I^\leq \iff$ the value of f_i should be decreased to the given aspiration level L_i;
3. $i \in I^= \iff f_i$ is satisfactory at the moment;
4. $i \in I^> \iff f_i$ is allowed to increase to a given upper bound $U_i > f_i(\boldsymbol{x}^c)$;
5. $i \in I^0 \iff f_i$ is allowed to change freely.

This means that the decision maker wishes to improve the values of the functions in $I^<$ and $I^\leq$ from their current level at the expense of the functions in $I^>$ and I^0.

According to this classification and to the values L_i, U_i given by the decision maker, we form an auxiliary problem:

$$\begin{aligned} &\text{minimize} \max_{i \in I^<,\ j \in I^\leq} \left\{ f_i(\boldsymbol{x}) - z_i^*,\ \max\left[f_j(\boldsymbol{x}) - L_j, 0 \right] \right\} \\ &\text{subject to } f_i(\boldsymbol{x}) \leq f_i(\boldsymbol{x}^c), \quad i \in I^< \cup I^\leq \cup I^=, \\ &\qquad f_i(\boldsymbol{x}) \leq U_i, \quad i \in I^>, \\ &\qquad \boldsymbol{x} \in \mathcal{U}. \end{aligned} \tag{4.31}$$

This is a scalar and nonsmooth problem regardless of the smoothness of the original problem. It can be solved by a proximal bundle method designed for nondifferentiable optimization [MN92]. It is known [MM00] that if $I^< \neq \emptyset$, then the solution of problem (4.31) is a weakly Pareto optimal solution of (4.26).

It is clear that the decision maker must be ready to allow the values of some objective functions to increase in order to achieve improvement for some other objective functions when moving around the weakly Pareto optimal set. The search procedure stops when the decision maker does not want to improve the value of any objective function.

The NIMBUS method has been successfully applied to problems of optimal control and design (see [Mie94], [MMM96], [MMM98]). Moreover, a Web-based implementation of the method is available online at http://nimbus.mit.jyu.fi/.

4.4.4 Genetic algorithms for multiobjective optimization problems

We could, of course, transform the multiobjective optimization problem into a scalar one using the techniques presented in Subsection 4.4.2, and after that apply scalar GAs. One approach in multiobjective problems is to give as many as possible of the Pareto optimal solutions to the decision maker for the selection of the most suitable one. For this reason we shall modify the basic GAs to handle multiple objective functions directly without any scalarization and to let GAs generate a population whose individuals are all approximations of *different* Pareto optimal points.

There are several variants of GAs for multiobjective optimization problems. In what follows we describe a modification of the nondominated sorting GA (NSGA) of Srinivas and Deb [SD95]. The NSGA differs from a scalar GA only in the way the fitness of an individual is defined and the parents are chosen. The crossover and mutation operations remain the same as in the scalar GA.

A point $\boldsymbol{x}^{(1)}$ is said to *dominate* another point $\boldsymbol{x}^{(2)}$ if the following conditions hold:

- $f_j(\boldsymbol{x}^{(1)}) \leq f_j(\boldsymbol{x}^{(2)}) \quad \forall j = 1, \ldots, k;$
- $f_\ell(\boldsymbol{x}^{(1)}) < f_\ell(\boldsymbol{x}^{(2)})$ for at least one $1 \leq \ell \leq k$.

Let $\mathcal{P} \equiv \{\boldsymbol{x}^{(i)}\}_{i=1}^N$ be a population of N points. A subset of all points in $\mathcal{P}$ that is not dominated by any other point of $\mathcal{P}$ is called the set of *nondominated* points. If a point $\boldsymbol{x}^{(\ell)} \in \mathcal{P}$ is Pareto optimal, it is also a nondominated point.

If $\boldsymbol{x}^{(1)}$ and $\boldsymbol{x}^{(2)}$ are two Pareto optimal points, then they are considered to be equally good and it is natural to assign the same fitness value to both of them. Therefore it is natural to assign the largest fitness value to the nondominated points and smaller values to

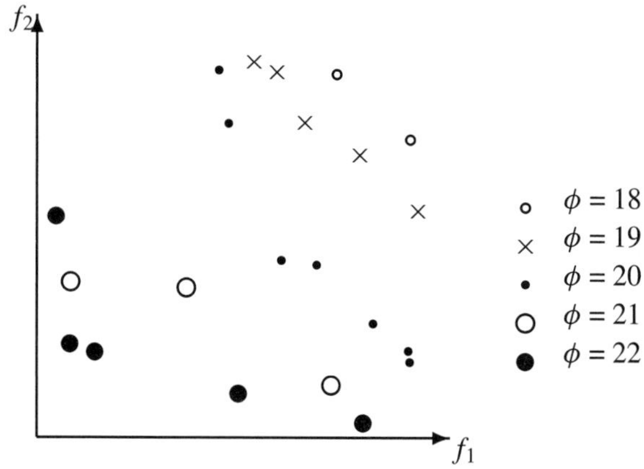

Figure 4.3. *Fitness values ϕ assigned to a population of 22 random pairs (f_1, f_2) of objective function values.*

the remaining ones. In an NSGA the fitness values of individuals are set using the following nondominated sorting procedure:

procedure Evaluate_fitness(**in:** $N, \mathcal{P}$, **out:** $\mathcal{F}$)
set $\phi := N$;
repeat
  Mark all individuals whose fitness value is not set as "nondominated";
  do $i = 1, \dots, N$, fitness value of $\boldsymbol{x}^{(i)}$ not set
    do $j = 1, \dots, N$; $j \neq i$, fitness value of $\boldsymbol{x}^{(j)}$ not set
      if $\boldsymbol{x}^{(j)}$ dominates $\boldsymbol{x}^{(i)}$ **then** mark $\boldsymbol{x}^{(i)}$ as "dominated";
    end do
  end do
  set the fitness $\mathcal{F}$ of all individuals not marked "dominated" equal to ϕ;
  set $\phi := \phi - 1$;
until fitness value of all individuals is set;

An example of the fitness values obtained by using the previous procedure is shown in Figure 4.3.

To select individuals becoming a parent in breeding we could again use the tournament selection rule. Unfortunately, if there are no modifications of the standard tournament selection, the population may converge toward one point on the set of Pareto optimal solutions. Since our aim is to obtain several points from the Pareto set a mechanism is needed in order to maintain *diversity* in the population.

In [MTP99] the following *tournament slot sharing approach* was employed to preserve the diversity of the population. A *sharing function* is defined by

$$\mathrm{Sh}(d_{ij}) = \begin{cases} 1 - \left(\frac{d_{ij}}{d_s}\right)^2 & \text{if } d_{ij} < d_s, \\ 0 & \text{otherwise,} \end{cases}$$

where d_{ij} is the genotypic distance between the individuals $\boldsymbol{x}^{(i)}$ and $\boldsymbol{x}^{(j)}$, the Euclidean one in our case. The parameter $d_s > 0$ is the maximum sharing distance for a tournament slot. The same sharing function $\mathrm{Sh}(d_{ij})$ is also used in the original NSGA. The probability of the individual $\boldsymbol{x}^{(i)}$ entering a tournament is now computed using the formula

$$p_i = \frac{1/\sum_{j=1}^{N} \mathrm{Sh}(d_{ij})}{\sum_{k=1}^{N}(1/\sum_{j=1}^{N} \mathrm{Sh}(d_{kj}))}. \tag{4.32}$$

Each individual's probability of entering the tournament is proportional to the inverse of the sum of all sharing functions associated with this individual. Therefore the probability of a point located in a "cluster" of points becoming a parent is small.

The parent selection process can be summarized as the following procedure:

procedure Choose_parent(**in:** $\mathcal{P}, \mathcal{F}, N, n_T$, **out:** $\widetilde{\boldsymbol{x}}$)
select $k \in \{1, \dots, N\}$ at random using probabilities in (4.32);
do $i = 2, \dots, n_T$
  select $m \in \{1, \dots, N\}$ at random using probabilities in (4.32);
  if $F_m > F_k$ **then** set $k := m$;
end do
set $\widetilde{\boldsymbol{x}} := \boldsymbol{x}^{(k)}$;

A modification of the scalar GA to the multiobjective case now consists of replacing the procedures *Evaluate_fitness* and *Choose_parent* in Algorithm 4.5 by the modified versions presented above.

Again an elitism mechanism can be added to the algorithm. One possibility is to copy from the old population to the new all individuals that would be nondominated in the new population. In this way the number of copied individuals will vary from one generation to another.

Problems

Problem 4.1. Let $\boldsymbol{x}^* \in \mathbb{R}^n$ be a local minimizer of a function $f \in C^2(\mathbb{R}^n)$. Further, let

(i) the Hessian $\boldsymbol{H}(\boldsymbol{x}^*)$ be positive definite;

(ii) $\boldsymbol{H}$ be Lipschitz continuous in a neighborhood of $\boldsymbol{x}^*$; i.e., $\exists q > 0:$

$$\|\boldsymbol{H}(\boldsymbol{x}) - \boldsymbol{H}(\boldsymbol{y})\| \leq q\|\boldsymbol{x} - \boldsymbol{y}\| \quad \forall \boldsymbol{x}, \boldsymbol{y} \in B_\delta(\boldsymbol{x}^*), \quad \delta > 0.$$

Show that, if the kth iteration $\boldsymbol{x}_k$ is sufficiently close to $\boldsymbol{x}^*$, Newton's method is well defined for any k and converges quadratically.

Hint: denote $\boldsymbol{h}_k = \boldsymbol{x}_k - \boldsymbol{x}^*$, $\boldsymbol{g}(\boldsymbol{x}) = \nabla f(\boldsymbol{x})$. From the Taylor expansion

$$\boldsymbol{0} = \boldsymbol{g}(\boldsymbol{x}^*) = \boldsymbol{g}(\boldsymbol{x}_k) - \boldsymbol{H}(\boldsymbol{x}_k)\boldsymbol{h}_k + O(\|\boldsymbol{h}_k\|^2)$$

it follows that $\boldsymbol{h}_{k+1} = O(\|\boldsymbol{h}_k^2\|)$ provided that $\boldsymbol{x}_k$ is close enough to $\boldsymbol{x}^*$.

Problem 4.2. Let $\boldsymbol{x}_k^*$ be a solution to penalized problem $(\mathbb{P}_{\varepsilon_k})$, $\varepsilon_k \to 0+$ as $k \to \infty$. Show that any convergent subsequence of $\{\boldsymbol{x}_k^*\}$ tends to a solution of $(\mathbb{P})$.

PROBLEM 4.3. Consider the minimization problem

$$\min_{x \in \mathcal{U}} \| \boldsymbol{f}(\boldsymbol{x}) - z^* \|_p, \tag{4.33}$$

where $1 \leq p < \infty$, $\boldsymbol{f}(\boldsymbol{x}) = (f_1(\boldsymbol{x}), \ldots, f_k(\boldsymbol{x}))$, and $z^* \in \mathbb{R}^k$ is an ideal criterion vector. Prove (by contradiction) that every solution of (4.33) is Pareto optimal.

PROBLEM 4.4. Consider minimization of the weighted sum of k continuous objective functions

$$\min_{x \in \mathcal{U}} \sum_{i=1}^{k} w_i f_i(\boldsymbol{x}). \tag{4.34}$$

Assume that $w_i > 0 \; \forall i = 1, \ldots, k$. Show (by contradiction) that if $\boldsymbol{x}^*$ is a solution of (4.34), then $\boldsymbol{x}^*$ is a Pareto optimal point of the corresponding multiobjective optimization problem.

Chapter 5

On Automatic Differentiation of Computer Programs

Using the techniques presented in Chapter 3, one can program the algebraic sensitivity analysis in optimal sizing and shape design problems with a reasonable amount of work. In the papers [HM92], [MT94], and [TMH98], algebraic sensitivity analysis was performed by hand for quite complicated shape optimization problems. However, full hand coding of derivatives requires a lot of time. This is not acceptable when solving industrial design problems.

Automatic differentiation (AD) is a technique for augmenting computer programs with derivative computations [Gri89]. It exploits the fact that every computer program executes a sequence of elementary arithmetic operations. By applying the chain rule of differential calculus repeatedly to these operations, accurate derivatives of arbitrary order can be computed automatically.

AD is now under active study and is used in many applications in engineering [BCG+92], [MMRS96] and mathematical physics [KL91], [IG00].

5.1 Introduction to automatic differentiation of programs

AD of computer programs is still sometimes confused with finite difference approximations of derivatives or symbolic differentiation of a single expression using packages like Mathematica [Wol99].

It is always possible to approximate partial derivatives of a smooth function $f : \mathbb{R}^n \to \mathbb{R}$ by using simple divided difference approximations such as the forward difference approximation

$$\frac{\partial f(\boldsymbol{x}_0)}{\partial x_i} \approx \frac{f(\boldsymbol{x}_0 + \delta \boldsymbol{e}^{(i)}) - f(\boldsymbol{x}_0)}{\delta}. \tag{5.1}$$

Here $\boldsymbol{e}^{(i)}$ is the ith unit vector and $\delta > 0$ is the step length. It is well known that this approach leads to the first order approximation of the partial derivatives. Computation of approximate derivatives in this way has the advantage that one needs f only as a "black box." There are, however, two disadvantages to this approach: first, the accuracy of this approximation is difficult to estimate. Second, difference approximations are expensive when the evaluation of f needs a solution of a (possibly nonlinear) state problem. A small step size δ is needed to minimize the truncation error due to neglect of higher order terms in the Taylor expansion of f in (5.1). On the other hand, if δ is too small, the subtraction of nearly equal floating-point numbers may lead to a significant cancellation error. Therefore very conservative (large) values of δ have to be used. As a rule of thumb one can say that if f is calculated with d-digit precision, then the partial derivative calculated using (5.1) has only $d/2$-digit precision. More accurate higher order approximations are practically useless in sizing and shape optimization because they require too many function evaluations.

Symbolic manipulation packages are, in general, unable to deal with whole computer programs containing subroutines, loops, and branches. It takes a lot of human effort to differentiate a large computer program in small pieces using a symbolic manipulator and to reassemble the resulting pieces of the derivative code into a new program. Because the Fortran code generated by the manipulators is usually unreadable, the process is also very error prone.

Most problems of numerical analysis (approximate solutions of a nonlinear system of algebraic equations, numerical integration, etc.) can be viewed as the evaluation of a nonlinear vector function $\Phi : \mathbb{R}^n \to \mathbb{R}^m$. A computer program evaluating Φ at $\boldsymbol{\xi} \in \mathbb{R}^n$ can be viewed as a sequence of K scalar assignment statements:

$$
\begin{aligned}
&x_k = \xi_k, \quad k = 1, \dots, n; \\
&\textbf{do } i = n+1, \dots, K \\
&\qquad x_i = \varphi_i(x_1, \dots, x_{i-1}) \\
&\textbf{end do}.
\end{aligned}
\tag{5.2}
$$

The computer program has K variables $x_1, x_2, \dots, x_K$. The first n of them are called *input* or *independent variables*. The last m variables are *output* variables. The rest of the variables are *temporary* variables. The elementary functions φ_i may depend on all known variables $x_1, \dots, x_{i-1}$. If φ_i is a function of n_i variables x_{i_k}, $k = 1, \dots, n_i$, only, then we define $\mathcal{I}_i = \{i_1, \dots, i_{n_i}\}$ and denote $\boldsymbol{x}_{\mathcal{I}_i} = \{x_{i_1}, \dots, x_{i_{n_i}}\}$. In practice, each φ_i is a unary or binary arithmetic operation ($\pm, *, /$) or a univariate transcendental function (sin, exp, etc). In this case the length $|\mathcal{I}_i| \le 2$.

The dependencies among variables in the computer program can be visualized using a directed acyclic graph. The nodes of the graph represent the variables $x_1, \dots, x_K$ and the edges represent the dependencies. An arc (x_j, x_i) exists iff $j \in \mathcal{I}_i$, i.e., the variable x_i depends directly on x_j. If a node corresponds to an independent variable, no arc enters it. Similarly no arc leaves a node corresponding to an output variable.

Example 5.1. Let us consider the evaluation of the simple function

$$f(x_1, x_2) = x_1 x_2 + \sin(x_1 - x_2) \tag{5.3}$$

at a point (ξ_1, ξ_2). In this case $n = 2$ and $m = 1$. The corresponding computer program and the associated graph are shown in Figure 5.1. We have used several temporary variables to make the right-hand side of each assignment statement contain a binary arithmetic operation or evaluation of a univariate standard function.

$$
\begin{aligned}
x_1 &= \xi_1\\
x_2 &= \xi_2\\
x_3 &= x_1 \cdot x_2\\
x_4 &= x_1 - x_2\\
x_5 &= \sin(x_4)\\
x_6 &= x_3 + x_5\\
f &= x_6
\end{aligned}
$$

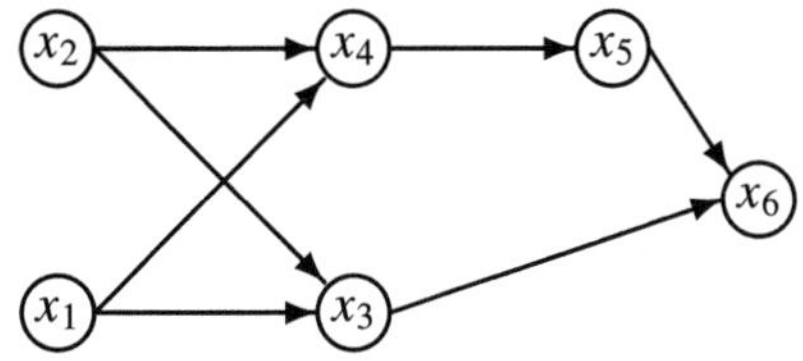

Figure 5.1. *Computer program for the evaluation of f and the associated acyclic directed graph.*

Our aim is to differentiate the output variables with respect to the independent ones. A variable $x_i,\ i > n$, is said to become *active* when an independent or an active one is assigned to it. Let $\varphi_1, \dots, \varphi_n$ denote the functions corresponding to the first n assignment statements in (5.2). Application of the chain rule to both sides of the composite functions in (5.2) yields

$$
\frac{\partial x_i}{\partial x_j} = \begin{cases} \delta_{ij} + \displaystyle\sum_{k=1}^{i-1} \frac{\partial \varphi_i}{\partial x_k}\frac{\partial x_k}{\partial x_j}, & j \le i \le K, \\ 0, & j > i. \end{cases} \tag{5.4}
$$

Defining the matrices

$$
\boldsymbol{D\varphi} := \left\{\frac{\partial \varphi_i}{\partial x_j}\right\}_{i,j=1}^{K} = \begin{pmatrix} 0 & \cdots & & \\ \partial\varphi_2/\partial x_1 & 0 & \cdots & \\ \partial\varphi_3/\partial x_1 & \partial\varphi_3/\partial x_2 & 0 & \cdots \\ \vdots & & & \end{pmatrix},
$$

$$
\boldsymbol{Dx} := \left\{\frac{\partial x_i}{\partial x_j}\right\}_{i,j=1}^{K} = \begin{pmatrix} 1 & 0 & \cdots & \\ \partial x_2/\partial x_1 & 1 & \cdots & \\ \partial x_3/\partial x_1 & \partial x_3/\partial x_2 & 1 & \cdots \\ \vdots & & & \end{pmatrix},
$$

(5.4) can be written in the compact matrix form

$$
\boldsymbol{Dx} = \boldsymbol{I} + (\boldsymbol{D\varphi})(\boldsymbol{Dx}).
$$

This can be further expressed in the form

$$
(\boldsymbol{I} - \boldsymbol{D\varphi})\,\boldsymbol{Dx} = \boldsymbol{I}. \tag{5.5}
$$

Using standard manipulations (5.5) can be expressed as follows:

$$\begin{aligned} \boldsymbol{Dx} &= (\boldsymbol{I} - \boldsymbol{D\varphi})^{-1}, \\ \boldsymbol{Dx}(\boldsymbol{I} - \boldsymbol{D\varphi}) &= \boldsymbol{I}, \\ (\boldsymbol{I} - \boldsymbol{D\varphi})^{\mathrm{T}}(\boldsymbol{Dx})^{\mathrm{T}} &= \boldsymbol{I}. \end{aligned} \tag{5.6}$$

From the lower triangular system (5.5), $\partial x_i/\partial x_j,\ i \neq j$, can be computed by using the *forward* substitutions

do $i = 2, \ldots, K$

  do $j = 1, \ldots, i-1$

$$\frac{\partial x_i}{\partial x_j} = \sum_{k=j}^{i-1} \frac{\partial \varphi_i}{\partial x_k} \frac{\partial x_k}{\partial x_j}. \tag{5.7}$$

  end do

end do

Similarly the upper triangular system (5.6) can be solved by using the *backward* substitutions

do $j = K-1, K-2, \ldots, 1$

  do $i = K, K-1, \ldots, j+1$

$$\frac{\partial x_i}{\partial x_j} = \sum_{k=j+1}^{i} \frac{\partial \varphi_k}{\partial x_j} \frac{\partial x_i}{\partial x_k}. \tag{5.8}$$

  end do

end do

Equations (5.5) and (5.6) represent the *forward* and *reverse* methods, respectively, of AD. In practice, because $K \gg n$ and $|\mathcal{I}_i| \leq 2$, it follows that the matrix $\boldsymbol{D\varphi}$ is very sparse. Therefore the practical implementation of AD is not based on the direct application of (5.5), (5.6).

5.1.1 Evaluation of the gradient using the forward and reverse methods

Consider now the program shown on the left in Figure 5.2 evaluating a real-valued function $f : \mathbb{R}^n \to \mathbb{R}$ at $\boldsymbol{\xi} \in \mathbb{R}^n$. Let us complete this program by gradient computations applying the forward method (5.7). In this case we need to store only the partial derivatives of variables x_i with respect to the input variables. The augmented program is shown on the right in Figure 5.2.

Following [Gri89] we briefly mention the complexity of the straightforward implementation of the forward and reverse modes in terms of arithmetic operations and memory usage. For more detailed analysis and more sophisticated implementations we refer to [Gri00]. Let work(f) and work$(f, \nabla f)$ denote the work (in terms of arithmetic operations) done in the original and augmented codes (5.2), respectively. Similarly, let mem(f) and

```
do i = 1, ..., n
    x_i = ξ_i
end do
do i = n + 1, ..., K
    x_i = φ_i(x_{I_i})
end do
f = x_K
```

```
do i = 1, ..., n
    x_i = ξ_i
    ∇x_i = e^(i)
end do
do i = n + 1, ..., K
    x_i = φ(x_{I_i})
    ∇x_i = Σ_{j∈I_i} (∂φ_i/∂x_j) ∇x_j
end do
f = x_K
∇f = ∇x_K
```

Figure 5.2. *Original program (left) and augmented program using the forward mode of AD (right).*

$\mathrm{mem}(f, \nabla f)$ denote the number of memory locations used in the original and augmented programs, respectively.

Suppose that every function φ_i is restricted to the elementary arithmetic operations and standard univariate library functions. Moreover, suppose that every φ_i requires at most cn_i arithmetic operations, where c is a fixed positive constant and $n_i = |\mathcal{I}_i|$. Then one can show (see [Gri89]) that

$$\begin{aligned} \mathrm{work}(f, \nabla f) &\geq \left(1 + \frac{n}{c}\right) \cdot \mathrm{work}(f), \\ \mathrm{mem}(f, \nabla f) &= (1 + n) \cdot \mathrm{mem}(f); \end{aligned}$$

i.e., the computational work and memory requirement both grow linearly in n.

Example 5.2. Let us evaluate the gradient of the function f defined in (5.3) at a point (ξ_1, ξ_2) using the forward method:

$$\begin{aligned}
&x_1 = \xi_1, \quad \nabla x_1 = (1, 0)^{\mathrm{T}}, \\
&x_2 = \xi_2, \quad \nabla x_2 = (0, 1)^{\mathrm{T}}, \\
&x_3 = \varphi_3(x_1, x_2) = x_1 \cdot x_2, \\
&\nabla x_3 = \tfrac{\partial \varphi_3}{\partial x_1} \nabla x_1 + \tfrac{\partial \varphi_3}{\partial x_2} \nabla x_2 = x_2 \nabla x_1 + x_1 \nabla x_2 = (x_2, x_1)^{\mathrm{T}}, \\
&x_4 = \varphi_4(x_1, x_2) = x_1 - x_2, \\
&\nabla x_4 = \tfrac{\partial \varphi_4}{\partial x_1} \nabla x_1 + \tfrac{\partial \varphi_4}{\partial x_2} \nabla x_2 = (1, -1)^{\mathrm{T}}, \\
&x_5 = \varphi_5(x_4) = \sin x_4, \\
&\nabla x_5 = \tfrac{\partial \varphi_5}{\partial x_4} \nabla x_4 = (\cos x_4, -\cos x_4)^{\mathrm{T}}, \\
&x_6 = \varphi_6(x_3, x_5) = x_3 + x_5, \\
&\nabla x_6 = \tfrac{\partial \varphi_6}{\partial x_3} \nabla x_3 + \tfrac{\partial \varphi_6}{\partial x_5} \nabla x_5 = (x_2 + \cos x_4, x_1 - \cos x_4)^{\mathrm{T}}, \\
&\nabla f = \nabla x_6 = (x_2 + \cos x_4, x_1 - \cos x_4)^{\mathrm{T}}.
\end{aligned}$$

Consider now the reverse method. Instead of (5.6) it is interesting to present an alternative way of deriving the reverse method for calculating ∇f. One may interpret the input variables $x_1, \dots, x_n$ as "control variables" and $x_{n+1}, \dots, x_K$ as "state variables" and treat the assignment statements involving state variables on the left-hand side as state constraints ("state equation"). Then the problem of adding gradient computations to the program can be written in a form that is already familiar to us from Chapter 3, i.e., to evaluate the gradient of the function

$$\tilde{f}(x_{n+1}, \dots, x_K) = x_K \tag{5.9}$$

subject to the constraints

$$\begin{cases} r_{n+1}(\boldsymbol{x}) := x_{n+1} - \varphi_{n+1}(\boldsymbol{x}_{\mathcal{I}_{n+1}}) = 0, \\ \quad\vdots \\ r_K(\boldsymbol{x}) := x_K - \varphi_K(\boldsymbol{x}_{\mathcal{I}_K}) = 0. \end{cases} \tag{5.10}$$

The partial derivatives of $\tilde{f}$ with respect to the control variables $x_1, \dots, x_n$ are given by

$$\frac{\partial \tilde{f}}{\partial x_i} = \sum_{j=n+1}^{K} p_j \frac{\partial r_j(\boldsymbol{x})}{\partial x_i}, \qquad i = 1, \dots, n, \tag{5.11}$$

where $\boldsymbol{p} := (p_{n+1}, \dots, p_K)^{\mathrm{T}}$ is the solution of the "adjoint equation"

$$(\boldsymbol{J}(\boldsymbol{x}))^{\mathrm{T}} \boldsymbol{p} = \left(\frac{\partial \tilde{f}}{\partial x_{n+1}}, \dots, \frac{\partial \tilde{f}}{\partial x_K} \right)^{\mathrm{T}}. \tag{5.12}$$

Here

$$\boldsymbol{J}(\boldsymbol{x}) = \begin{pmatrix} 1 & 0 & \cdots & \\ \frac{\partial r_{n+2}}{\partial x_{n+1}} & 1 & \cdots & \\ \frac{\partial r_{n+3}}{\partial x_{n+1}} & \frac{\partial r_{n+3}}{\partial x_{n+2}} & 1 & \cdots \\ \cdots & & & \end{pmatrix} \in \mathbb{R}^{(K-n)\times(K-n)}$$

is the Jacobian corresponding to the nonlinear system (5.10). As $\boldsymbol{J}$ is a lower triangular matrix with ones at the diagonal, the upper diagonal system (5.12) is readily solved. Due to this analogy with optimal control the reverse method is often termed the *adjoint mode* of AD.

The previous calculations can be compactly augmented into the program shown on the left of Figure 5.3, resulting in the program shown on the right of the same figure.

For the reverse mode one may derive a quite surprising result:

$$\text{work}(f, \nabla f) \le 5 \cdot \text{work}(f), \tag{5.13}$$

$$\text{mem}(f, \nabla f) \le K + \text{mem}(f). \tag{5.14}$$

Because the program flow must be reversed in the reverse mode, one has to store the call graph (the index sets $\mathcal{I}_i$ in practice) and the partial derivatives $\partial \varphi_i / \partial x_j,\ j \in \mathcal{I}_i$, during the execution of the program. This implies that the memory required in the worst case may be equal to the number of assignment statements K in the program.

do $i = 1, \dots, n$
$\quad x_i = \xi_i$
end do
do $i = n+1, \dots, K$
$\quad x_i = \varphi_i(\boldsymbol{x}_{\mathcal{I}_i})$
end do
$f = x_K$

do $i = 1, \dots, n$
$\quad x_i = \xi_i$
end do
do $i = n+1, \dots, K$
$\quad x_i = \varphi_i(\boldsymbol{x}_{\mathcal{I}_i})$
end do
$f = x_K$
do $i = 1, \dots, K-1$
$\quad \overline{x}_i = 0$
end do
$\overline{x}_K = 1$
do $i = K, K-1, \dots, n+1$
$$\overline{x}_j := \overline{x}_j + \frac{\partial \varphi_i}{\partial x_j}\overline{x}_i \qquad \forall\, j \in \mathcal{I}_i$$
end do
$\nabla f = \{\overline{x}_i\}_{i=1}^n$

Figure 5.3. *Original program (left) and augmented program using the reverse mode of AD (right).*

Although the reverse mode seems to be superior in terms of computational work, it is more difficult to implement it efficiently. Moreover, storing and retrieving the call graph take time that is not included in (5.13).

If one wishes to calculate the gradient of several dependent variables $x_{K-m+1}, x_{K-m+2}, \dots, x_K$, then the forward mode is more efficient in terms of computational work provided $m > n$.

Example 5.3. Let us evaluate the gradient of the function f in (5.3) at a point (ξ_1, ξ_2) using the reverse method:

$$\begin{aligned}
&x_3 = \varphi_3(x_1, x_2) = x_1 \cdot x_2,\\
&x_4 = \varphi_4(x_1, x_2) = x_1 - x_2,\\
&x_5 = \varphi_5(x_4) = \sin x_4,\\
&x_6 = \varphi_6(x_3, x_5) = x_3 + x_5,\\
&f = x_6,\\
&\overline{x}_i = 0, \quad i = 1, \dots, 5,\\
&\overline{x}_6 = 1,\\
&\overline{x}_3 := \overline{x}_3 + \overline{x}_6 \partial\varphi_6/\partial x_3 = 1,\\
&\overline{x}_5 := \overline{x}_5 + \overline{x}_6 \partial\varphi_6/\partial x_5 = 1,\\
&\overline{x}_4 := \overline{x}_4 + \overline{x}_5 \partial\varphi_5/\partial x_4 = \cos x_4,\\
&\overline{x}_1 := \overline{x}_1 + \overline{x}_4 \partial\varphi_4/\partial x_1 = \cos x_4,\\
&\overline{x}_2 := \overline{x}_2 + \overline{x}_4 \partial\varphi_4/\partial x_2 = -\cos x_4,\\
&\overline{x}_1 := \overline{x}_1 + \overline{x}_3 \partial\varphi_3/\partial x_1 = x_2 + \cos x_4,\\
&\overline{x}_2 := \overline{x}_2 + \overline{x}_3 \partial\varphi_3/\partial x_2 = x_1 - \cos x_4,\\
&\nabla f = (\overline{x}_1, \overline{x}_2)^{\mathrm{T}} = (x_2 + \cos x_4, x_1 - \cos x_4)^{\mathrm{T}}.
\end{aligned}$$

5.2 Implementation of automatic differentiation

In this section we describe very briefly the main steps needed to implement AD. The pieces of computer code are in Fortran 90 with keywords typed in uppercase letters. A reader not familiar with Fortran 90 should consult [MR96], for example.

In the forward method, the gradient ∇x_i of each variable x_i with respect to the n independent variables $x_1, \ldots, x_n$ is stored after being calculated. Therefore we may define a new data type `dvar_t` for storing the pair $(z, \nabla z)$ as follows:

```
INTEGER, PARAMETER:: nmax=10
INTEGER:: n

TYPE dvar_t
  REAL:: val, der(nmax)
END type dvar_t
```

Here `n` and `nmax` are the actual number and maximum allowed number of independent variables, respectively.

If z is the ith independent variable, then its gradient must be initialized by $\nabla z = \boldsymbol{e}^{(i)}$. Next we will implement simple routines to declare some variables to be independent and to extract the value or the gradient as normal floating-point numbers:

```
!
! x will be the nth independent variable with an initial
! value x=val.
! grad(x) is set equal to the nth unit vector.
!
SUBROUTINE ad_declare_indep_var( x, val )
  REAL, INTENT(IN):: val
  TYPE(dvar_t):: x
  IF ( n < nmax ) THEN
     n = n + 1
  ELSE
     CALL error()
  END IF
  x%val = val
  x%der(1:nmax) = 0.0
  x%der(n) = 1.0
END SUBROUTINE ad_declare_indep_var

FUNCTION ad_value( x ) RESULT ( v )
  TYPE(dvar_t), INTENT(IN):: x
  REAL:: v
  v = x%val
END FUNCTION ad_value

FUNCTION ad_gradient( f ) RESULT ( df )
  TYPE(dvar_t), INTENT(IN):: f
  REAL:: df(n)
  df = f%der(1:n)
  n  = 0
END FUNCTION ad_gradient
```

Finally we must define the standard arithmetic operations and library functions for the new data type `dvar_t` using the standard rules of differential calculus. For example, the operators +, -, *, / and the library function `sin` for the new data type could be implemented as the following Fortran 90 functions:

```
FUNCTION add_dvar( a, b ) RESULT( c )
  TYPE(dvar_t), INTENT(in):: a, b
  TYPE(dvar_t):: c
  c%val = a%val + b%val
  c%der(1:n) = a%der(1:n) + b%der(1:n)
END FUNCTION add_dvar

FUNCTION sub_dvar( a, b ) RESULT( c )
  TYPE(dvar_t), INTENT(in):: a, b
  TYPE(dvar_t):: c
  c%val = a%val - b%val
  c%der(1:n) = a%der(1:n) - b%der(1:n)
END FUNCTION sub_dvar

FUNCTION mul_dvar( a, b ) RESULT( c )
  TYPE(dvar_t), INTENT(in):: a, b
  TYPE(dvar_t):: c
  c%val = a%val * b%val
  c%der(1:n) = a%der(1:n) * b%val  +  a%val * b%der(1:n)
END FUNCTION mul_dvar

FUNCTION div_dvar( a, b ) RESULT( c )
  TYPE(dvar_t), INTENT(in):: a, b
  TYPE(dvar_t):: c
  c%val = a%val / b%val
  c%der(1:n) = ( a%der(1:n)*b%val - a%val*b%der(1:n) ) /
               (b%val*b%val)
END FUNCTION div_dvar

FUNCTION sin_dvar( a ) RESULT( c )
  TYPE(dvar_t), INTENT(in):: a
  TYPE(dvar_t):: c
  c%val = SIN( a%val )
  c%der(1:n) = COS( a%val ) * a%der(1:n)
END FUNCTION sin_dvar
```

Let us consider once again the evaluation of the simple function of Example 5.1. The corresponding Fortran 90 subroutine is shown below:

```
SUBROUTINE myfunc( xi, f )
  REAL xi(2), f
  REAL x1, x2, x3, x4, x5, x6
  x1 = xi(1)
  x2 = xi(2)
  x3 = x1*x2
  x4 = x1 - x2
  x5 = SIN(x4)
```

```
    x6 = x3 + x5
    f = x6
END SUBROUTINE myfunc
```

Let us augment this code with the gradient computations. Now x1 and x2 are the independent variables and f is the output variable. We define all variables except xi and f to be of type dvar_t. Then the arithmetic operations and the call to library function sin must be replaced by calls to appropriate functions introduced above. In addition, the values of the independent variables and their derivatives are initialized using subroutine declare_indep_var. After computations, the values of f and ∇f as ordinary floating-point numbers are recovered by functions ad_value and ad_gradient. Let the definitions of new AD-related data types and subroutines be encapsulated in a Fortran 90 module ad_m. Then the augmented subroutine reads as follows:

```
SUBROUTINE myfunc2( xi, f, df )
  USE  ad_m
  REAL xi(2), f, df(2)
  TYPE(dvar_t) x1, x2, x3, x4, x5, x6
  CALL declare_indep_var( x1, xi(1) )
  CALL declare_indep_var( x2, xi(2) )
  x3 = mul_dvar( x1, x2 )
  x4 = sub_dvar( x1, x2 )
  x5 = sin_dvar( x4 )
  x6 = add_dvar( x3, x5 )
  f  = ad_value( x6 )
  df = ad_gradient( x6 )
END SUBROUTINE myfunc2
```

The transformation of the subroutine myfunc into myfunc2 was done by hand. To justify the term *automatic* differentiation, the transformation should be done by the computer without any human intervention. This can be achieved in two different ways either by using the *source transformation* (done by a preprocessor) or the *operator overloading*.

Automatic differentiation using a preprocessor

If a preprocessor like ADIFOR or Odyssée (see [RS93], [BCKM96]) is used, the source transformation is usually done in three steps:

1. In the *code canonicalization* step long right-hand sides of assignment statements are broken up into smaller pieces of code; expressions appearing as arguments of function calls are written as separate assignment statements using temporary variables. For example, the following piece of code:

```
y = SIN(a*b+c) + d*a
```

is transformed into, e.g., the following form:

```
t1 = a*b
t2 = t1 + c
t3 = SIN(t2)
t4 = d*a
y = t3 + t4
```

2. In the *variable nomination* step a derivative object is associated with every active variable whose value has an effect on the output variables. Savings in storage and computing time can be achieved by identifying those active variables that have no effect on the output variables.

3. After the code canonicalization and variable nomination steps the augmented source code is *generated* by allocating storage for additional derivative information and replacing the original assignment statements with the appropriate subroutine calls (or their inline counterparts). The user then compiles the augmented code using a standard Fortran or C compiler instead of the original one. We do not enter into further detail on the use of precompilers in AD.

Automatic differentiation using operator overloading

Modern programming languages such as Fortran 90, C++, and Ada make it possible to extend the meaning of an intrinsic operator or function to additional user-defined data types.

For each elementary operation and a standard function (+, *, `sin(),...`) we can define the meaning of the operation for variables of the new data type `dvar_t` introduced before. In Fortran 90 this requires special interface blocks to be added into the module containing AD-related definitions and subroutines. The compiler then replaces the operator by a call to a subroutine specified in the interface block. This step, however, is totally invisible to the user.

Finally, we can encapsulate all previous declarations and subroutines needed to implement AD into a single Fortran 90 module that can be used where needed:

```
MODULE ad_m
  INTEGER, PRIVATE:: n
  INTEGER, PRIVATE, PARAMETER:: nmax=10
  ! -- declaration of a new data type --
  TYPE dvar_t
    REAL:: val, der(nmax)
  END TYPE dvar_t
  ! -- interface blocks for overloaded operators
  INTERFACE OPERATOR(*)
    MODULE PROCEDURE mult_dvar
  END INTERFACE

  INTERFACE sin
    MODULE PROCEDURE sin_dvar
  END INTERFACE
   . . .

CONTAINS

  ! -- code of subroutines needed for declaring input and output
       variables
  SUBROUTINE ad_declare_indep_var( x, val )
   . . . ( code not shown )
```

```
    FUNCTION ad_value( x ) RESULT ( v )
     . . . ( code not shown )

    FUNCTION ad_gradient( f ) RESULT ( df )
     . . . ( code not shown )

    ! -- code implementing overloaded operators

    FUNCTION mul_dvar( a, b ) RESULT( c )
     . . . ( code not shown )

    FUNCTION sin_dvar( a ) RESULT( c )
     . . . ( code not shown )
  END MODULE ad_m
```

The advantage of the operator overloading is that it almost completely hides the AD tool from the user. The code used for the evaluation of f itself and the code used for the evaluation of both f and ∇f are identical except for the variable declaration and the identification of input and output variables. If the implementation of the AD tool is changed, the user-written source code needs no modifications.

EXAMPLE 5.4. Consider again the code we are familiar with:

```
SUBROUTINE myfunc( xi, f )
  REAL xi(2), f
  REAL x1, x2, x3, x4, x5, x6
  x1 = xi(1)
  x2 = xi(2)
  x3 = x1 * x2
  x4 = x1 - x2
  x5 = SIN(x4)
  x6 = x3 + x5
  f = x6
END SUBROUTINE myfunc
```

Let us assume that the module ad_m implementing AD using operator overloading is available. Then the augmented code reads as follows:

```
SUBROUTINE myfunc2( xi, f, df )
  USE  ad_m
  REAL xi(2), f, df(2)
  TYPE(dvar_t) x1, x2, x3, x4, x5, x6
  CALL declare_indep_var( x1, xi(1) )
  CALL declare_indep_var( x2, xi(2) )
  x3 = x1 * x2
  x4 = x1 - x2
  x5 = SIN( x4 )
  x6 = x3 + x5
  f  = ad_value( x6 )
  df = ad_gradient( x6 )
END SUBROUTINE myfunc2
```

Of course, this simple example does not fully show the advantage of operator overloading. If, however, the body of the original code had, e.g., 100 lines of Fortran 90 code, then there would still be only 5 additional lines and 1 modified line in the augmented code.

Note that there is no need to canonicalize the code by hand. The Fortran 90 compiler will introduce additional temporary variables of type `dvar_t` corresponding, e.g., to the statement `x5 = sin( x1 - x2 )`.

5.3 Application to sizing and shape optimization

In standard software used for numerical minimization the evaluation of an objective function and its gradient is done in a separate program unit. Usually, the user of the minimization software is responsible for writing a subroutine that evaluates the value of the objective function $\mathcal{J}$ and its gradient $\nabla\mathcal{J}$ at a point $\boldsymbol{\alpha} \in \mathbb{R}^d$ given by the (sub)program calling the user-written subroutine.

Consider the numerical solution of a simple shape optimization problem. Let us first assume that a computer program solves the direct problem on a domain described by the discrete design vector $\boldsymbol{\alpha} = (\alpha_1, \ldots, \alpha_d)$. Moreover, the program also evaluates a cost function (the compliance in our case) depending on the solution of the state problem. This program can be written as a Fortran subroutine as follows:

```
subroutine costfun( α, J )
    real α, J
    Generate a finite element mesh corresponding to α.
    Assemble the stiffness matrix K := K(α) and the force vector f := f(α).
    Solve the linear system Kq = f.
    J = f^T q.
end
```

This subroutine executes a finite sequence of basic arithmetic operations. Therefore any of the AD techniques presented in the previous sections could be directly applied to add gradient computations to this subroutine. However, the *efficient* use (in terms of memory and computing time) of AD for large-scale industrial shape optimization problems is not often so straightforward.

A direct application of AD can be too expensive due to the complexity of the state equation. If the forward mode of AD is used, the required computing time and the memory of one combined cost and gradient evaluation is approximately d times as expensive as the evaluation of the cost function only. Note that in `costfun` all entries of the stiffness matrix and all statements of the linear system solver are differentiated with respect to design variables.

For AD to be useful it is important that the problem structure be understood and used. In what follows we describe a "hybrid approach" in which both hand coding of derivatives and AD are used to produce a subroutine augmented with gradient computations.

Let us consider a system of (non)linear algebraic equations depending on a parameter $\boldsymbol{\alpha} \in \mathcal{U} \subset \mathbb{R}^d$:

$$\boldsymbol{K}(\boldsymbol{\alpha}, \boldsymbol{q}(\boldsymbol{\alpha}))\boldsymbol{q}(\boldsymbol{\alpha}) = \boldsymbol{f}(\boldsymbol{\alpha}, \boldsymbol{q}(\boldsymbol{\alpha})), \tag{5.15}$$

in which $\boldsymbol{K} \in C^1(\mathcal{U} \times \mathbb{R}^n; \mathbb{R}^{n\times n})$ and $\boldsymbol{f} \in C^1(\mathcal{U} \times \mathbb{R}^n; \mathbb{R}^n)$ are a (non)linear matrix and a vector function, respectively. Let $J : \mathcal{U} \times \mathbb{R}^n \to \mathbb{R}$ be a real function and define $\mathcal{J} : \mathcal{U} \to \mathbb{R}$ by

$$\mathcal{J}(\boldsymbol{\alpha}) = J(\boldsymbol{\alpha}, \boldsymbol{q}(\boldsymbol{\alpha})) \tag{5.16}$$

with $\boldsymbol{q}(\boldsymbol{\alpha}) \in \mathbb{R}^n$ the solution of (5.15).

Equation (5.15) can be equivalently stated as

$$\boldsymbol{r}(\boldsymbol{\alpha}, \boldsymbol{q}(\boldsymbol{\alpha})) := \boldsymbol{K}(\boldsymbol{\alpha}; \boldsymbol{q}(\boldsymbol{\alpha}))\, \boldsymbol{q}(\boldsymbol{\alpha}) - \boldsymbol{f}(\boldsymbol{\alpha}; \boldsymbol{q}(\boldsymbol{\alpha})) = \boldsymbol{0}, \tag{5.17}$$

where $\boldsymbol{r}(\boldsymbol{\alpha}, \boldsymbol{q})$ is the residual vector. Following the steps of Section 3.1 (see also Problem 3.1) we obtain

$$\boldsymbol{r}'(\boldsymbol{\alpha}, \boldsymbol{q}(\boldsymbol{\alpha}); \boldsymbol{\beta}) + \boldsymbol{J}(\boldsymbol{\alpha}, \boldsymbol{q}(\boldsymbol{\alpha}))\boldsymbol{q}'(\boldsymbol{\alpha}; \boldsymbol{\beta}) = \boldsymbol{0}, \quad \boldsymbol{\beta} \in \mathbb{R}^d, \tag{5.18}$$

where $\boldsymbol{r}'(\boldsymbol{\alpha}, \boldsymbol{q}; \boldsymbol{\beta})$ denotes the (partial) directional derivative of $\boldsymbol{r}$ at $(\boldsymbol{\alpha}, \boldsymbol{q})$ with respect to $\boldsymbol{\alpha}$ in the direction $\boldsymbol{\beta}$, $\boldsymbol{J}(\boldsymbol{\alpha}, \boldsymbol{q})$ is the (partial) Jacobian of $\boldsymbol{r}$ with respect to $\boldsymbol{q}$ at $(\boldsymbol{\alpha}, \boldsymbol{q})$, and $\boldsymbol{q}'(\boldsymbol{\alpha}; \boldsymbol{\beta})$ is the directional derivative of $\boldsymbol{q}$. Introducing the adjoint state system

$$\left(\boldsymbol{J}(\boldsymbol{\alpha}, \boldsymbol{q}(\boldsymbol{\alpha}))\right)^{\mathrm{T}} \boldsymbol{p}(\boldsymbol{\alpha}) = \nabla_{\boldsymbol{q}} J(\boldsymbol{\alpha}, \boldsymbol{q}(\boldsymbol{\alpha})) \tag{5.19}$$

one can express the directional derivative of $\mathcal{J}$ as follows:

$$\mathcal{J}'(\boldsymbol{\alpha}; \boldsymbol{\beta}) = (\nabla_{\boldsymbol{\alpha}} J(\boldsymbol{\alpha}, \boldsymbol{q}(\boldsymbol{\alpha})))^{\mathrm{T}}\boldsymbol{\beta} - \boldsymbol{p}(\boldsymbol{\alpha})^{\mathrm{T}}\boldsymbol{r}'(\boldsymbol{\alpha}, \boldsymbol{q}(\boldsymbol{\alpha}); \boldsymbol{\beta}), \tag{5.20}$$

making use of (5.18) and (5.19).

Assume now that the system (5.15) arises from a finite element discretization of a quasi-linear partial differential equation. In addition to the previous application of the chain rule of differential calculus we exploit the *sparsity of the partial gradient and Jacobian* with respect to $\boldsymbol{q}$. Let every finite element have p degrees of freedom. The residual vector $\boldsymbol{r}(\boldsymbol{\alpha}, \boldsymbol{q}) \in \mathbb{R}^n$ is obtained by using the standard assembly process:

$$\boldsymbol{r}(\boldsymbol{\alpha}, \boldsymbol{q}) = \sum_e (\boldsymbol{P}^e)^{\mathrm{T}}\boldsymbol{r}^e(\boldsymbol{\alpha}, \boldsymbol{q}^e), \tag{5.21}$$

where

$$\boldsymbol{r}^e(\boldsymbol{\alpha}, \boldsymbol{q}^e) = \boldsymbol{K}^e(\boldsymbol{\alpha}, \boldsymbol{q}^e)\boldsymbol{q}^e - \boldsymbol{f}^e(\boldsymbol{\alpha}, \boldsymbol{q}^e),$$

$\boldsymbol{q}^e \in \mathbb{R}^p$ is the vector of element degrees of freedom, and $\boldsymbol{P}^e$ are Boolean matrices (see (3.70)). Finally, let the objective function J be separable in the following sense:

$$J(\boldsymbol{\alpha}, \boldsymbol{q}) = \sum_e J_e(\boldsymbol{\alpha}, \boldsymbol{q}^e). \tag{5.22}$$

This assumption is true for most cost functions of practical interest.

In many practical shape optimization problems the order of magnitude of d and p is 10. On the other hand the number of global degrees of freedom n even in simple two-dimensional state problems may be over 10,000.

To obtain the partial derivatives needed in (5.19) and (5.20) we use AD. Taking into account (5.21) and (5.22) it is obvious that one needs to apply AD only to the *local* contributions $\boldsymbol{r}^e$, J_e in (5.21), (5.22), respectively, to get the partial derivatives with respect to $\boldsymbol{\alpha}$ and $\boldsymbol{q}$. Thus, one has to differentiate only scalars J_e and small vectors $\boldsymbol{r}^e$ with respect to a small number of independent variables $\alpha_1, \ldots, \alpha_d, q_1^e, \ldots, q_p^e$. The global terms in (5.19), (5.20) are then obtained using the standard assembly process, which involves no differentiation.

In shape optimization, differentiation with respect to $\boldsymbol{\alpha}$ requires differentiation of the mesh produced by the mesh generator. This is not a time-consuming task if the topology of the mesh is shape independent and the position of each node is given by a simple algebraic formula. This was the case presented in Section 2.2. More complicated mesh generation methods, such as advancing front and Voronoi methods (see [Geo91]), are iterative and the dependence of the nodal positions on design variables is highly implicit. In this case the differentiation of the mesh generation process can be the most time-consuming step.

The proposed hybrid method for shape design sensitivity analysis is both easy to program and efficient in terms of computing time and memory. It is efficient because the (non)linear state solver is not differentiated and thus the standard floating-point arithmetic can be used in solving (5.15), (5.19). This makes it possible to use standard software (LAPACK [ABB+99], for example) to solve the linear(ized) state problem. Computed sensitivities are very accurate provided that the *mesh topology remains fixed* and the (non)linear state equation is solved with *sufficiently high accuracy.* Our approach is general because it can also be applied in multidisciplinary shape optimization problems and it is not restricted to some specific type (such as linear triangular) of finite element.

Problems

PROBLEM 5.1. Consider the function $f : \mathbb{R}^3 \to \mathbb{R}$,

$$f(x) = \sqrt{1 + x_1^2 + x_2^2} - x_2 x_3.$$

(i) Write a subroutine for evaluating f using the elementary functions φ_i of (5.2) and draw the associated directed graph.

(ii) Count the number of arithmetic operations needed to evaluate ∇f with the algorithms that use the forward and reverse modes shown on the right of Figures 5.2 and 5.3.

PROBLEM 5.2. Complete the simple implementation of the forward mode discussed in Section 5.2 to handle the basic arithmetic operations $\pm, \cdot, /$ and the following library routines: sin, cos, exp, sqrt. If it is more convenient, C++ may be used instead of Fortran 90.

Chapter 6

Fictitious Domain Methods in Shape Optimization

Sizing and shape optimization problems are typical bilevel problems. The upper level consists of minimizing a cost functional by using appropriate mathematical programming methods. Some of these have been presented in Chapter 4. The lower level provides solutions of discretized state problems needed to evaluate cost and constraint functions and their derivatives at the upper level. Typically, discrete state problems are given by large-scale systems of algebraic equations arising from finite element approximations of state relations. Usually the lower level is run many times. Thus it is not surprising that the efficiency of solving discrete state problems is one of the decisive factors of the whole computational process. This chapter deals with a type of method for the numerical realization of linear elliptic state equations that is based on the so-called fictitious domain formulation. A common feature of all these methods is that all computations are carried out in an auxiliary simply shaped domain $\widehat{\Omega}$ (called *fictitious*) in which the original domain Ω representing the shape of a structure is embedded. There are different ways to link the solution of the original problem to the solution of the problem solved in the fictitious domain. Here we present a class of methods based on the use of boundary Lagrange (BL) and distributed Lagrange (DL) multipliers. Just the fact that the new problem is solved in a domain with a simple shape (e.g., a rectangle) enables us to use *uniform* or *"almost" uniform meshes* for constructing finite element spaces yielding a special structure of the resulting stiffness matrix. Systems with such matrices can be solved by special fast algorithms and special preconditioning techniques. In this chapter we confine ourselves to fictitious domain solvers that use the so-called nonfitted meshes, i.e., meshes not respecting the geometry of Ω. Besides the advantages we have already mentioned there is yet another one, which makes this type of state solver interesting: programming shape optimization problems is easier and "user friendly." To see this let us recall the classical approach based on the so-called boundary variation technique widely used in practice. Suppose that a gradient type minimization method at the upper level and a classical finite element approach at the lower level are used. The program realizes a minimizing sequence $\{\Omega^{(k)}\}$ of domains, i.e., a sequence decreasing the value of a cost functional. Each new term $\Omega^{(k+1)}$ of this sequence is obtained from $\Omega^{(k)}$ by an appropriate change in $\partial\Omega^{(k)}$. If the classical finite element method is used and the state problem is linear we have to

(i) *remesh the new domain,*

(ii) *assemble the new stiffness matrix and the right-hand side of the system,*

(iii) *solve the new system of linear equations.*

As we have already mentioned, all these steps are repeated many times. Fictitious domain methods with nonfitted meshes completely avoid step (i) and partially avoid step (ii) since the stiffness matrix remains the same for any admissible shape.

6.1 Fictitious domain formulations based on boundary and distributed Lagrange multipliers

Let us consider a homogeneous Dirichlet boundary value problem in a bounded domain $\Omega \subset \mathbb{R}^n$ with the Lipschitz boundary $\partial\Omega$:

$$\begin{cases} -\Delta u = f & \text{in } \Omega, \quad f \in L^2(\Omega), \\ \quad u = 0 & \text{on } \partial\Omega \end{cases} \tag{$\mathcal{P}'$}$$

or, in a weak form:

$$\begin{cases} \text{Find } u \in H_0^1(\Omega) \text{ such that} \\ \displaystyle\int_\Omega \nabla u \cdot \nabla v \, dx = \int_\Omega f v \, dx \quad \forall v \in H_0^1(\Omega). \end{cases} \tag{$\mathcal{P}$}$$

Let $\widehat{\Omega} \subset \mathbb{R}^n$ be another domain with a simple shape, containing Ω in its interior; see Figure 6.1.

On $\widehat{\Omega}$ we shall formulate the following boundary value problem:

$$\begin{cases} \text{Find } \widehat{u} \in H_0^1(\widehat{\Omega}, \partial\Omega) \text{ such that} \\ \displaystyle\int_{\widehat{\Omega}} \nabla \widehat{u} \cdot \nabla v \, dx = \int_{\widehat{\Omega}} \widetilde{f} v \, dx \quad \forall v \in H_0^1(\widehat{\Omega}, \partial\Omega), \end{cases} \tag{$\widehat{\mathcal{P}}$}$$

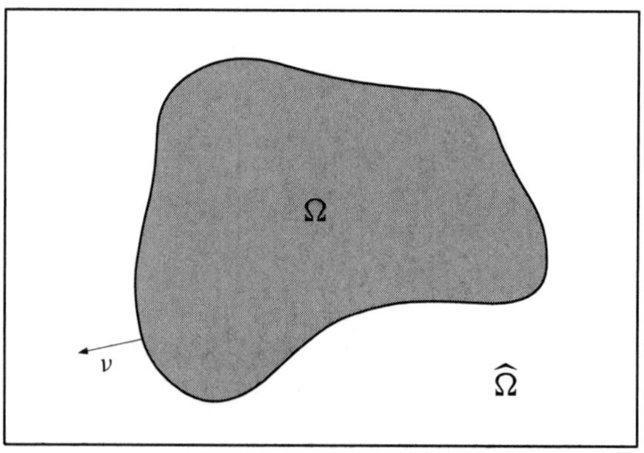

Figure 6.1. *Fictitious domain method.*

where $H_0^1(\widehat{\Omega}, \partial\Omega)$ is the subspace of $H_0^1(\widehat{\Omega})$ containing all functions vanishing on $\partial\Omega$:

$$H_0^1(\widehat{\Omega}, \partial\Omega) = \{v \in H_0^1(\widehat{\Omega}) \mid v = 0 \text{ on } \partial\Omega\}, \tag{6.1}$$

and $\widetilde{f} \in L^2(\widehat{\Omega})$ is an extension of f from Ω to $\widehat{\Omega}$.

It is left as an easy exercise to show that $(\widehat{\mathcal{P}})$ has a unique solution $\widehat{u}$ (see Problem 6.1). In addition, $u := \widehat{u}_{|\Omega}$ solves the original problem $(\mathcal{P})$. Next we reformulate problem $(\widehat{\mathcal{P}})$ by using a *mixed variational formulation.*

Before doing that we introduce more notation and recall well-known facts that will be used later on. Denote by $H^{\frac{1}{2}}(\partial\Omega)$ the space of all traces on $\partial\Omega$ of functions belonging to $H^1(\Omega)$:

$$H^{\frac{1}{2}}(\partial\Omega) = \{\varphi \in L^2(\partial\Omega) \mid \exists v \in H^1(\Omega) : \quad v = \varphi \text{ on } \partial\Omega\}.$$

It can be shown (see [GR79]) that $H^{\frac{1}{2}}(\partial\Omega)$ endowed with the norm

$$\|\varphi\|_{\frac{1}{2},\partial\Omega} := \inf_{\substack{v\in H^1(\Omega)\\ v=\varphi \text{ on } \partial\Omega}} \|v\|_{1,\Omega} \tag{6.2}$$

is a Banach space. It is easy to show that the infimum in (6.2) is realized by a function $u(\varphi) \in H^1(\Omega)$ that is the unique solution of

$$\begin{cases} -\Delta u(\varphi) + u(\varphi) = 0 & \text{in } \Omega, \\ u(\varphi) = \varphi & \text{on } \partial\Omega; \end{cases} \tag{6.3}$$

i.e.,

$$\|\varphi\|_{\frac{1}{2},\partial\Omega} = \|u(\varphi)\|_{1,\Omega}. \tag{6.4}$$

Further, let $H^{-\frac{1}{2}}(\partial\Omega)$ be the dual of $H^{\frac{1}{2}}(\partial\Omega)$ equipped with the standard dual norm

$$\|\mu\|_{-\frac{1}{2},\partial\Omega} := \sup_{\substack{\varphi\in H^{\frac{1}{2}}(\partial\Omega)\\ \varphi\neq 0}} \frac{\langle \mu, \varphi\rangle}{\|\varphi\|_{\frac{1}{2},\partial\Omega}},$$

where $\langle \cdot, \cdot \rangle$ is the duality pairing between $H^{-\frac{1}{2}}(\partial\Omega)$ and $H^{\frac{1}{2}}(\partial\Omega)$. One can show again that the previous supremum is attained at $u(\mu) \in H^1(\Omega)$ solving the nonhomogeneous Neumann problem

$$\begin{cases} -\Delta u(\mu) + u(\mu) = 0 & \text{in } \Omega, \\ \dfrac{\partial u(\mu)}{\partial \nu} = \mu & \text{on } \partial\Omega; \end{cases} \tag{6.5}$$

i.e.,

$$\|\mu\|_{-\frac{1}{2},\partial\Omega} = \|u(\mu)\|_{1,\Omega} \tag{6.6}$$

(see [GR79]).

The fact that a function v belongs to $H_0^1(\widehat{\Omega}, \partial\Omega)$ can be equivalently expressed as follows:

$$v \in H_0^1(\widehat{\Omega}, \partial\Omega) \iff v \in H_0^1(\widehat{\Omega}) \text{ and } \langle \mu, v \rangle = 0 \quad \forall \mu \in H^{-\frac{1}{2}}(\partial\Omega). \tag{6.7}$$

For simplicity of notation we use the same symbol for functions and their traces.

Instead of $(\widehat{\mathcal{P}})$ we shall consider the following problem also formulated in $\widehat{\Omega}$:

$$\begin{cases} \text{Find } (w, \lambda) \in H_0^1(\widehat{\Omega}) \times H^{-\frac{1}{2}}(\partial\Omega) \text{ such that} \\ \displaystyle\int_{\widehat{\Omega}} \nabla w \cdot \nabla v \, dx = \int_{\widehat{\Omega}} \widetilde{f} v \, dx + \langle \lambda, v \rangle \quad \forall v \in H_0^1(\widehat{\Omega}), \\ \langle \mu, w \rangle = 0 \quad \forall \mu \in H^{-\frac{1}{2}}(\partial\Omega). \end{cases} \tag{$\widehat{\mathcal{M}}$}$$

The relation between $(\widehat{\mathcal{P}})$ and $(\widehat{\mathcal{M}})$ is established in the following lemma.

Lemma 6.1. *Problem* $(\widehat{\mathcal{M}})$ *has a unique solution* (w, λ). *In addition,* $(w, \lambda) = (\widehat{u}, [\partial\widehat{u}/\partial\nu])$, *where* $\widehat{u}$ *solves* $(\widehat{\mathcal{P}})$; $[\partial\widehat{u}/\partial\nu]$ *is the jump of the normal derivative of* $\widehat{u}$ *across* $\partial\Omega$; *and* ν *is oriented as shown in Figure* 6.1.

Proof. Suppose that the solution $\widehat{u}$ of $(\widehat{\mathcal{P}})$ is sufficiently smooth (for a general case see Remark 6.1). From Green's theorem it follows that

$$\begin{aligned} \int_{\widehat{\Omega}} \nabla\widehat{u} \cdot \nabla v \, dx &= -\int_{\widehat{\Omega}} \Delta\widehat{u} v \, dx + \int_{\partial\Omega} \left[\frac{\partial\widehat{u}}{\partial\nu}\right] v \, ds \\ &= \int_{\widehat{\Omega}} \widetilde{f} v \, dx + \int_{\partial\Omega} \left[\frac{\partial\widehat{u}}{\partial\nu}\right] v \, ds \quad \forall v \in H_0^1(\widehat{\Omega}), \end{aligned}$$

making use of the fact that $-\Delta\widehat{u} = \widetilde{f}$ in $\widehat{\Omega}$. From this we see that $(\widehat{\mathcal{M}})_1$ is satisfied with $w = \widehat{u}$ and $\lambda = [\partial\widehat{u}/\partial\nu]$. Since $\widehat{u} = 0$ on $\partial\Omega$, $(\widehat{\mathcal{M}})_2$ is satisfied as well. Therefore $(\widehat{u}, [\partial\widehat{u}/\partial\nu])$ solves $(\widehat{\mathcal{M}})$. On the contrary if (w, λ) is a solution to $(\widehat{\mathcal{M}})$, then using a similar approach one can show that $(w, \lambda) = (\widehat{u}, [\partial\widehat{u}/\partial\nu])$ with $\widehat{u}$ the solution to $(\widehat{\mathcal{P}})$. The uniqueness of (w, λ) follows from the uniqueness of $\widehat{u}$. □

Remark 6.1. The proof of Lemma 6.1 can be done without any smoothness assumption on $\widehat{u}$. We only need that $\widetilde{f} \in L^2(\widehat{\Omega})$. Then the jump $[\partial\widehat{u}/\partial\nu]$ has to be interpreted as an element of $H^{-\frac{1}{2}}(\partial\Omega)$ and the integrals over $\partial\Omega$ have to be replaced with the corresponding duality pairing $\langle\cdot, \cdot\rangle$.

Convention: *In view of Lemma* 6.1, *the solution to* $(\widehat{\mathcal{M}})$ *will be denoted by* $(\widehat{u}, \lambda)$ *in what follows.*

Remark 6.2. The couple $(\widehat{u}, \lambda)$ being the solution to $(\widehat{\mathcal{M}})$ can be equivalently characterized as a unique saddle point of the Lagrange function $\mathcal{L} : H_0^1(\widehat{\Omega}) \times H^{-\frac{1}{2}}(\partial\Omega) \to \mathbb{R}$:

$$\begin{cases} (\widehat{u}, \lambda) \in H_0^1(\widehat{\Omega}) \times H^{-\frac{1}{2}}(\partial\Omega) \quad \text{such that} \\ \mathcal{L}(\widehat{u}, \mu) \le \mathcal{L}(\widehat{u}, \lambda) \le \mathcal{L}(v, \lambda) \quad \forall (v, \mu) \in H_0^1(\widehat{\Omega}) \times H^{-\frac{1}{2}}(\partial\Omega), \end{cases}$$

where

$$\mathcal{L}(v, \mu) = \frac{1}{2}\int_{\widehat{\Omega}} |\nabla v|^2\, dx - \int_{\widehat{\Omega}} \widetilde{f} v\, dx - \langle \mu, v \rangle.$$

The second component λ of the saddle point represents the Lagrange multiplier associated with the constraint $v = 0$ on $\partial\Omega$ satisfied by $v \in H_0^1(\widehat{\Omega}, \partial\Omega)$.

Remark 6.3. (*Very important.*) The function $\widetilde{f}$ on the right-hand side of $(\widehat{\mathcal{P}})$ and $(\widehat{\mathcal{M}})$ is an arbitrary $L^2(\widehat{\Omega})$-extension of f. It turns out that the zero extension of f to $\Xi := \widehat{\Omega} \setminus \overline{\Omega}$, i.e.,

$$\widetilde{f} = \begin{cases} f & \text{in } \Omega, \\ 0 & \text{in } \Xi, \end{cases}$$

is of particular importance. Indeed, for such a special extension it holds that $\widehat{u} = 0$ in Ξ, implying that $\lambda = [\partial\widehat{u}/\partial\nu] = \partial\widehat{u}/\partial\nu$ on $\partial\Omega$. Since $u := \widehat{u}_{|\Omega}$ solves $(\mathcal{P})$, the Lagrange multiplier λ is equal to the normal derivative of u on $\partial\Omega$.

Problem $(\widehat{\mathcal{M}})$ is a fictitious domain formulation of $(\mathcal{P})$ based on the BL multiplier technique. As we have already mentioned, $(\widehat{\mathcal{M}})$ is a particular example of a mixed variational formulation. Since such formulations are frequently used for the numerical realization of partial differential equations and will be employed for introducing another type of fictitious domain method, let us recall their abstract setting and main existence, uniqueness, and convergence results (for a detailed analysis we refer to [BF91]).

Let V, Q be two real Hilbert spaces equipped with norms denoted by $\|\cdot\|_V$, $\|\cdot\|_Q$, respectively; V', Q' be the corresponding dual spaces; and $\langle\cdot,\cdot\rangle_{V'\times V}$, $[\cdot,\cdot]_{Q'\times Q}$ be the duality pairings on $V' \times V$, $Q' \times Q$, respectively. Let $a : V \times V \to \mathbb{R}$, $b : V \times Q \to \mathbb{R}$ be two bounded bilinear forms:

$$\exists M = \text{const.} > 0 : \quad |a(u, v)| \le M\|u\|_V\|v\|_V \quad \forall u, v \in V; \tag{6.8}$$

$$\exists m = \text{const.} > 0 : \quad |b(v, q)| \le m\|v\|_V\|q\|_Q \quad \forall (v, q) \in V \times Q. \tag{6.9}$$

Finally, let $f \in V'$, $g \in Q'$ be given. By a *mixed variational formulation* determined by $\{V, Q, a, b, f, g\}$ we mean the following problem:

$$\begin{cases} \text{Find } (u, \lambda) \in V \times Q \quad \text{such that} \\ a(u, v) + b(v, \lambda) = \langle f, v\rangle_{V'\times V} \quad \forall v \in V, \\ b(u, q) = [g, q]_{Q'\times Q} \quad \forall q \in Q. \end{cases} \tag{$\widehat{\mathcal{P}}_m$}$$

To guarantee the existence and uniqueness of (u, λ) solving $(\widehat{\mathcal{P}}_m)$ for any $(f, g) \in V' \times Q'$ the following two assumptions are needed (in fact, (6.10) can be weakened):

$$\exists \alpha = \text{const.} > 0 : \quad a(v, v) \ge \alpha\|v\|^2 \quad \forall v \in V; \tag{6.10}$$

$$\exists k = \text{const.} > 0 : \quad \sup_{\substack{v\in V\\ v\neq 0}} \frac{b(v, q)}{\|v\|_V} \ge k\|q\|_Q \quad \forall q \in Q. \tag{6.11}$$

Then we have the following result.

Theorem 6.1. *Let* (6.8)–(6.11) *be satisfied. Then* $(\widehat{\mathcal{P}}_m)$ *has a unique solution* (u, λ) *for any* $(f, g) \in V' \times Q'$.

Proof. See [BF91]. □

Our problem $(\widehat{\mathcal{M}})$ is a special case of $(\widehat{\mathcal{P}}_m)$ with the following choice of data:

$$V = H_0^1(\widehat{\Omega}), \quad Q = H^{-\frac{1}{2}}(\partial\Omega), \quad a(u, v) = \int_{\widehat{\Omega}} \nabla u \cdot \nabla v \, dx,$$

$$b(v, q) = -\langle q, v\rangle, \quad u, v \in H_0^1(\widehat{\Omega}), \quad q \in H^{-\frac{1}{2}}(\partial\Omega),$$

$$\langle f, v\rangle_{V' \times V} = \int_{\widehat{\Omega}} \widetilde{f} v \, dx, \quad g = 0$$

(recall that $\langle \cdot, \cdot \rangle$ stands for the duality pairing on $H^{-\frac{1}{2}}(\partial\Omega) \times H^{\frac{1}{2}}(\partial\Omega)$). The existence and uniqueness of a solution to $(\widehat{\mathcal{M}})$ can be obtained directly from Theorem 6.1 (see Problem 6.3).

For a discretization of $(\widehat{\mathcal{P}}_m)$ we use a Galerkin type method. Let $\{V_h\}$ and $\{Q_h\}$, $h \to 0+$, be two systems of finite dimensional subspaces of V and Q, respectively. The discretization of $(\widehat{\mathcal{P}}_m)$ on $V_h \times Q_h$ reads as follows:

$$\begin{cases} \text{Find } (u_h, \lambda_h) \in V_h \times Q_h \quad \text{such that} \\ a(u_h, v_h) + b(v_h, \lambda_h) = \langle f, v_h\rangle_{V' \times V} \quad \forall v_h \in V_h, \\ b(u_h, q_h) = [g, q_h]_{Q' \times Q} \quad \forall q_h \in Q_h. \end{cases} \qquad (\widehat{\mathcal{P}}_m^h)$$

To ensure the existence and uniqueness of (u_h, λ_h) solving $(\widehat{\mathcal{P}}_m^h)$ the following assumption will be needed:

$$b(v_h, q_h) = 0 \quad \forall v_h \in V_h \implies q_h = 0. \qquad (6.12)$$

From (6.12) we see that there is a constant $k_h > 0$, *generally depending* on the discretization parameter h, such that

$$\sup_{\substack{v_h \in V_h \\ v_h \neq 0}} \frac{b(v_h, q_h)}{\|v_h\|_V} \geq k_h \|q_h\|_Q \quad \forall q_h \in Q_h. \qquad (6.13)$$

To get convergence of approximate solutions one needs a stronger assumption, namely the so-called Ladyzhenskaya–Babuska–Brezzi (LBB) condition: there exists a positive constant $\overline{k}$ such that

$$\sup_{\substack{v_h \in V_h \\ v_h \neq 0}} \frac{b(v_h, q_h)}{\|v_h\|_V} \geq \overline{k} \|q_h\|_Q \quad \forall q_h \in Q_h \; \forall h \to 0+; \qquad (6.14)$$

i.e., the constant k_h in (6.13) can be bounded from below by a positive constant $\overline{k}$ independent of h.

The convergence result follows from Theorem 6.2.

Theorem 6.2. *Let* (6.8)–(6.11) *and* (6.14) *be satisfied. In addition, suppose that systems* $\{V_h\}$, $\{Q_h\}$, $h \to 0+$, *are dense in* V, Q, *respectively. Then the sequence* $\{(u_h, \lambda_h)\}$ *of approximate solutions to* $(\widehat{\mathcal{P}}_m^h)$, $h \to 0+$, *tends to the unique solution* (u, λ) *of* $(\widehat{\mathcal{P}}_m)$*:*

$$
\begin{aligned}
u_h &\to u \quad \text{in } V,\\
\lambda_h &\to \lambda \quad \text{in } Q, \quad h \to 0+.
\end{aligned}
$$

We now give an algebraic form of $(\widehat{\mathcal{P}}_m^h)$ for a fixed value of the discretization parameter h. We first choose bases of V_h and Q_h: $V_h = \{\varphi_1, \dots, \varphi_n\}$, $Q_h = \{\psi_1, \dots, \psi_m\}$, $n = \dim V_h$, $m = \dim Q_h$. Then $(\widehat{\mathcal{P}}_m^h)$ leads to the following system of linear algebraic equations:

$$
\begin{aligned}
\boldsymbol{A}\boldsymbol{u} + \boldsymbol{B}^{\mathrm{T}}\boldsymbol{\lambda} &= \boldsymbol{f},\\
\boldsymbol{B}\boldsymbol{u} &= \boldsymbol{g},
\end{aligned}
\tag{6.15}
$$

where $\boldsymbol{u}, \boldsymbol{\lambda}$ are the coordinates of u_h, λ_h with respect to $\{\varphi_i\}_{i=1}^n$, $\{\psi_i\}_{i=1}^m$, respectively. Furthermore $\boldsymbol{A} \in \mathbb{R}^{n\times n}$, $\boldsymbol{B} \in \mathbb{R}^{m\times n}$ are matrices whose elements are given by

$$
a_{ij} = a(\varphi_j, \varphi_i), \quad b_{kl} = b(\varphi_l, \psi_k), \quad i, j, l = 1, \dots, n; \; k = 1, \dots, m,
$$

respectively, and $\boldsymbol{f} \in \mathbb{R}^n$, $\boldsymbol{g} \in \mathbb{R}^m$ are vectors with the coordinates

$$
f_i = \langle f, \varphi_i \rangle_{V'\times V}, \; i = 1, \dots, n, \quad g_j = [g, \psi_j]_{Q'\times Q}, \quad j = 1, \dots, m,
$$

respectively.

After this short excursion to mixed variational formulations and their approximations let us return to the homogeneous Dirichlet boundary value problem formulated at the beginning of this section. We now present another fictitious domain formulation of this problem using *DL multipliers*.

Denote by $V(\Xi)$ the space of restrictions to $\Xi := \widehat{\Omega} \setminus \overline{\Omega}$ of all functions belonging to $H_0^1(\widehat{\Omega})$:

$$
V(\Xi) = \{w \in H^1(\Xi) \mid \exists\, v \in H_0^1(\widehat{\Omega}) : \; w = v_{|\Xi}\}.
$$

Further, let $V'(\Xi)$ be the dual of $V(\Xi)$ with the duality pairing denoted by $\langle\!\langle \cdot, \cdot \rangle\!\rangle$. Next we shall consider the following mixed type variational formulation:

$$
\left\{
\begin{aligned}
&\text{Find } (\widehat{u}, \lambda) \in H_0^1(\widehat{\Omega}) \times V'(\Xi) \quad \text{such that}\\
&\int_{\widehat{\Omega}} \nabla \widehat{u} \cdot \nabla v \, dx = \int_{\widehat{\Omega}} \widetilde{f} v \, dx + \langle\!\langle \lambda, v \rangle\!\rangle \quad \forall v \in H_0^1(\widehat{\Omega}),\\
&\langle\!\langle \mu, \widehat{u} \rangle\!\rangle = 0 \quad \forall \mu \in V'(\Xi).
\end{aligned}
\right.
\tag{$\widehat{\mathcal{N}}$}
$$

Again, $\widetilde{f}$ denotes an $L^2(\widehat{\Omega})$-extension of f to $\widehat{\Omega}$. Before proving the existence and uniqueness of $(\widehat{u}, \lambda)$ solving $(\widehat{\mathcal{N}})$ let us show that $u := \widehat{u}_{|\Omega}$ solves $(\mathcal{P})$. Indeed, from $(\widehat{\mathcal{N}})_2$ it follows that $\widehat{u} = 0$ in Ξ and, consequently, $u \in H_0^1(\Omega)$. Restricting to functions $v \in H_0^1(\widehat{\Omega})$ with $\operatorname{supp} v \subset \overline{\Omega}$ in $(\widehat{\mathcal{N}})_1$ we have

$$\int_\Omega \nabla u \cdot \nabla v \, dx = \int_\Omega f v \, dx;$$

i.e., $-\Delta u = f$ in Ω, implying that u solves $(\mathcal{P})$.

To prove the existence and uniqueness of a solution to $(\widehat{\mathcal{N}})$ we use Theorem 6.1.

Lemma 6.2. *Problem $(\widehat{\mathcal{N}})$ has a unique solution $(\widehat{u}, \lambda)$.*

Proof. Setting $V = H_0^1(\widehat{\Omega})$, $Q = V'(\Xi)$, $b(v, g) = -\langle\langle g, v\rangle\rangle$ in abstract formulation $(\widehat{\mathcal{P}}_m)$ we see that (6.8)–(6.10) are satisfied. It remains to verify (6.11). Let $w \in V(\Xi)$ be arbitrary. Then the trace of w on $\partial\Omega$ can be continuously extended from $\partial\Omega$ into Ω; i.e., one can construct a function $\widetilde{w} \in H_0^1(\widehat{\Omega})$ such that $\widetilde{w}_{|\Xi} = w$ in Ξ and

$$\exists \beta = \text{const.} > 0 : \quad \|\widetilde{w}\|_{1,\widehat{\Omega}} \le \beta \|w\|_{1,\Xi}, \tag{6.16}$$

where β does not depend on w. Denote by $\widetilde{V}(\Xi)$ the subset of $H_0^1(\widehat{\Omega})$ whose elements are the continuous extensions of all $v \in V(\Xi)$ from Ξ to Ω constructed above. Then from (6.16) it follows that

$$\sup_{\substack{v \in H_0^1(\widehat{\Omega}) \\ v \neq 0}} \frac{\langle\langle \mu, v\rangle\rangle}{\|v\|_{1,\widehat{\Omega}}} \ge \sup_{\substack{\widetilde{w} \in \widetilde{V}(\Xi) \\ \widetilde{w} \neq 0}} \frac{\langle\langle \mu, \widetilde{w}\rangle\rangle}{\|\widetilde{w}\|_{1,\widehat{\Omega}}} \ge \frac{1}{\beta} \sup_{\substack{w \in V(\Xi) \\ w \neq 0}} \frac{\langle\langle \mu, w\rangle\rangle}{\|w\|_{1,\Xi}} := \frac{1}{\beta}\|\mu\|_*,$$

where $\|\cdot\|_*$ stands for the dual norm of $\mu \in V'(\Xi)$. □

Remark 6.4. Denote

$$H_0^1(\widehat{\Omega}, \Xi) = \{v \in H_0^1(\widehat{\Omega}) \mid v = 0 \text{ in } \Xi\}.$$

From Lemma 6.2 it follows that the first component $\widehat{u}$ of the solution to $(\widehat{\mathcal{N}})$ belongs to $H_0^1(\widehat{\Omega}, \Xi)$. The second component λ is the Lagrange multiplier releasing the constraint $v = 0$ in Ξ satisfied by elements of $H_0^1(\widehat{\Omega}, \Xi)$.

Let us now pass to discretizations of both fictitious domain formulations of the homogeneous Dirichlet boundary value problem. For the sake of simplicity of our presentation we shall consider Ω to be a *plane, polygonal* domain. We start with a discretization of the BL multiplier approach.

Let $\widehat{\Omega}$ be a rectangular domain containing Ω in its interior and $\{\widehat{\mathcal{T}}_h\}$, $h \to 0+$, be a family of *uniform triangulations* of $\widehat{\Omega}$ constructed as follows: we first subdivide $\widehat{\Omega}$ by a uniform square mesh of size h and then divide each square into two triangles along the same diagonal. With any such $\widehat{\mathcal{T}}_h$ the following space of continuous piecewise linear functions will be associated:

$$V_h = \{v_h \in C(\overline{\widehat{\Omega}}) \mid v_{h|T} \in P_1(T)\ \forall T \in \widehat{\mathcal{T}}_h,\ v_h = 0 \text{ on } \partial\widehat{\Omega}\}. \tag{6.17}$$

Next we discretize the space $H^{-\frac{1}{2}}(\partial\Omega)$ of the Lagrange multipliers on $\partial\Omega$. Let $\{\mathcal{T}_H\}$, $H \to 0+$, be another regular system of partitions of $\partial\Omega$ into a finite number $m := m(H)$ of straight segments S of length not exceeding H. On any $\mathcal{T}_H$ we construct the space Λ_H of *piecewise constant* functions:

$$\Lambda_H = \{\mu_H \in L^2(\partial\Omega) \mid \mu_{H|S} \in P_0(S)\ \forall S \in \mathcal{T}_H\}.$$

The discretization of $(\widehat{\mathcal{M}})$ is as follows:

$$\left\{\begin{array}{l}\text{Find } (\widehat{u}_h, \lambda_H) \in V_h \times \Lambda_H \text{ such that}\\ \displaystyle\int_{\widehat{\Omega}} \nabla\widehat{u}_h \cdot \nabla v_h\, dx = \int_{\widehat{\Omega}} \widetilde{f} v_h\, dx + \int_{\partial\Omega} \lambda_H v_h\, ds \quad \forall v_h \in V_h,\\ \displaystyle\int_{\partial\Omega} \widehat{u}_h \mu_H\, ds = 0 \quad \forall \mu_H \in \Lambda_H.\end{array}\right. \qquad (\widehat{\mathcal{M}}_h^H)$$

As we already know, the LBB condition (6.14) is the fundamental property in the convergence analysis. In [GG95] it has been shown that the LBB condition is satisfied provided the ratio H/h is *sufficiently large*, i.e., the partition $\mathcal{T}_H$ is *coarser* than the triangulation $\widehat{\mathcal{T}}_h$. More precisely Lemma 6.3 holds.

Lemma 6.3. *Let the ratio H/h satisfy*

$$3 \le H/h \le L,$$

where $L > 0$ is a fixed positive number. Then there exists a positive constant $\overline{k}$ independent of h, H such that

$$\sup_{\substack{v_h \in V_h\\ v_h \neq 0}} \frac{\int_{\partial\Omega} \mu_H v_h\, ds}{\|v_h\|_{1,\widehat{\Omega}}} \ge \overline{k}\|\mu_H\|_{-\frac{1}{2},\partial\Omega}.$$

With this result at hand and from abstract error estimates (see [BF91]) we arrive at the following result.

Theorem 6.3. *Let the mesh sizes h, H satisfy the assumptions of Lemma 6.3 and let $(\widehat{u}, \lambda)$ be a solution to $(\widehat{\mathcal{M}})$. Then there exists a unique solution $(\widehat{u}_h, \lambda_H)$ to $(\widehat{\mathcal{M}}_h^H)$ and*

$$\|\widehat{u} - \widehat{u}_h\|_{1,\widehat{\Omega}} + \|\lambda - \lambda_H\|_{-\frac{1}{2},\partial\Omega} = O(h^{\frac{1}{2}-\varepsilon}), \quad h \to 0+, \tag{6.18}$$

holds for any $\varepsilon > 0$ provided that the restrictions of $\widehat{u}$ to Ω, Ξ belong to $H^2(\Omega)$, $H^2(\Xi)$, respectively.

For the proof we refer to [GG95].

Remark 6.5. We see that the error estimate (6.18) is not optimal in view of the fact that the solution $\widehat{u}$ is not generally smooth in the whole domain $\widehat{\Omega}$. What one can expect is that $\widehat{u} \in H^{\frac{3}{2}-\varepsilon}(\widehat{\Omega})$ for any $\varepsilon > 0$, explaining the rate of convergence in (6.18). (For the definition of Sobolev spaces with noninteger exponents we refer to [Neč67], [LM68].) The normal derivative of $\widehat{u}$ may have a nonzero jump across $\partial\Omega$. From (6.18) we also see that *both* components of $(\widehat{u}, \lambda)$ are approximated. In particular if $\widetilde{f} = 0$ in Ξ, then λ_H approximates $\partial u/\partial \nu$ on $\partial\Omega$, where $u \in H_0^1(\Omega)$ solves the homogeneous Dirichlet boundary value problem $(\mathcal{P})$ (see Remark 6.3).

Let h and H be fixed. Then $(\widehat{\mathcal{M}}_h^H)$ leads to a linear system of algebraic equations of type (6.15) with $\boldsymbol{A} \in \mathbb{R}^{n\times n}$, $\boldsymbol{B} \in \mathbb{R}^{m\times n}$ given by

$$a_{ij} = \int_{\widehat{\Omega}} \nabla\varphi_i \cdot \nabla\varphi_j \, dx, \quad i, j = 1, \dots, n; \; n = \dim V_h, \tag{6.19}$$

$$b_{kl} = -\int_{S_k} \varphi_l \, ds, \quad S_k \in \mathcal{T}_H, \; k = 1, \dots, m, \; l = 1, \dots, n, \; m = \dim \Lambda_H, \tag{6.20}$$

where $\{\varphi_i\}_{i=1}^n$ are the Courant basis functions of V_h.

Remark 6.6. From the second equation in $(\widehat{\mathcal{M}}_h^H)$ it follows that the homogeneous Dirichlet boundary condition on $\partial\Omega$ is satisfied by $\widehat{u}_h$ in an integral average sense:

$$\int_S \widehat{u}_h \, ds = 0 \quad \forall S \in \mathcal{T}_H.$$

Next we briefly describe how to discretize the second formulation based on the DL multipliers. As before, we divide $\widehat{\Omega}$ into a square mesh of size h. Then each square will be divided by its diagonals into four isoceles right triangles (for reasons see Remark 6.7). The system of all such triangles creates a uniform "union jack" mesh of $\widehat{\Omega}$ denoted by $\widehat{\mathcal{T}}_h$. Let V_h be defined by (6.17) using the triangulation $\widehat{\mathcal{T}}_h$. The space $V'(\Xi)$ of the DL multipliers will be discretized by the finite dimensional space Λ_h realized by the restrictions of all $v_h \in V_h$ to Ξ; i.e.,

$$\Lambda_h = V_{h|\Xi}. \tag{6.21}$$

The discretization of $(\widehat{\mathcal{N}})$ now reads as follows:

$$\left\{\begin{array}{l} \text{Find } (\widehat{u}_h, \lambda_h) \in V_h \times \Lambda_h \text{ such that} \\ \displaystyle\int_{\widehat{\Omega}} \nabla\widehat{u}_h \cdot \nabla v_h \, dx = \int_{\widehat{\Omega}} \widetilde{f} v_h \, dx + \int_{\Xi} \lambda_h v_h \, dx \quad \forall v_h \in V_h, \\ \displaystyle\int_{\Xi} \widehat{u}_h \mu_h \, dx = 0 \quad \forall \mu_h \in \Lambda_h. \end{array}\right. \tag{$\widehat{\mathcal{N}}_h$}$$

It is readily seen that condition (6.12) is automatically satisfied in this case owing to the definition of Λ_h; i.e.,

$$\int_{\Xi} \mu_h v_h \, dx = 0 \quad \forall v_h \in V_h \implies \mu_h = 0 \text{ in } \Xi.$$

Therefore problem $(\widehat{\mathcal{N}}_h)$ has a unique solution $(\widehat{u}_h, \lambda_h)$. From the last equation in $(\widehat{\mathcal{N}}_h)$ we also see that $\widehat{u}_h = 0$ in Ξ. Since $\widehat{u}_h$ is piecewise linear on $\widehat{\mathcal{T}}_h$ it vanishes not only in Ξ but in

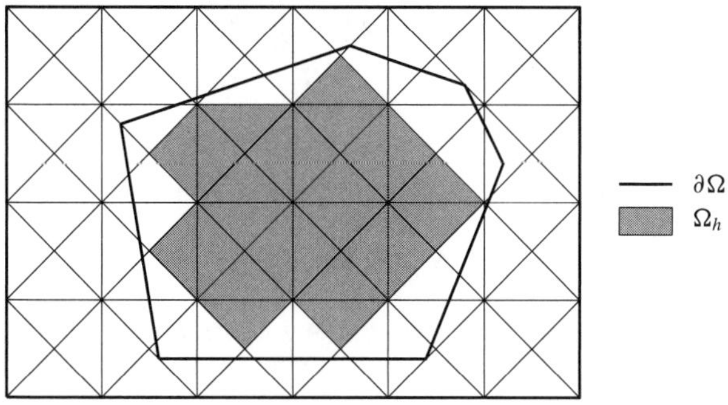

Figure 6.2. *Polygonal domain Ω and its inner approximation Ω_h.*

a larger set $\Xi_h \supset \Xi$ that is the union of all triangles from $\widehat{\mathcal{T}}_h$ having a nonempty intersection with the interior of Ξ (see Figure 6.2). This phenomenon is known as a *locking effect*.

Denote $\Omega_h := \widehat{\Omega} \setminus \overline{\Xi}_h$. Then the restriction $u_h := \widehat{u}_{h|\Omega_h}$ belongs to $H_0^1(\Omega_h)$, and from the first equation in $(\widehat{\mathcal{N}}_h)$ it follows that u_h solves the following discrete homogeneous Dirichlet boundary value problem in Ω_h:

$$u_h \in V_h(\Omega_h) : \quad \int_{\Omega_h} \nabla u_h \cdot \nabla v_h \, dx = \int_{\Omega_h} f v_h \, dx \quad v_h \in V_h(\Omega_h), \tag{6.22}$$

where

$$V_h(\Omega_h) = V_{h|\Omega_h} \cap H_0^1(\Omega_h). \tag{6.23}$$

Unlike with the BL multiplier approach, this time the homogeneous Dirichlet boundary condition is satisfied *exactly* on $\partial\Omega$.

Remark 6.7. The assumption that $\widehat{\mathcal{T}}_h$ is a uniform "union jack" mesh is needed in the error analysis. In [HMT01] it has been shown that for such $\widehat{\mathcal{T}}_h$ the following error estimate holds:

$$\|u - u_h\|_{1,\Omega_h} \le c h^{\frac{1}{2}-\varepsilon}, \quad h \to 0+,$$

provided that Ω is convex; $f \in L^p(\Omega)$ with $p > 2$, where $\varepsilon > 0$ is an arbitrary number; c is a positive constant that does not depend on h; and $u \in H_0^1(\Omega)$ is the solution to $(\mathcal{P})$.

Remark 6.8. The space Λ_h defined by (6.21) uses the same triangulation $\widehat{\mathcal{T}}_h$ as V_h. But this is not necessary. Suppose that $\widehat{\mathcal{T}}_H$ is another triangulation of $\overline{\widehat{\Omega}}$ such that $\widehat{\mathcal{T}}_h$ is its refinement; i.e., each $K \in \widehat{\mathcal{T}}_H$ is a union of a finite number of triangles from $\widehat{\mathcal{T}}_h$. Denote by V_H and Λ_H the following finite dimensional spaces:

$$V_H = \left\{ v_H \in C(\overline{\widehat{\Omega}}) \mid v_{H|T} \in P_1(T) \; \forall T \in \widehat{\mathcal{T}}_H, \; v_H = 0 \text{ on } \partial\widehat{\Omega} \right\},$$

$$\Lambda_H = V_{H|\Xi}.$$

Since $\Lambda_H \subset V_{h|\Xi}$ we see that condition (6.12) remains valid:

$$\mu_H \in \Lambda_H : \quad \int_\Xi \mu_H v_h \, dx = 0 \quad \forall v_h \in V_h \implies \mu_H = 0 \text{ in } \Xi.$$

Therefore the following problem:

$$\begin{cases} \text{Find } (\widehat{u}_h, \lambda_H) \in V_h \times \Lambda_H \text{ such that} \\ \displaystyle\int_{\widehat{\Omega}} \nabla \widehat{u}_h \cdot \nabla v_h \, dx = \int_{\widehat{\Omega}} \widetilde{f} v_h \, dx + \int_\Xi \lambda_H v_h \, dx \quad \forall v_h \in V_h, \\ \displaystyle\int_\Xi \widehat{u}_h \mu_H \, dx = 0 \quad \forall \mu_H \in \Lambda_H \end{cases} \qquad (\widehat{\mathcal{N}}_h^H)$$

still has a unique solution $(\widehat{u}_h, \lambda_H)$. In this way one can reduce the dimension of the space of Lagrange multipliers. On the other hand the homogeneous Dirichlet boundary condition is satisfied only in the following integral sense:

$$\int_\Xi \widehat{u}_h \mu_H \, dx = 0 \quad \forall \mu_H \in \Lambda_H.$$

From this it does *not* follow that $\widehat{u}_h = 0$ in Ξ and, consequently, by using a coarser partition $\widehat{\mathcal{T}}_H$ *one can avoid the locking effect.*

The algebraic form of $(\widehat{\mathcal{N}}_h)$ leads again to the linear system (6.15). This time the elements of the matrix $\boldsymbol{B}$ are given by

$$b_{kl} = -\int_\Xi \varphi_k \varphi_l \, dx, \quad l = 1, \dots, n; \quad n = \dim V_h, \; k \in \mathcal{I}, \tag{6.24}$$

where $\{\varphi_i\}_{i=1}^n$ are the Courant basis functions of V_h and $\mathcal{I}$ is the set containing indices of all φ_i whose support has a nonempty intersection with the interior of Ξ. Let us observe that the system $\{\varphi_i\}_{i\in\mathcal{I}}$ forms a basis of Λ_h. If two different partitions $\widehat{\mathcal{T}}_h$, $\widehat{\mathcal{T}}_H$ are used, appropriate modifications are necessary.

In the next section we use the previous fictitious domain techniques for the numerical realization of state problems in shape optimization. To this end let us check how data of the linear system (6.15) depend on the geometry of the original domain Ω. Since the previous discretizations use the so-called nonfitted meshes, meaning that the partitions $\widehat{\mathcal{T}}_h$ of $\widehat{\Omega}$ are constructed independently of Ω, the stiffness matrix $\boldsymbol{A}$ *does not depend* on Ω. The information on the geometry of Ω *appears solely* in matrix $\boldsymbol{B}$ (see (6.20), (6.24)) and eventually in the right-hand side vectors $\boldsymbol{f}$ and $\boldsymbol{g}$. The fact that $\boldsymbol{A}$ is independent of Ω is of practical significance: it can be computed once and stored forever. In addition, due to the particular construction of $\widehat{\mathcal{T}}_h$, the matrix $\boldsymbol{A}$ possesses a special structure enabling us to use fast direct solvers for large and sparse systems of linear algebraic equations.

Remark 6.9. There is another class of finite element discretizations of the fictitious domain formulations using partitions $\widehat{\mathcal{T}}_h$ that take into account the geometry of Ω (locally fitted meshes) and at the same time still preserve the good properties of $\boldsymbol{A}$. In this case the stiffness matrix $\boldsymbol{A}$ *depends* on the shape of Ω so that the main advantage of the previous approach is lost. For this reason we shall confine ourselves to finite element discretizations with nonfitted meshes in the rest of this chapter.

Remark 6.10. Consider now a nonhomogeneous Dirichlet boundary condition $u = g$ on $\partial\Omega$, $g \in H^{\frac{1}{2}}(\partial\Omega)$. Then the second equations in the mixed finite element formulation $(\widehat{\mathcal{M}}_h^H)$, $(\widehat{\mathcal{N}}_h)$ have to be replaced by

$$\int_{\partial\Omega} \widehat{u}_h \mu_H \, ds = \int_{\partial\Omega} g \mu_H \, ds \quad \forall \mu_H \in \Lambda_H,$$
$$\int_{\Xi} \widehat{u}_h v_h \, dx = \int_{\Xi} \widetilde{g} v_h \, dx \quad \forall v_h \in \Lambda_h,$$

respectively, where $\widetilde{g}$ is an extension of g from $\partial\Omega$ to Ξ. Fictitious domain approaches can be used not only for solving Dirichlet boundary value problems, but also for the case of Neumann boundary conditions analyzed in [HMT01], [HK00].

6.2 Fictitious domain formulations of state problems in shape optimization

As we already know, a discretization of any shape optimization problem consists of two steps: First, a family of admissible domains is replaced by another one containing domains whose shapes are fully characterized by a finite number of parameters. Second, a state problem is discretized by using an appropriate numerical method. We have used a classical finite element method in all examples presented up to now. To overcome some of the drawbacks of this approach, which have been mentioned in the introduction to this chapter, we now use fictitious domain methods for the numerical realization of state problems. Special attention will be paid to sensitivity analysis. It turns out that fictitious domain solvers with nonfitted meshes *reduce* the smoothness of the respective control state mapping and consequently may lead to nondifferentiable optimization problems.

To illustrate the whole matter let us consider the optimal shape design problem with the family $\mathcal{O}$ consisting of domains $\Omega(\alpha)$ whose shapes, captured in Figure 2.2, are described by functions α from U^{ad} defined by

$$U^{ad} = \big\{\alpha \in C^{0,1}([0,1]) \mid 0 < \alpha_{\min} \le \alpha \le \alpha_{\max} \text{ in } [0,1],$$
$$|\alpha'| \le L_0 \text{ almost everywhere in }]0,1[\big\}, \qquad (6.25)$$

where $\alpha_{\min}$, $\alpha_{\max}$, and L_0 are given positive parameters (this time the constant volume constraint in the definition of U^{ad} is omitted).

On any $\Omega(\alpha)$, $\alpha \in U^{ad}$, we consider the following homogeneous Dirichlet boundary value state problem:

$$\begin{cases} -\Delta u(\alpha) = f & \text{in } \Omega(\alpha), \quad f \in L^2_{\mathrm{loc}}(\mathbb{R}^2), \\ u(\alpha) = 0 & \text{on } \partial\Omega(\alpha). \end{cases} \qquad (\mathcal{P}(\alpha))$$

Let the cost functional J be given by (2.99) and define the following problem:

$$\begin{cases} \text{Find } \alpha^* \in U^{ad} \quad \text{such that} \\ J(u(\alpha^*)) \le J(u(\alpha)) \quad \forall \alpha \in U^{ad}, \end{cases} \tag{$\mathbb{P}$}$$

where $u(\alpha) \in H_0^1(\Omega(\alpha))$ is the weak solution of $(\mathcal{P}(\alpha))$.

To discretize $\mathcal{O}$ we use exactly the same approach as in Section 2.2: the moving part of the boundary is shaped by piecewise quadratic Bézier functions $s_\varkappa$. The set of all such $s_\varkappa$ will be denoted by $U_\varkappa^{ad}$. Let us observe that, because of the absence of the constant volume constraint, $U_\varkappa^{ad}$ is a subset of U^{ad}. Denote $\mathcal{O}_\varkappa = \{\Omega(s_\varkappa) \mid s_\varkappa \in U_\varkappa^{ad}\}$.

First we choose a fictitious domain $\widehat{\Omega}$ in which all computations will be performed and such that $\widehat{\Omega} \supset \Omega\ \forall \Omega \in \mathcal{O}_\varkappa$ (one may take $\widehat{\Omega} =]0, 2\alpha_{\max}[\times]0,1[$, for example). We use the BL variant of the fictitious domain formulation of $(\mathcal{P}(\alpha))$. Let $\widehat{\mathcal{T}}_h$ be a uniform triangulation of $\widehat{\Omega}$ and V_h be the space of all continuous piecewise linear functions over $\widehat{\mathcal{T}}_h$ defined by (6.17). The space of the BL multipliers will be discretized by piecewise constant functions. To this end let $\widehat{\mathcal{T}}_H$ be a regular partition of the vertical side $\widehat{\Gamma} = \{0\}\times]0,1[$ into segments Δ_i, $i = 1, \dots, m(H)$, of length not exceeding H. With any $\widehat{\mathcal{T}}_H$ we associate the space Λ_H defined as follows:

$$\Lambda_H = \left\{\mu_H \in L^2(\widehat{\Gamma}) \mid \mu_{H|\Delta} \in P_0(\Delta)\ \forall \Delta \in \widehat{\mathcal{T}}_H\right\}.$$

The BL variant of the fictitious domain formulation of $(\mathcal{P}(\alpha))$ reads as follows:

$$\begin{cases} \text{Find } (\widehat{u}_h, \lambda_H) := (\widehat{u}_h(s_\varkappa), \lambda_H(s_\varkappa)) \in V_h \times \Lambda_H \quad \text{such that} \\ \displaystyle\int_{\widehat{\Omega}} \nabla \widehat{u}_h \cdot \nabla v_h\, dx = \int_{\widehat{\Omega}} \widetilde{f} v_h\, dx + \langle \lambda_H, v_h \rangle_{r_H s_\varkappa} \quad \forall v_h \in V_h, \\ \langle \mu_H, \widehat{u}_h \rangle_{r_H s_\varkappa} = 0 \quad \forall \mu_H \in \Lambda_H, \end{cases} \tag{$\mathcal{P}_h^H(s_\varkappa)$}$$

where

$$\langle \mu_H, v_h \rangle_{r_H s_\varkappa} := \int_0^1 \mu_H(x_2) v_h(r_H s_\varkappa(x_2), x_2)\, dx_2, \quad \mu_H \in \Lambda_H, v_h \in V_h,$$

and $r_H s_\varkappa$ is the piecewise linear Lagrange interpolant of $s_\varkappa$ on $\widehat{\mathcal{T}}_H$ (see Figure 6.3).

Comments 6.1.

(i) In the original setting of the BL variant of the fictitious domain method the space Λ_H is defined on the variable part of the boundary. Therefore it depends on the design variable, too. In addition the respective duality pairing is realized (for regular functions) by the curvilinear integral on $\Gamma(\alpha)$:

$$\langle \mu, v \rangle = \int_{\Gamma(\alpha)} \mu v\, ds, \quad \mu \in L^2(\Gamma(\alpha)), v \in H^1(\Omega(\alpha)).$$

Because of the special parametrization of shapes of $\Omega \in \mathcal{O}$, the integral along $\Gamma(\alpha)$ can be transformed into the integral on $\widehat{\Gamma}$ in a standard way:

$$\int_{\Gamma(\alpha)} \mu v\, ds = \int_0^1 \mu \circ \alpha\, v \circ \alpha \sqrt{1 + (\alpha')^2}\, dx_2,$$

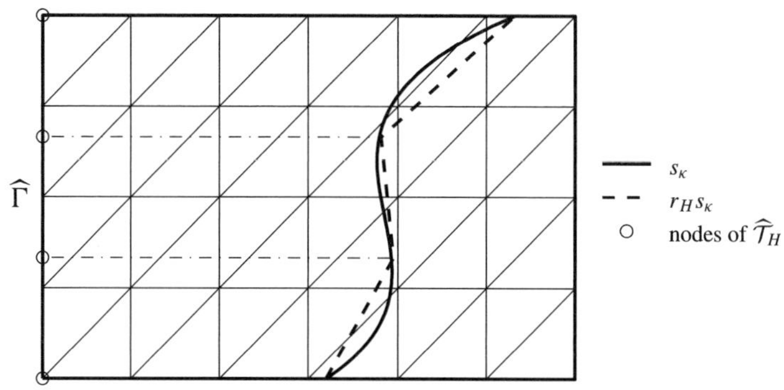

Figure 6.3. *Partitions defining V_h and Λ_H.*

where $\Gamma(\alpha)$ is the graph of $\alpha \in U^{ad}$; $\mu \circ \alpha(x_2) := \mu(\alpha(x_2), x_2)$ and $v \circ \alpha$ is defined similarly. Suppose now that $\alpha : \widehat{\Gamma} \to \mathbb{R}$ is a *continuous, piecewise linear* function over the partition $\widehat{\mathcal{T}}_H$ of $\widehat{\Gamma}$ and $\mu : \Gamma(\alpha) \to \mathbb{R}$ is a function that is *constant* on any straight segment S of $\Gamma(\alpha)$. Then the function $\mu \circ \alpha\sqrt{1 + (\alpha')^2} : \widehat{\Gamma} \to \mathbb{R}$ remains piecewise constant on $\widehat{\mathcal{T}}_H$. This makes it possible to introduce Λ_H as a space of functions defined in $\widehat{\Gamma}$ that does not depend on the design variable. In addition the Riemann integral on $\widehat{\Gamma}$ without the term $\sqrt{1 + [(r_H s_\varkappa)']^2}$ can be used in place of the duality term in $(\mathcal{P}_h^H(s_\varkappa))$.

(ii) As before, the symbol $\widetilde{f}$ stands for an $L^2(\widehat{\Omega})$-extension of f from $\Omega(r_H s_\varkappa)$ to $\widehat{\Omega}$. Since f is defined in the whole plane, the simplest way of defining $\widetilde{f}$ is to take $\widetilde{f} = f_{|\widehat{\Omega}}$. This extension *does not depend* on the geometry of $\Omega \in \mathcal{O}_\varkappa$. On the other hand, we know from Remark 6.3 that the zero extension of f is of particular importance since the Lagrange multiplier λ is related to $\partial u/\partial \nu$ on the variable part of the boundary of the original domain. In this case $\widetilde{f}$ *depends* on the shape of $\Omega \in \mathcal{O}_\varkappa$.

(iii) Suppose now that the zero extension of f to $\widehat{\Omega}$ is used in $(\mathcal{P}_h^H(s_\varkappa))$. Then, in view of the definition of the duality term $\langle \cdot, \cdot \rangle_{r_H s_\varkappa}$, the second component λ_H approximates the product $\partial u/\partial \nu\sqrt{1 + [(r_H s_\varkappa)']^2}$ on $\Gamma(r_H s_\varkappa)$! This form of the duality term makes the forthcoming convergence analysis easier. In practical computations, however, the original curvilinear integral $\int_{\Gamma(r_H s_\varkappa)} \mu_H v_h \, ds$ may be used, in which case λ_H directly approximates $\partial u/\partial \nu$ on $\Gamma(r_H s_\varkappa)$.

Three different discretization parameters appear in the definition of $(\mathcal{P}_h^H(s_\varkappa))$: the parameter $\varkappa$ is related to the approximation of the geometry of admissible domains, while h and H denote the norms of $\widehat{\mathcal{T}}_h$ and $\widehat{\mathcal{T}}_H$ used for constructing V_h and Λ_H, respectively. To ensure the existence and uniqueness of solutions to $(\mathcal{P}_h^H(s_\varkappa))$ we shall suppose that the condition

$$\langle \mu_H, v_h \rangle_{r_H s_\varkappa} = 0 \quad \forall v_h \in V_h \implies \mu_H = 0 \quad \text{on } \widehat{\Gamma}$$

is satisfied $\forall s_\varkappa \in U_\varkappa^{ad}$. From the results of the previous section we know this is true whenever the ratio H/h is sufficiently large. Let us observe that we do not require the LBB condition to be satisfied. To keep the notation simple we shall suppose there is a one-to-one relation between h and H enabling us to label the discretized shape optimization problem by only two parameters: h and $\varkappa$. Finally we shall suppose that all three parameters *simultaneously* tend to zero:

$$h \to 0+ \iff H \to 0+ \iff \varkappa \to 0+ .$$

The discretized optimal shape design problem now reads as follows:

$$\begin{cases} \text{Find } s_\varkappa^* \in U_\varkappa^{ad} \quad \text{such that} \\ J(\widehat{u}_h(s_\varkappa^*)) \le J(\widehat{u}_h(s_\varkappa)) \quad \forall s_\varkappa \in U_\varkappa^{ad}, \end{cases} \tag{$\mathbb{P}_{h\varkappa}$}$$

where $\widehat{u}_h(s_\varkappa) \in V_h$ is the first component of the solution to $(\mathcal{P}_h^H(s_\varkappa))$ and $J(y) = \frac{1}{2}\|y - z_d\|_{0,D}^2$.

Remark 6.11. In a classical finite element method, approximated solutions of state problems are computed in their own computational domain. On the contrary if fictitious domain methods are used, then solutions are available on a larger domain $\widehat{\Omega}$ so that their appropriate restrictions have to be inserted into the cost functionals. In the previous example the situation is simple: the target domain D is chosen in such a way that $D \subset \Omega$ for any $\Omega \in \mathcal{O}$, and thus one can always use the same restriction of $\widehat{u}_h(s_\varkappa)$ to D. Consider now the following cost functional: $J(\alpha, y) = \frac{1}{2}\|y - z_d\|_{0,\Omega(\alpha)}^2$, $z_d \in L^2(\widehat{\Omega})$, which depends explicitly on the design variable α. If fictitious domain methods were used, the discretized shape optimization problem would be as follows:

$$\begin{cases} \text{Find } s_\varkappa^* \in U_\varkappa^{ad} \quad \text{such that} \\ J(r_H s_\varkappa^*, \widehat{u}_h(s_\varkappa^*)_{|\Omega(r_H s_\varkappa^*)}) \le J(r_H s_\varkappa, \widehat{u}_h(s_\varkappa)_{|\Omega(r_H s_\varkappa)}) \quad \forall s_\varkappa \in U_\varkappa^{ad}. \end{cases} \tag{$\mathbb{P}_{h\varkappa}$}$$

This time the computed solution $\widehat{u}_h(s_\varkappa)$ is restricted to the polygonal approximation $\Omega(r_H s_\varkappa)$ of $\Omega(s_\varkappa)$; i.e., $\Omega(r_H s_\varkappa)$ plays the role of the computational domain.

It is left as an exercise to prove Theorem 6.4.

Theorem 6.4. *Problem* $(\mathbb{P}_{h\varkappa})$ *has a solution for any* h *and* $\varkappa > 0$.

As usual we shall show that solutions to $(\mathbb{P}_{h\varkappa})$ and $(\mathbb{P})$ are close in the sense of subsequences for $h, \varkappa \to 0+$. But first we prove the following auxiliary statement.

Lemma 6.4. *Let* $\{\widehat{u}_h(s_\varkappa)\}$ *be a sequence of solutions to* $(\mathcal{P}_h^H(s_\varkappa))$, $h \to 0+$, *where* $s_\varkappa \in U_\varkappa^{ad}$ *are such that*

$$s_\varkappa \rightrightarrows \alpha \quad \text{in } [0,1] \text{ as } \varkappa \to 0+ . \tag{6.26}$$

Then

$$\widehat{u}_h(s_\varkappa) \to \widehat{u} \quad \text{in } H_0^1(\widehat{\Omega}) \text{ as } h, \varkappa \to 0+,$$

and $u(\alpha) := \widehat{u}_{|\Omega(\alpha)}$ *is the solution of the homogeneous Dirichlet boundary value problem in* $\Omega(\alpha)$.

Proof. Inserting $v_h := \widehat{u}_h(s_\varkappa)$ into the first equation in $(\mathcal{P}_h^H(s_\varkappa))$ we have

$$\int_{\widehat{\Omega}} |\nabla \widehat{u}_h(s_\varkappa)|^2\, dx = \int_{\widehat{\Omega}} \widetilde{f}\widehat{u}_h(s_\varkappa)\, dx, \tag{6.27}$$

making use of $\langle \lambda_H, \widehat{u}_h(s_\varkappa)\rangle_{r_H s_\varkappa} = 0$. From (6.27) and the Friedrichs inequality in $H_0^1(\widehat{\Omega})$ we see that $\{\widehat{u}_h(s_\varkappa)\}$ is bounded: there exists a positive constant c such that

$$\|\widehat{u}_h(s_\varkappa)\|_{1,\widehat{\Omega}} \le c \quad \forall h, \varkappa > 0.$$

Thus one can pass to a subsequence of $\{\widehat{u}_h(s_\varkappa)\}$ such that

$$\widehat{u}_h(s_\varkappa) \rightharpoonup \widehat{u} \quad \text{in } H_0^1(\widehat{\Omega}), \quad h, \varkappa \to 0+, \tag{6.28}$$

for some element $\widehat{u} \in H_0^1(\widehat{\Omega})$. We now prove that $\widehat{u} = 0$ on $\Gamma(\alpha)$. This is equivalent to showing that

$$\int_0^1 \mu \widehat{u} \circ \alpha\, dx_2 = 0 \quad \forall \mu \in L^2(\widehat{\Gamma}).$$

Let $\overline{\mu} \in L^2(\widehat{\Gamma})$ be given. Then there exists a sequence $\{\overline{\mu}_H\}$, $\overline{\mu}_H \in \Lambda_H$, such that

$$\overline{\mu}_H \to \overline{\mu} \quad \text{in } L^2(\widehat{\Gamma}) \text{ as } H \to 0+. \tag{6.29}$$

From the second equation in $(\mathcal{P}_h^H(s_\varkappa))$ we have

$$\langle \overline{\mu}_H, \widehat{u}_h(s_\varkappa)\rangle_{r_H s_\varkappa} = 0. \tag{6.30}$$

From Lemma 2.21, the definition of $\langle \cdot, \cdot \rangle_{r_H s_\varkappa}$, (6.26), (6.28), (6.29), and (6.30) one has

$$\langle \overline{\mu}_H, \widehat{u}_h(s_\varkappa)\rangle_{r_H s_\varkappa} \to \int_0^1 \overline{\mu}\, \widehat{u} \circ \alpha\, dx_2 = 0, \quad h, \varkappa \to 0+, \tag{6.31}$$

proving that $\widehat{u}$ belongs to the space $H_0^1(\widehat{\Omega}, \partial\Omega(\alpha))$ defined by (6.1) and in particular $\widehat{u} = 0$ on $\Gamma(\alpha)$. Next we show that $\widehat{u}$ solves problem $(\widehat{\mathcal{P}})$ formulated at the beginning of Section 6.1.

Denote

$$\mathcal{W}(\alpha) = \left\{ v \in C_0^\infty(\overline{\widehat{\Omega}}) \mid \operatorname{supp} v \subset \Omega(\alpha) \cup \Xi(\alpha) \right\}, \quad \Xi(\alpha) = \widehat{\Omega} \setminus \overline{\Omega(\alpha)}.$$

Let $\overline{v} \in \mathcal{W}(\alpha)$ be given and $\{r_h\overline{v}\}$, $r_h\overline{v} \in V_h$, be a sequence of the piecewise linear Lagrange interpolants of $\overline{v}$. Then

$$r_h\overline{v} \to \overline{v} \quad \text{in } H_0^1(\widehat{\Omega}) \text{ as } h \to 0+ \tag{6.32}$$

and, in addition, there exists a δ-neighborhood $B_\delta(\Gamma(\alpha))$ of $\Gamma(\alpha)$ such that $\operatorname{supp} r_h\overline{v} \cap B_\delta(\Gamma(\alpha)) = \emptyset$ for any h sufficiently small, as follows from the definition of $\mathcal{W}(\alpha)$. From (6.26) and the definition of $U_\varkappa^{ad}$ it also follows that the sequence $\{r_H s_\varkappa\}$ tends to α uniformly in $[0, 1]$, implying that the graph of $r_H s_\varkappa$ has an empty intersection with $\operatorname{supp} r_h\overline{v}$ for H and $\varkappa$ small enough, too. Therefore $\langle \lambda_H, r_h\overline{v}\rangle_{r_H s_\varkappa} = 0$ and the first equation in $(\mathcal{P}_h^H(s_\varkappa))$ takes the form

$$\int_{\widehat{\Omega}} \nabla \widehat{u}_h \cdot \nabla r_h\overline{v}\, dx = \int_{\widehat{\Omega}} \widetilde{f} r_h\overline{v}\, dx.$$

Passing here to the limit with $h, \varkappa \to 0+$ and using (6.28), (6.32) we obtain

$$\int_{\widehat{\Omega}} \nabla \widehat{u} \cdot \nabla \overline{v}\, dx = \int_{\widehat{\Omega}} \widetilde{f} \overline{v}\, dx. \tag{6.33}$$

Since $\mathcal{W}(\alpha)$ is dense in $H_0^1(\widehat{\Omega}, \partial\Omega(\alpha))$ in the $H_0^1(\widehat{\Omega})$-norm we may conclude from (6.33) that $\widehat{u}$ solves $(\widehat{\mathcal{P}})$ and consequently $\widehat{u}_{\Omega(\alpha)}$ solves $(\mathcal{P}(\alpha))$. Problem $(\widehat{\mathcal{P}})$ has a unique solution; therefore the whole sequence $\{\widehat{u}_h\}$ tends weakly to $\widehat{u}$. Strong convergence of $\{\widehat{u}_h\}$ to $\widehat{u}$ in the $H_0^1(\widehat{\Omega})$-norm can be proved in a standard way. □

Remark 6.12. Since the cost functional J contains only the primal variable $\widehat{u}_h$, we did not pay attention to the behavior of $\{\lambda_H\}$. This is why we did not require the LBB condition to be satisfied. If on the other hand knowledge of the Lagrange multiplier λ were needed in computations (as a part of a cost functional, for example; see the end of this chapter), satisfaction of the LBB condition would be necessary. The mathematical analysis of this case is rather technical and therefore is omitted.

On the basis of the previous lemma we are now able to prove the following convergence result.

Theorem 6.5. *For any sequence $\{(s_\varkappa^*, \widehat{u}_h(s_\varkappa^*))\}$ of optimal pairs of $(\mathbb{P}_{h\varkappa})$, $h \to 0+$, there exists its subsequence such that*

$$\begin{cases} s_\varkappa^* \rightrightarrows \alpha^* & \text{in } [0,1], \\ \widehat{u}_h(s_\varkappa^*) \to \widehat{u}^* & \text{in } H_0^1(\widehat{\Omega}) \text{ as } h, \varkappa \to 0+. \end{cases} \tag{6.34}$$

In addition, $(\alpha^*, \widehat{u}^*_{|\Omega(\alpha^*)})$ is an optimal pair of $(\mathbb{P})$. Any accumulation point of $\{(s_\varkappa^*, \widehat{u}_h(s_\varkappa^*))\}$ in the sense of (6.34) *possesses this characteristic.*

Proof. Since U^{ad} is compact, one can pass to a subsequence of $\{s_\varkappa^*\}$ such that (6.34) holds for some elements $\alpha^* \in U^{ad}$ and $\widehat{u}^* \in H_0^1(\widehat{\Omega})$. In addition $u(\alpha^*) := \widehat{u}^*_{|\Omega(\alpha^*)}$ solves $(\mathcal{P}(\alpha^*))$, as follows from Lemma 6.4. It remains to show that α^* solves $(\mathbb{P})$. Let $\overline{\alpha} \in U^{ad}$ be given. Then from (iii) of Lemma 2.10 we know that there exists a sequence $\{\overline{s}_\varkappa\}$, $\overline{s}_\varkappa \in U_\varkappa^{ad}$, such that

$$\overline{s}_\varkappa \rightrightarrows \overline{\alpha} \quad \text{in } [0,1], \quad \varkappa \to 0+, \tag{6.35}$$

and also

$$\widehat{u}_h(\overline{s}_\varkappa) \to \widehat{u} \quad \text{in } H_0^1(\widehat{\Omega}), \quad h, \varkappa \to 0+, \tag{6.36}$$

where $u(\overline{\alpha}) := \widehat{u}_{|\Omega(\overline{\alpha})}$ solves $(\mathcal{P}(\overline{\alpha}))$, making use of Lemma 6.4 once again. Passing to the limit with $h, \varkappa \to 0+$ in

$$J(\widehat{u}_h(s_\varkappa^*)) \le J(\widehat{u}_h(\overline{s}_\varkappa)),$$

using $(6.34)_2$, (6.36), and the continuity of J, we arrive at

$$J(u(\alpha^*)) \le J(u(\overline{\alpha})). \quad \square$$

Remark 6.13. The same convergence result can also be established for the DL variant of the fictitious domain method, taking for example $\Lambda_h = V_{h|\Xi(r_H s_\varkappa)}$, where $\Xi(r_H s_\varkappa) = \Omega \setminus \overline{\Omega(r_H s_\varkappa)}$; i.e., both spaces use the same triangulation (see Problem 6.6).

The remainder of this section will be devoted to sensitivity analysis in discretized optimal shape design problems using fictitious domain solvers with nonfitted meshes. As stated in the previous section, the algebraic form based on the fictitious domain formulations of the state problems leads to the following linear algebraic system:

$$\begin{cases} \text{Find } (\boldsymbol{u}(\boldsymbol{\alpha}), \boldsymbol{\lambda}(\boldsymbol{\alpha})) \in \mathbb{R}^n \times \mathbb{R}^m \quad \text{such that} \\ \boldsymbol{A}\,\boldsymbol{u}(\boldsymbol{\alpha}) + \boldsymbol{B}(\boldsymbol{\alpha})^{\mathrm{T}}\boldsymbol{\lambda}(\boldsymbol{\alpha}) = \boldsymbol{f}(\boldsymbol{\alpha}), \\ \boldsymbol{B}(\boldsymbol{\alpha})\boldsymbol{u}(\boldsymbol{\alpha}) = \boldsymbol{g}(\boldsymbol{\alpha}), \end{cases} \tag{6.37}$$

where $\boldsymbol{\alpha} \in \mathcal{U} \subset \mathbb{R}^d$ is a discrete design variable characterizing the shape of $\Omega \in \mathcal{O}_\varkappa$. Recall once again that only the matrix $\boldsymbol{B}$ and the right-hand side of (6.37) depend on $\boldsymbol{\alpha}$. Let us check the differentiability of the mapping $\Phi : \mathcal{U} \to \mathbb{R}^n \times \mathbb{R}^m$, $\Phi(\boldsymbol{\alpha}) = (\boldsymbol{u}(\boldsymbol{\alpha}), \boldsymbol{\lambda}(\boldsymbol{\alpha})) \in \mathbb{R}^n \times \mathbb{R}^m$, $\boldsymbol{\alpha} \in \mathcal{U}$. A common way of proving this property is to apply the implicit function theorem to the mapping $\Psi : \mathcal{U} \times \mathbb{R}^n \times \mathbb{R}^m \to \mathbb{R}^n \times \mathbb{R}^m$, where

$$\Psi(\boldsymbol{\alpha}, \boldsymbol{u}, \boldsymbol{\lambda}) = \begin{Bmatrix} \boldsymbol{A}\,\boldsymbol{u} + \boldsymbol{B}(\boldsymbol{\alpha})^{\mathrm{T}}\boldsymbol{\lambda} - \boldsymbol{f}(\boldsymbol{\alpha}) \\ \boldsymbol{B}(\boldsymbol{\alpha})\,\boldsymbol{u} - \boldsymbol{g}(\boldsymbol{\alpha}) \end{Bmatrix}.$$

As we shall see, one of the assumptions of this theorem, namely the differentiability of Ψ with respect to $\boldsymbol{\alpha}$, is not satisfied.

Let us start with the BL variant of the fictitious domain approach. For the simplicity of our presentation we shall suppose that the variable part of the boundary of discrete design domains is realized by continuous, *piecewise linear* functions α_H over the partition $\widehat{\mathcal{T}}_H$ of $\widehat{\Gamma}$. If this is so then the discrete design variable $\boldsymbol{\alpha}$ can be identified with the values of α_H at the nodes of $\widehat{\mathcal{T}}_H$ (see Figure 6.4). Let us recall that the elements of $\boldsymbol{B}(\boldsymbol{\alpha})$ are given by

$$b_{kl}(\boldsymbol{\alpha}) := b_{kl}(\alpha_H) = -\int_{\Delta_k} \varphi_l \circ \alpha_H \, dx_2,$$

where φ_l is the lth Courant basis function of V_h. Denote by S_k the graph of α_H over $\Delta_k \in \widehat{\mathcal{T}}_H$.

Let β_H be another continuous, piecewise linear function over $\widehat{\mathcal{T}}_H$ and $\boldsymbol{\beta}$ be the respective discrete design variable. Compute the directional derivative

$$b'_{kl}(\boldsymbol{\alpha}; \boldsymbol{\beta}) := b'_{kl}(\alpha_H; \beta_H) = -\frac{d}{dt}\left(\int_{\Delta_k} \varphi_l \circ (\alpha_H + t\beta_H)\, dx_2\right)\Bigg|_{t=0+}.$$

The formal derivation under the sign of the integral yields

$$b'_{kl}(\boldsymbol{\alpha}; \boldsymbol{\beta}) = -\int_{\Delta_k} \left(\frac{\partial \varphi_l}{\partial x_1}\right) \circ \alpha_H \beta_H \, dx_2. \tag{6.38}$$

From this expression we see that if $(\partial \varphi_l / \partial x_1) \circ \alpha_H$ were not defined in a subset of Δ_k whose one-dimensional Lebesgue measure is positive, then it could happen that

$$b'_{kl}(\alpha_H; \beta_H) \neq -b'_{kl}(\alpha_H; -\beta_H);$$

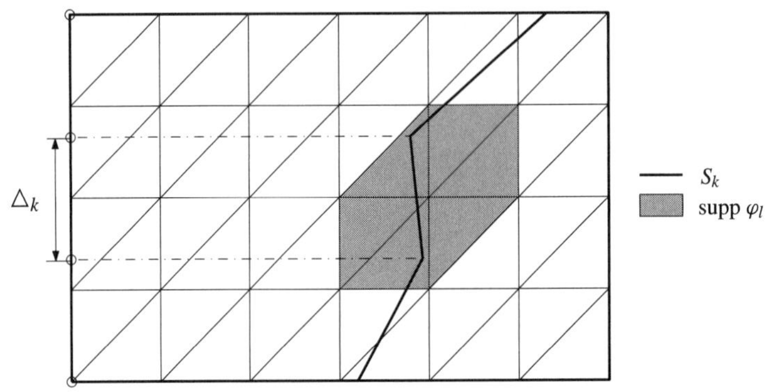

Figure 6.4. *Computation of $b_{kl}(\boldsymbol{\alpha})$.*

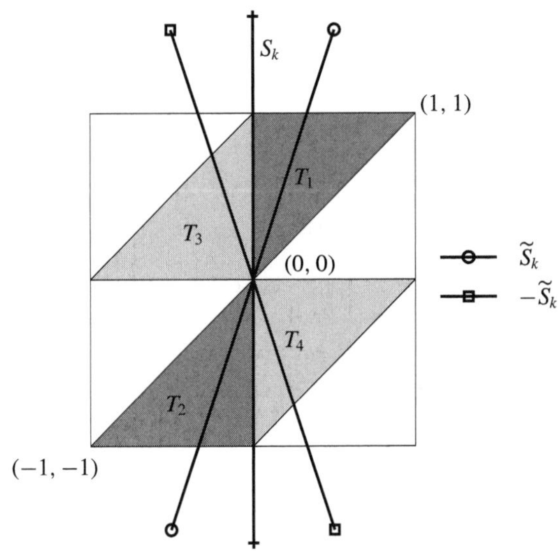

Figure 6.5. *Nondifferentiability of $b_{kl}(\boldsymbol{\alpha})$.*

i.e., b_{kl} is not continuously differentiable as a function of $\boldsymbol{\alpha}$. To see this consider the segment S_k having a nonempty intersection with an interelement boundary whose one-dimensional Lebesgue measure is positive (see Figure 6.5, where the function φ_l is associated with the node (0, 0) of the unit square grid in $\mathbb{R}^2$). Denote by $\varphi_l^{(i)}$ the restriction of φ_l to the triangle T_i, $i = 1, \ldots, 4$. It is readily seen that

$$\varphi_l^{(1)} = 1 - x_2, \quad \varphi_l^{(2)} = 1 + x_2,$$
$$\varphi_l^{(3)} = 1 + x_1 - x_2, \quad \varphi_l^{(4)} = 1 - x_1 + x_2.$$

Further, let the vertical interelement boundary separating T_2, T_3 from T_1, T_4 be a part of S_k. We want to compute $b'_{kl}(\alpha_H; \beta_H)$, where β_H is such that its graph $\widetilde{S}_k$ over Δ_k passes through the point (0, 0) as shown in Figure 6.5. It is readily seen that the integral in (6.38) is always equal to zero. On the other hand $b'_{kl}(\alpha_H; -\beta_H)$ equals $2\int_0^1 \beta_H \, dx_2$ (observe that the direction $-\beta_H$ corresponds to the segment $-\widetilde{S}_k$). From this elementary example it follows that the mapping $\boldsymbol{B} : \boldsymbol{\alpha} \mapsto \boldsymbol{B}(\boldsymbol{\alpha})$ cannot be differentiable in general. Since one of the basic assumptions of the implicit function theorem is not fulfilled, there is no reason to expect that the mapping Φ assigning to any $\boldsymbol{\alpha} \in \mathcal{U}$ the unique solution $(\boldsymbol{u}(\boldsymbol{\alpha}), \boldsymbol{\lambda}(\boldsymbol{\alpha}))$ of (6.37) should be continuously differentiable. On the other hand one can prove that Φ is continuous in $\mathcal{U}$ (see Problem 6.7).

Let us now consider the DL variant of the fictitious domain method. In this case the situation is even worse! Suppose that the space Λ_h uses the same partition $\widehat{\mathcal{T}}_h$ as the space V_h; i.e., $\Lambda_h = V_{h|\Xi(r_H s_\varkappa)}$ (see Remark 6.13). Then owing to the locking effect the fictitious domain solution $\widehat{u}_h(s_\varkappa)$ solves a discrete homogeneous Dirichlet boundary value problem (6.22) in Ω_h and vanishes outside of Ω_h, where Ω_h is the approximation of the polygonal domain $\Omega(r_H s_\varkappa)$ from inside made of all triangles of $\widehat{\mathcal{T}}_h$ lying in the interior of $\Omega(r_H s_\varkappa)$. From this it follows that any variation of the designed part $\Gamma(r_H s_\varkappa)$, $s_\varkappa \in U_\varkappa^{ad}$, that does not change Ω_h still leads to the *same* solution of (6.22). Since there is only a *finite number* of different Ω_h for $s_\varkappa \in U_\varkappa^{ad}$ the same holds for the respective solutions $\widehat{u}_h(s_\varkappa)$, $h, \varkappa$ fixed. In other words, these solutions, after being substituted into a cost functional, entail its *discontinuity* (see Figure 6.8). This phenomenon can be weakened by using a coarser triangulation $\widehat{\mathcal{T}}_H$ to construct the discrete space of the DL multipliers.

From all that has been said above we may conclude that fictitious domain solvers used at the lower level of shape optimization problems reduce the smoothness of minimized functions regardless of the fact that the original problem is smooth. This fact has a practical consequence: gradient type methods may give unsatisfactory results, especially if too coarse meshes are used. Global optimization methods based only on function evaluations represent a possible way to overcome this difficulty. Some of these have been mentioned in Chapter 4.

We end this subsection by illustrating how the previous variants of the fictitious domain methods can be used in shape optimization. We start with the DL variant.

Let the family $\mathcal{O}$ of admissible domains be of the form

$$\mathcal{O} = \{\Omega \subset \mathbb{R}^2 \mid \exists Q \in \mathcal{O}^* : \Omega = Q \setminus \overline{B}\},$$

where B is the unit open disk with center at the point $C = (4, 4)$ and $\mathcal{O}^*$ is another system of simply connected bounded domains $Q \subset \mathbb{R}^2$ with Lipschitz boundary, containing B in their interior, and contained in the square $]0.5, 7.5[\times]0.5, 7.5[$. In other words $\mathcal{O}$ consists of double connected domains with fixed inner boundary ∂B and variable outer part given by ∂Q, $Q \in \mathcal{O}^*$, which characterizes the shape of $\Omega \in \mathcal{O}$ (see Figure 6.6). On any $\Omega \in \mathcal{O}$ we consider the following state problem:

$$\begin{cases} -\Delta u(\Omega) = f & \text{in } \Omega, \\ u(\Omega) = 0 & \text{on } \partial\Omega = \partial B \cup \partial Q, \end{cases} \tag{6.39}$$

with $f = -\Delta u_d$ in $\mathbb{R}^2$, where

$$u_d = \left((x-4)^2 + (y-4)^2 - 1\right)\left(1 - \frac{1}{9}(x-4)^2 - \frac{1}{4}(y-4)^2\right). \tag{6.40}$$

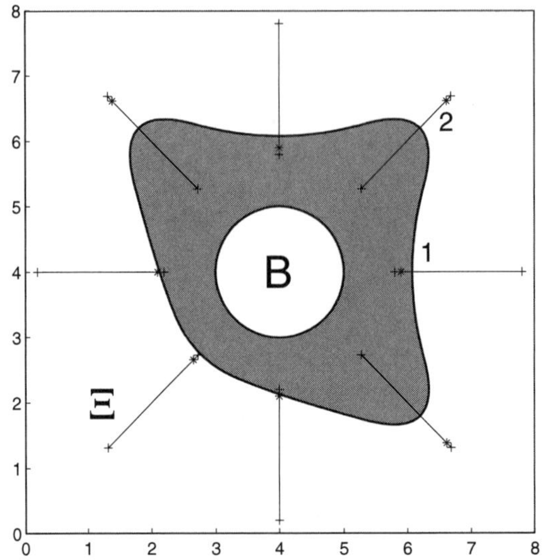

Figure 6.6. *Shape of admissible domains.*

The optimal shape design problem is defined as follows:

$$\begin{cases} \text{Find } \Omega^* \in \mathcal{O} \quad \text{such that} \\ J(\Omega^*, u(\Omega^*)) \le J(\Omega, u(\Omega)) \quad \forall \Omega \in \mathcal{O}, \end{cases} \tag{6.41}$$

where

$$J(\Omega, u(\Omega)) = \frac{1}{2}\|u(\Omega) - u_d\|^2_{0,\Omega},$$

$u(\Omega)$ solves (6.39) in $\Omega \in \mathcal{O}$, and u_d is as in (6.40).

Denote by $\widetilde{\Omega} \in \mathcal{O}$ the domain whose outer boundary is realized by the ellipse

$$\frac{1}{9}(x-4)^2 + \frac{1}{4}(y-4)^2 = 1. \tag{6.42}$$

From (6.39) and (6.40) it follows that $u(\widetilde{\Omega}) = u_d$ in $\widetilde{\Omega}$ so that $J(\widetilde{\Omega}, u(\widetilde{\Omega})) = 0$. Therefore $\widetilde{\Omega}$ is a solution to (6.41). We try to identify $\widetilde{\Omega}$ by solving (6.41).

Any $\Omega \in \mathcal{O}$ will be embedded into the fictitious domain $\widehat{\Omega} =]0, 8[\times]0, 8[$. Denote by $V(B) = H_0^1(\widehat{\Omega})_{|B}$ and $V(\Xi) = H_0^1(\widehat{\Omega})_{|\Xi}$ the spaces of restrictions to two components B and Ξ of $\widehat{\Omega} \setminus \overline{\Omega}$. Further let $V'(B)$, $V'(\Xi)$ be the dual spaces with duality pairings denoted by $\langle \cdot, \cdot \rangle_B$, $\langle \cdot, \cdot \rangle_\Xi$, respectively.

The fictitious domain formulation of (6.39) in $\Omega \in \mathcal{O}$ reads as follows:

$$\begin{cases} \text{Find } (\widehat{u}(\Omega), \lambda_1(\Omega), \lambda_2(\Omega)) \in H_0^1(\widehat{\Omega}) \times V'(B) \times V'(\Xi) \quad \text{such that} \\ \displaystyle\int_{\widehat{\Omega}} \nabla \widehat{u}(\Omega) \cdot \nabla v \, dx = \int_{\widehat{\Omega}} f v \, dx + \langle \lambda_1(\Omega), v \rangle_B + \langle \lambda_2(\Omega), v \rangle_\Xi \quad \forall v \in H_0^1(\widehat{\Omega}), \\ \langle \mu_1, \widehat{u}(\Omega) \rangle_B + \langle \mu_2, \widehat{u}(\Omega) \rangle_\Xi = 0 \quad \forall (\mu_1, \mu_2) \in V'(B) \times V'(\Xi). \end{cases} \tag{6.43}$$

Remark 6.14. Problem (6.43) and its solution depend on $\Omega \in \mathcal{O}$ through the duality term $\langle \cdot, \cdot \rangle_{\Xi}$.

It is left as an exercise to show that (6.43) has a unique solution and, in addition, $\widehat{u}(\Omega)_{|\Omega}$ solves the original problem (6.39) in Ω.

We now describe the approximation of (6.41). The outer part of the boundary of any $\Omega \in \mathcal{O}$ will be approximated by piecewise quadratic Bézier curves. In our example the number of Bézier segments is equal to eight. The set of all such domains will be denoted by $\mathcal{O}_{\varkappa}$. The control points of Bézier curves whose positions define the boundary of any $\Omega_{\varkappa} \in \mathcal{O}_{\varkappa}$ are subject to constraints ensuring that $\mathcal{O}_{\varkappa} \subset \mathcal{O}$. For discretizations of $H_0^1(\widehat{\Omega})$, $V'(B)$, $V'(\Xi)$ we use *piecewise bilinear* functions over a uniform partition $\widehat{\mathcal{R}}_h$ of $\widehat{\Omega}$ into squares of size h. The use of bilinear functions over rectangles makes programming easier and the locking effect more "visible." We define

$$V_h = \left\{ v_h \in C(\overline{\widehat{\Omega}}) \mid v_{h|R} \in Q_1(R)\ \forall R \in \widehat{\mathcal{R}}_h,\ v_h = 0 \text{ on } \partial\widehat{\Omega} \right\},$$

$$\Lambda_{1h}(B) = V_{h|B}, \quad \Lambda_{2h}(\Xi_{\varkappa}) = V_{h|\Xi_{\varkappa}},$$

where $Q_1(R)$ stands for the space of bilinear functions defined in $R \in \widehat{\mathcal{R}}_h$ and $\Xi_{\varkappa}$ is the outer component of $\Omega_{\varkappa} \in \mathcal{O}_{\varkappa}$ in $\widehat{\Omega}$. All these spaces use the *same* mesh $\widehat{\mathcal{R}}_h$. With any $\Omega_{\varkappa} \in \mathcal{O}_{\varkappa}$ we associate the following mixed finite element problem:

$$\left\{\begin{array}{l} \text{Find } (\widehat{u}_h(\Omega_{\varkappa}), \lambda_{1h}(\Omega_{\varkappa}), \lambda_{2h}(\Omega_{\varkappa})) \in V_h \times \Lambda_{1h}(B) \times \Lambda_{2h}(\Xi_{\varkappa}) \text{ such that} \\ \displaystyle\int_{\widehat{\Omega}} \nabla\widehat{u}_h(\Omega_{\varkappa}) \cdot \nabla v_h\, dx = \int_{\widehat{\Omega}} f v_h\, dx + \int_B \lambda_{1h}(\Omega_{\varkappa}) v_h\, dx \\ \qquad\qquad \displaystyle + \int_{\Xi_{\varkappa}} \lambda_{2h}(\Omega_{\varkappa}) v_h\, dx \quad \forall v_h \in V_h, \\ \displaystyle\int_B \mu_{1h}\widehat{u}_h(\Omega_{\varkappa})\, dx + \int_{\Xi_h} \mu_{2h}\widehat{u}_h(\Omega_{\varkappa})\, dx = 0 \\ \qquad\qquad \forall (\mu_{1h}, \mu_{2h}) \in \Lambda_{1h}(B) \times \Lambda_{2h}(\Xi_{\varkappa}). \end{array}\right. \tag{6.44}$$

We already know that (6.44) has a unique solution. Let $\Omega_{\varkappa}^h$ be the union of all squares $R \in \widehat{\mathcal{R}}_h$ having an empty intersection with the interior of B and $\Xi_{\varkappa}$; i.e., $\Omega_{\varkappa}^h$ is the "checkerboard" approximation of $\Omega_{\varkappa}$ from inside. Due to the locking effect the solution $\widehat{u}_h(\Omega_{\varkappa})$ vanishes outside of $\Omega_{\varkappa}^h$ and the function $u_h(\Omega_{\varkappa}) := \widehat{u}_h(\Omega_{\varkappa})_{|\Omega_{\varkappa}^h}$ solves the following discrete homogeneous Dirichlet boundary value problem:

$$\left\{\begin{array}{l} \text{Find } u_h(\Omega_{\varkappa}) \in V_h(\Omega_{\varkappa}^h) \quad \text{such that} \\ \displaystyle\int_{\Omega_{\varkappa}^h} \nabla u_h(\Omega_{\varkappa}) \cdot \nabla v_h\, dx = \int_{\Omega_{\varkappa}^h} f v_h\, dx \quad \forall v_h \in V_h(\Omega_{\varkappa}^h), \end{array}\right.$$

where $V_h(\Omega_{\varkappa}^h) = V_{h|\Omega_{\varkappa}^h} \cap H_0^1(\Omega_{\varkappa}^h)$ (see also (6.22) and (6.23)). The discretization of (6.41) using (6.44) as a discrete state solver reads as follows:

$$\left\{\begin{array}{l} \text{Find } \Omega_{\varkappa}^* \in \mathcal{O}_{\varkappa} \quad \text{such that} \\ J(\Omega_{\varkappa}^{h*}, \widehat{u}_h(\Omega_{\varkappa}^*)_{|\Omega_{\varkappa}^{h*}}) \le J(\Omega_{\varkappa}^h, \widehat{u}_h(\Omega_{\varkappa})_{|\Omega_{\varkappa}^h}) \quad \forall \Omega_{\varkappa} \in \mathcal{O}_{\varkappa}, \end{array}\right. \tag{6.45}$$

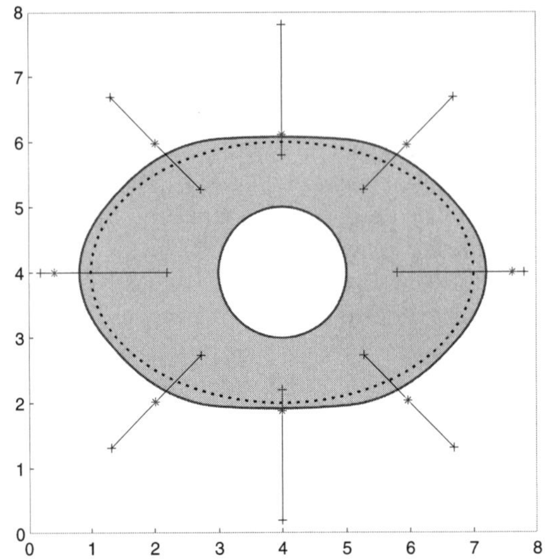

Figure 6.7. *Found solution.*

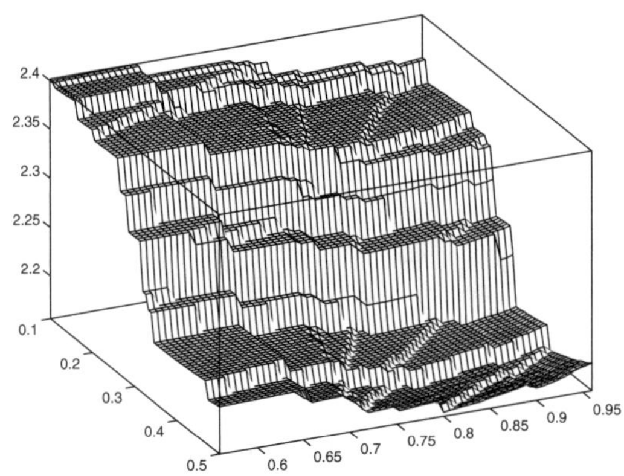

Figure 6.8. *Graph of the cost functional.*

where $\widehat{u}_h(\Omega_\varkappa)$ solves (6.44). From the definition of (6.45) we see that the cost functional J is evaluated using the checkerboard approximation $\Omega_\varkappa^h$ of $\Omega_\varkappa$ and the restriction of $\widehat{u}_h(\Omega_\varkappa)$ to $\Omega_\varkappa^h$. Due to the locking effect there is no reason to expect that the minimized function in (6.45) is continuous. To illustrate its character let us fix all control points defining the shape of $\Omega_\varkappa \in \mathcal{O}_\varkappa$ except two of them, denoted by 1, 2 in Figure 6.6, which are allowed to move along the marked directions. The graph of J as a function of these two design variables is depicted in Figure 6.8. We see that J is stairwise; i.e., there is no hope that gradient type minimization methods could be successful in this case. For this reason the modified controlled random search (MCRS) method described in Chapter 4 was used. The final result is shown in Figure 6.7. The dotted line denotes the exact shape defined by the ellipse (6.42);

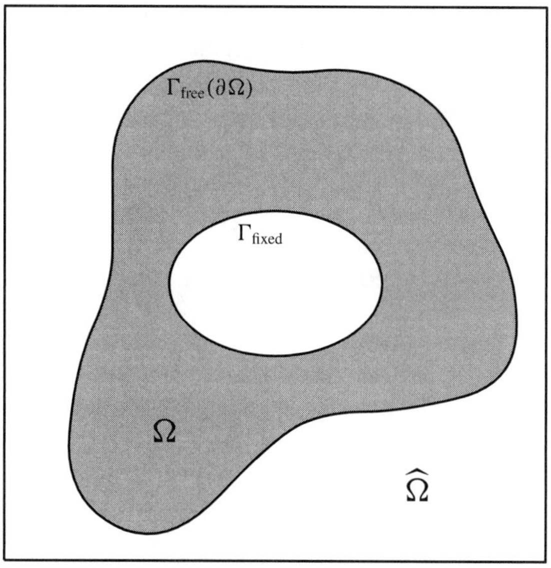

Figure 6.9. *External Bernoulli problem.*

the found shape after 2000 function evaluations of the MCRS method is represented by the solid line. Problem (6.44) was solved for the mesh size $h = 8/32$.

In the remainder of this chapter we show how the BL variant of the fictitious domain method combined with the shape optimization approach can be used for the numerical realization of a class of *free boundary value problems* arising in optimal insulation and electrochemistry modeling [Ack81], [Fas92]. These problems are known in the literature as Bernoulli problems.

Let $\Omega \subset \mathbb{R}^2$ be a double connected domain with the Lipschitz boundary $\partial\Omega = \Gamma_{\text{fixed}} \cup \Gamma_{\text{free}}(\partial\Omega)$, where Γ_{fixed} is a given component and $\Gamma_{\text{free}}(\partial\Omega)$ (exterior to Γ_{fixed}) is a searched part of $\partial\Omega$ (see Figure 6.9). The system of all such domains will be denoted by $\mathcal{O}$. We formulate the following problem:

$$\begin{cases} \text{Find } \Omega^* \in \mathcal{O} \text{ and } u^* : \Omega^* \to \mathbb{R} \text{ such that} \\ -\Delta u^* = 0 \quad \text{in } \Omega^*, \\ u^* = 1 \quad \text{on } \Gamma_{\text{fixed}}, \\ u^* = 0, \ \dfrac{\partial u^*}{\partial \nu} = Q \quad \text{on } \Gamma_{\text{free}}(\partial\Omega), \end{cases} \tag{6.46}$$

where $Q < 0$ is a given constant.

Remark 6.15. Problem (6.46) is the so-called external Bernoulli problem, unlike the *internal* one, in which $\Gamma_{\text{free}}(\partial\Omega)$ is interior to Γ_{fixed}.

First observe that $\Omega^* \in \mathcal{O}$ is one of the unknowns in (6.46). Indeed, for $\Omega \in \mathcal{O}$ given a priori, problem (6.46) is not well posed owing to the fact that the boundary conditions to be satisfied on $\Gamma_{\text{free}}(\partial\Omega)$ form an overdetermined system; i.e., they cannot be satisfied simultaneously, in general. To overcome this difficulty we shall consider the shape of $\Gamma_{\text{free}}(\partial\Omega)$ to be a design variable in an appropriate shape optimization problem. The redundant boundary condition in (6.46) will be included in a cost functional and will be satisfied by its minimization over $\mathcal{O}$, while the remaining one will be part of a (now well-posed) boundary value problem. Thus instead of (6.46) we shall consider the following optimal shape design problem:

$$\begin{cases} \text{Find } \Omega^* \in \mathcal{O} \quad \text{such that} \\ J(\Omega^*) \le J(\Omega) \quad \forall \Omega \in \mathcal{O}, \end{cases} \tag{6.47}$$

where

$$J(\Omega) = \frac{1}{2}\left\| \frac{\partial u(\Omega)}{\partial \nu} - Q \right\|^2_{-\frac{1}{2},\Gamma_{\text{free}}(\partial\Omega)} \tag{6.48}$$

is the cost functional and $u(\Omega)$ solves the following nonhomogeneous Dirichlet boundary value problem in Ω:

$$\begin{cases} -\Delta u(\Omega) = 0 & \text{in } \Omega, \\ u(\Omega) = 1 & \text{on } \Gamma_{\text{fixed}}, \\ u(\Omega) = 0 & \text{on } \Gamma_{\text{free}}(\partial\Omega). \end{cases} \tag{6.49}$$

The mutual relation between (6.46) and (6.47) is readily seen: $\Omega^* \in \mathcal{O}$ *solves* (6.46) *iff* Ω^* *solves* (6.47) *and* $J(\Omega^*) = 0$.

REMARK 6.16. The system of admissible domains has to be sufficiently large to guarantee that Ω^* (solution of (6.46)) belongs to $\mathcal{O}$.

Let $\widehat{\Omega} \subset \mathbb{R}^2$ be a box such that $\overline{\Omega} \subset \widehat{\Omega}\ \forall \Omega \in \mathcal{O}$. Further let $\Lambda_1 = H^{-1/2}(\Gamma_{\text{fixed}})$, $\Lambda_2(\partial\Omega) = H^{-1/2}(\Gamma_{\text{free}}(\partial\Omega))$ be two spaces of the Lagrange multipliers defined on Γ_{fixed}, $\Gamma_{\text{free}}(\partial\Omega)$, respectively, and $V = H^1_0(\widehat{\Omega})$. The fictitious domain formulation of (6.49) in $\widehat{\Omega}$ is as follows:

$$\begin{cases} \text{Find } (\widehat{u}, \lambda_1, \lambda_2) := (\widehat{u}(\Omega), \lambda_1(\Omega), \lambda_2(\Omega)) \in V \times \Lambda_1 \times \Lambda_2(\partial\Omega) \text{ such that} \\ \displaystyle\int_{\widehat{\Omega}} \nabla \widehat{u} \cdot \nabla v \, dx = \langle \lambda_1, v \rangle_{\Gamma_{\text{fixed}}} + \langle \lambda_2, v \rangle_{\Gamma_{\text{free}}(\partial\Omega)} \quad \forall v \in V, \\ \langle \mu_1, \widehat{u} \rangle_{\Gamma_{\text{fixed}}} + \langle \mu_2, \widehat{u} \rangle_{\Gamma_{\text{free}}(\partial\Omega)} = \langle \mu_1, 1 \rangle_{\Gamma_{\text{fixed}}}\ \forall (\mu_1, \mu_2) \in \Lambda_1 \times \Lambda_2(\partial\Omega), \end{cases} \tag{6.50}$$

where $\langle \cdot, \cdot \rangle_{\Gamma_{\text{fixed}}}$ and $\langle \cdot, \cdot \rangle_{\Gamma_{\text{free}}(\partial\Omega)}$ stand for the duality pairings between Λ_1 and $H^{1/2}(\Gamma_{\text{fixed}})$ and $\Lambda_2(\partial\Omega)$ and $H^{1/2}(\Gamma_{\text{free}}(\partial\Omega))$, respectively. It is left as an easy exercise to show that (6.50) has a unique solution, $u := \widehat{u}_{|\Omega}$ solves (6.49), $\lambda_1 = [\partial\widehat{u}/\partial\nu]$ is the jump of the normal derivative of $\widehat{u}$ across Γ_{fixed}, and $\lambda_2 = \partial u/\partial\nu$ is the normal derivative of u (the solution

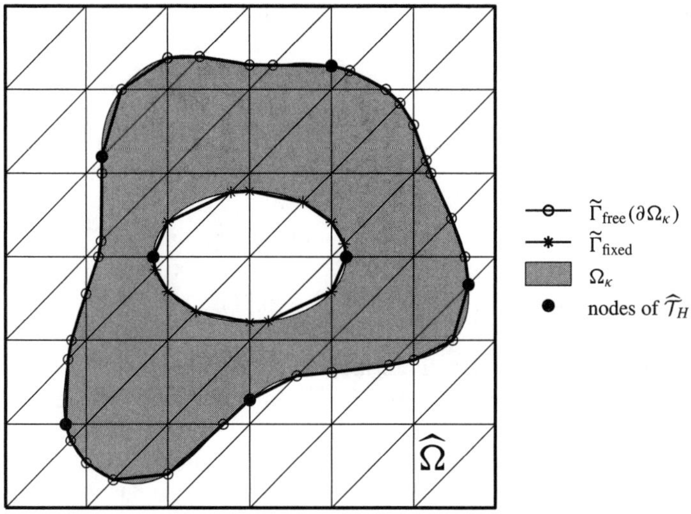

Figure 6.10. *Partitions defining finite element spaces.*

to (6.49)!) on $\Gamma_{\text{free}}(\partial\Omega)$ (see Problem 6.9). Therefore the cost functional (6.48) can be equivalently expressed as

$$J(\Omega) = \frac{1}{2}\|\lambda_2 - Q\|^2_{-1/2,\Gamma_{\text{free}}(\partial\Omega)}.$$

We shall now pass to a discretization and numerical realization of (6.47). The original family of $\mathcal{O}$ will be replaced by $\mathcal{O}_\varkappa$ made of all double connected domains $\Omega_\varkappa$, where both parts of the boundaries, Γ_{fixed} and $\Gamma_{\text{free}}(\partial\Omega_\varkappa)$, are realized by piecewise second degree Bézier curves. The fictitious domain formulation (6.50) will now be discretized the way we did in our model example at the beginning of this chapter: $V_h \subset H_0^1(\widehat{\Omega})$ denotes the space of continuous piecewise linear functions over a uniform triangulation $\widehat{\mathcal{T}}_h$ of $\overline{\widehat{\Omega}}$ and Λ_{H_1}, $\Lambda_{H_2}(\partial\Omega_\varkappa)$ are the spaces of piecewise constant functions defined on polygonal approximations $\widetilde{\Gamma}_{\text{fixed}}$, $\widetilde{\Gamma}_{\text{free}}(\partial\Omega_\varkappa)$ of Γ_{fixed}, $\Gamma_{\text{free}}(\partial\Omega_\varkappa)$, respectively (see Figure 6.10). With any $\Omega_\varkappa \in \mathcal{O}_\varkappa$ we associate the following mixed finite element problem:

$$\left\{\begin{aligned}
&\text{Find } (\widehat{u}_h, \lambda_{H_1}, \lambda_{H_2}) \\
&\qquad := (\widehat{u}_h(\Omega_\varkappa), \lambda_{H_1}(\Omega_\varkappa), \lambda_{H_2}(\Omega_\varkappa)) \in V_h \times \Lambda_{H_1} \times \Lambda_{H_2}(\partial\Omega_\varkappa) \text{ such that} \\
&\int_{\widehat{\Omega}} \nabla\widehat{u}_h \cdot \nabla v_h\, dx = \int_{\widetilde{\Gamma}_{\text{fixed}}} \lambda_{H_1} v_h\, ds + \int_{\widetilde{\Gamma}_{\text{free}}(\partial\Omega_\varkappa)} \lambda_{H_2} v_h\, ds \quad \forall v_h \in V_h; \\
&\int_{\widetilde{\Gamma}_{\text{fixed}}} \mu_{H_1}\widehat{u}_h\, ds + \int_{\widetilde{\Gamma}_{\text{free}}(\partial\Omega_\varkappa)} \mu_{H_2}\widehat{u}_h\, ds \\
&\qquad = \int_{\widetilde{\Gamma}_{\text{fixed}}} \mu_{H_1} ds \quad \forall(\mu_{H_1}, \mu_{H_2}) \in \Lambda_{H_1} \times \Lambda_{H_2}(\partial\Omega_\varkappa).
\end{aligned}\right. \tag{6.51}$$

The approximation of (6.47) using (6.51) as a state solver now reads as follows:

$$\begin{cases} \text{Find } \Omega_{\varkappa}^* \in \mathcal{O}_{\varkappa} \quad \text{such that} \\ J_{hH}(\Omega_{\varkappa}^*) \le J_{hH}(\Omega_{\varkappa}) \quad \forall \Omega_{\varkappa} \in \mathcal{O}_{\varkappa}, \end{cases} \qquad (\mathbb{P}_{\varkappa})$$

where

$$J_{hH}(\Omega_{\varkappa}) = \frac{1}{2}\|\lambda_{H_2} - Q\|^2_{0,\widetilde{\Gamma}_{\text{free}}(\partial\Omega_{\varkappa})} \qquad (6.52)$$

with λ_{H_2} being the last component of the solution to (6.51).

Remark 6.17. The mixed finite element formulation (6.51) yields the convergence of both sequences of the Lagrange multipliers in the $H^{-1/2}$-norm provided that the LBB condition is satisfied. For this reason one should use (to be correct) the dual instead of the L^2-norm in (6.52). Since the evaluation of the $H^{-1/2}$-norm is not so straightforward we decided on the formal use of the L^2-norm, which can be easily computed. On the other hand the use of (6.52) is justified if the classical mixed finite element method is replaced by its regularized form (see [IKP00]). The modification consists of adding an appropriate regularizing term to the mixed formulation, and it ensures the convergence of the Lagrange multipliers in the norm of a more regular space.

To illustrate this approach we present numerical results of a model example. Let us consider the external Bernoulli problem with Γ_{fixed} being L shaped. The fictitious domain $\widehat{\Omega} =]0,8[\times]0,8[$ is divided into small squares of size $h = 8/64$ and then each square is divided by its diagonal into two triangles. The designed part $\Gamma_{\text{free}}(\partial\Omega_{\varkappa})$ is realized by 10 Bézier segments of order 2. Let $m_1 = \dim \Lambda_{H_1}$, $m_2 = \dim \Lambda_{H_2}(\partial\Omega_{\varkappa})$ be the

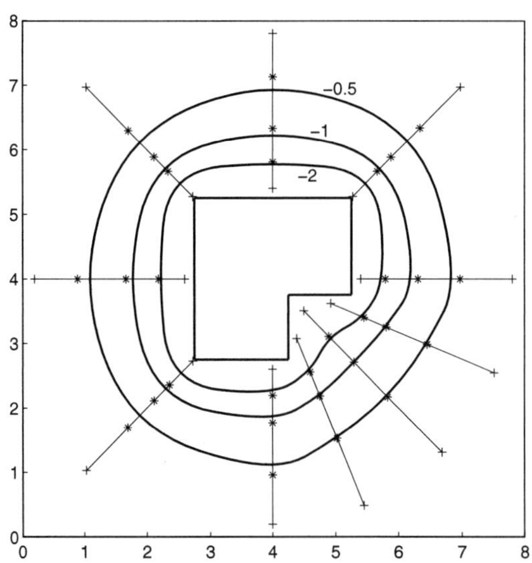

Figure 6.11. *Found solutions.*

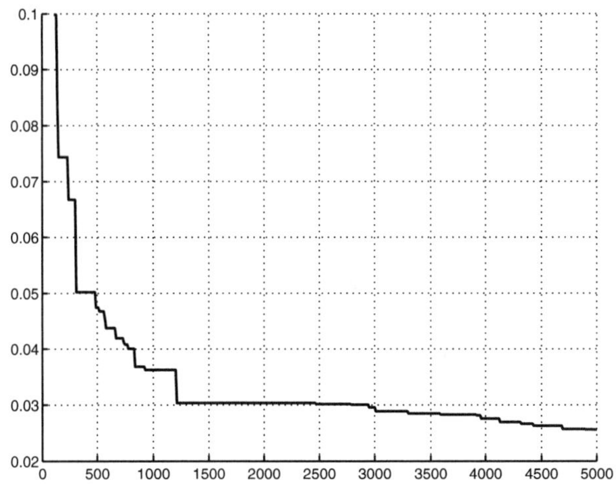

Figure 6.12. *Minimization history.*

number of constant Lagrange multipliers. In our case $m_1 = m_2 = 64$. In order to define $\Lambda_{H_1}, \Lambda_{H_2}(\partial\Omega_\varkappa)$ we first construct the polygonal approximation $\widetilde{\Gamma}_{\text{free}}(\partial\Omega_\varkappa)$ of $\Gamma_{\text{free}}(\partial\Omega_\varkappa)$ as depicted in Figure 6.10: the vertices of $\widetilde{\Gamma}_{\text{free}}(\partial\Omega_\varkappa)$ are the intersections of $\Gamma_{\text{free}}(\partial\Omega_\varkappa)$ with the interelement boundaries of $\widehat{\mathcal{T}}_h$. Γ_{fixed} and $\widetilde{\Gamma}_{\text{free}}(\partial\Omega_\varkappa)$ are then divided into connected parts S by nodes placed on interelement boundaries in such a way that the norm H of the respective partition $\widehat{\mathcal{T}}_H$ is such that the ratio H/h is as close as possible to 3. On each S, functions from $\Lambda_{H_1}, \Lambda_{H_2}(\partial\Omega_\varkappa)$ are constant.

The matrix formulation of (6.51) leads to a system of algebraic equations of type (6.15). Eliminating the primal variable $\boldsymbol{u}$ we obtain the *dual* system for the vector of Lagrange multipliers. This system was solved by the classical conjugate gradient method. The numerical minimization of the cost functional was performed again by the MCRS algorithm. The stopping criterion was given by the maximal number of function evaluations. This number was equal to 5000. In Figure 6.11 the free boundaries found after 2000 function evaluations are shown for $Q \in \{-2, -1, -0.5\}$. The behavior of the free boundary with respect to Q confirms known theoretical results (see [FR97]). A typical minimization history is shown in Figure 6.12.

Problems

PROBLEM 6.1. Prove that problem $(\widehat{\mathcal{P}})$ from Section 6.1 has a unique solution $\widehat{u} \in H_0^1(\widehat{\Omega}, \partial\Omega)$.

PROBLEM 6.2. Show that

$$\|\varphi\|_{\frac{1}{2},\partial\Omega} = \|u(\varphi)\|_{1,\Omega}, \quad \varphi \in H^{\frac{1}{2}}(\partial\Omega),$$

$$\|\mu\|_{-\frac{1}{2},\partial\Omega} = \|u(\mu)\|_{1,\Omega}, \quad \mu \in H^{-\frac{1}{2}}(\partial\Omega),$$

where $u(\varphi)$, $u(\mu)$ are unique solutions of (6.3), (6.5), respectively.

Problem 6.3. Prove that problem ($\widehat{\mathcal{M}}$) from Section 6.1 has a unique solution by using Theorem 6.1.

Problem 6.4. Prove Theorem 6.4.

Problem 6.5. Prove (6.31) provided that (6.26), (6.28), and (6.29) hold.

Problem 6.6. Consider optimal shape design problem ($\mathbb{P}$) using the DL variant of the fictitious domain method. Prove the counterpart of Theorem 6.5 with Λ_h as in Remark 6.13.

Problem 6.7. Prove that the mapping $\Phi : \mathcal{U} \to \mathbb{R}^n \times \mathbb{R}^m$, where $\Phi(\boldsymbol{\alpha}) = (\boldsymbol{u}(\boldsymbol{\alpha}), \boldsymbol{\lambda}(\boldsymbol{\alpha}))$, $\boldsymbol{\alpha} \in \mathcal{U}$, solves (6.37), is continuous in $\mathcal{U}$.

Problem 6.8. Prove that problem (6.43) has a unique solution $\widehat{u}$ and $\widehat{u}_{|\Omega}$ solves (6.39) on Ω.

Problem 6.9. Prove that problem (6.50) has a unique solution $(\widehat{u}, \lambda_1, \lambda_2)$ and interpret each component of it.

Part III

Applications

Chapter 7

Applications in Elasticity

7.1 Multicriteria optimization of a beam

In Section 2.1 the thickness optimization of an elastic beam loaded by a vertical force was analyzed. We looked for a thickness distribution minimizing the compliance of the beam. In practice, however, it is usually important to optimize structures subject to different types of loads. This section deals with a simple prototype of such problems. We shall study multiobjective thickness optimization of an elastic beam. In addition to the compliance cost functional we introduce another two functionals involving the smallest eigenvalues of two generalized eigenvalue problems. Eigenvalues represent natural frequencies of free oscillations and buckling loads of the beam and depend on the thickness distribution e. Our goal is to find a thickness minimizing the compliance of the perpendicularly loaded beam, maximizing the minimal natural frequency (i.e., the beam is stiffer under slowly varying dynamic forces), and maximizing the minimal buckling load (i.e., the beam does not lose its stability easily under the compressive load).

7.1.1 Setting of the problem

Let the beam of varying thickness e and represented by an interval $I = [0, \ell]$, $\ell > 0$, be clamped at $x = 0$ and simply supported at $x = \ell$, yielding the following boundary conditions (b.c.):

$$u(0) = u'(0) = u(\ell) = u''(\ell) = 0. \tag{7.1}$$

The deflection u of the beam under a vertical force f solves the following boundary value problem:

$$\mathcal{A}(e)u := (\beta e^3 u'')'' = f \quad \text{in }]0, \ell[\quad \text{with b.c. (7.1).} \tag{7.2}$$

The meaning of all symbols is the same as in Section 2.1 and $e \in U^{ad}$, where U^{ad} is the class of admissible thicknesses defined by (2.3).

The *equation of vibration* is represented by the following generalized eigenvalue problem:

$$\mathcal{A}(e)u = \lambda\mathcal{B}(e)u \quad \text{in }]0, \ell[\quad \text{with b.c. (7.1)}, \tag{7.3}$$

where $\mathcal{B}(e)u := \rho e u$ and $\rho \in [\rho_0, \rho_1]$ is a mass density with $\rho_1 \geq \rho_0$ positive numbers.

Finally, the *equation of buckling* is given by another eigenvalue problem:

$$\mathcal{A}(e)u = \mu\mathcal{D}u \quad \text{in }]0, \ell[\quad \text{with b.c. (7.1)}, \tag{7.4}$$

where $\mathcal{D}u := -u''$. Buckling of the beam may occur when a compressive force is greater than a critical load represented by the smallest eigenvalue of (7.4).

To give the variational formulation of the previous problems we introduce the space V:

$$V = \{v \in H^2(I) \mid v(0) = v'(0) = v(\ell) = 0\}$$

and the following bilinear forms defined on $V \times V$:

$$a_e(u, v) = \int_I \beta e^3 u'' v'' \, dx, \tag{7.5}$$

$$b_e(u, v) = \int_I \rho e u v \, dx, \tag{7.6}$$

$$d(u, v) = \int_I u' v' \, dx, \quad u, v \in V, \ e \in U^{ad}, \tag{7.7}$$

related to the differential operators $\mathcal{A}(e)$, $\mathcal{B}(e)$, and $\mathcal{D}$, respectively.

The weak formulations of (7.2), (7.3), and (7.4) read as follows:

$$\begin{cases} \text{Find } u := u(e) \in V \quad \text{such that} \\ a_e(u, v) = \displaystyle\int_I f v \, dx \quad \forall v \in V. \end{cases} \tag{$\mathcal{P}_1(e)$}$$

$$\begin{cases} \text{Find } (u, \lambda) := (u(e), \lambda(e)) \in V \times \mathbb{R}, \ u \neq 0, \quad \text{such that} \\ a_e(u, v) = \lambda b_e(u, v) \quad \forall v \in V. \end{cases} \tag{$\mathcal{P}_2(e)$}$$

$$\begin{cases} \text{Find } (u, \mu) := (u(e), \mu(e)) \in V \times \mathbb{R}, \ u \neq 0, \quad \text{such that} \\ a_e(u, v) = \mu d(u, v) \quad \forall v \in V. \end{cases} \tag{$\mathcal{P}_3(e)$}$$

The thickness optimization of the beam mentioned in the introduction to this section can be formulated as the following multicriteria optimization problem:

$$\begin{cases} \text{Find a Pareto optimal } e^* \in U^{ad} \quad \text{for the problem} \\ \displaystyle\min_{e \in U^{ad}} \{J_1(e), J_2(e), J_3(e)\}. \end{cases} \tag{$\mathbb{P}$}$$

The individual cost functionals are defined by

$$J_1(e) = \int_I f u(e)\,dx, \quad J_2(e) = 1/\lambda_1(e), \quad J_3(e) = 1/\mu_1(e),$$

where $u(e)$ solves $(\mathcal{P}_1(e))$ and $\lambda_1(e),\ \mu_1(e)$ are the smallest eigenvalues in $(\mathcal{P}_2(e))$, $(\mathcal{P}_3(e))$, respectively. Before we prove that the set of Pareto optimal solutions is nonempty, we will show that $(\mathbb{P})$ is well posed. It is known (see Appendix A) that $\lambda_1(e),\ \mu_1(e)$ are the minimizers of the Rayleigh quotients:

$$\lambda_1(e) = \min_{\substack{v\in V\\ v\neq 0}} \frac{a_e(v,v)}{b_e(v,v)}, \qquad \mu_1(e) = \min_{\substack{v\in V\\ v\neq 0}} \frac{a_e(v,v)}{d(v,v)}.$$

It is easy to prove that $\lambda_1(e), \mu_1(e)$ are bounded from above and below *uniformly* in U^{ad}: there exist positive constants $m \le M$ such that

$$m \le \lambda_1(e), \mu_1(e) \le M \quad \forall e \in U^{ad}. \tag{7.8}$$

Indeed,

$$\frac{e^3_{\min}\beta_0}{\rho_1 e_{\max}} \frac{|v|^2_{2,I}}{\|v\|^2_{0,I}} \le \frac{a_e(v,v)}{b_e(v,v)} \le \frac{e^3_{\max}\|\beta\|_{L^\infty(I)}}{\rho_0 e_{\min}} \frac{|v|^2_{2,I}}{\|v\|^2_{0,I}}.$$

Therefore,

$$\frac{e^3_{\min}\beta_0}{\rho_1 e_{\max}}\overline{\lambda}_1 \le \lambda_1(e) \le \frac{e^3_{\max}\|\beta\|_{L^\infty(I)}}{\rho_0 e_{\min}}\overline{\lambda}_1,$$

where $\overline{\lambda}_1 > 0$ is the smallest eigenvalue of the auxiliary problem

$$u'''' = \lambda u \quad \text{in }]0,\ell[\quad \text{with b.c. (7.1).} \tag{7.9}$$

Similar bounds hold for $\mu_1(e)$. Thus the functionals J_2 and J_3 are well defined for any $e \in U^{ad}$.

The multicriteria optimization problem will be realized by the *weighting method* described in Subsection 4.4.2. Problem $(\mathbb{P})$ will be replaced as follows:

$$\begin{cases} \text{Find } e^* \in U^{ad} \quad \text{such that} \\ J_0(e^*) \le J_0(e) \quad \forall e \in U^{ad}, \end{cases} \tag{$\mathbb{P}_w$}$$

where $J_0(e) = \sum_{i=1}^3 w_i J_i(e)$ and $w_i > 0,\ i = 1,2,3$, are given positive weights. To prove that $(\mathbb{P}_w)$ has a solution we need the following lemma.

Lemma 7.1. *The functional J_0 is continuous in U^{ad}:*

$$e_n \rightrightarrows e \text{ in } I, \quad e_n, e \in U^{ad} \implies J_0(e_n) \to J_0(e), \quad n \to \infty.$$

Proof. We already know that J_1 is continuous in U^{ad}. Let us prove the same for J_2. Since $\{\lambda_1(e_n)\}$ is bounded, as follows from (7.8), one can pass to a convergent subsequence such that

$$\lambda_1(e_n) \to \lambda, \quad n \to \infty, \tag{7.10}$$

for some $\lambda > 0$. Let $u_n := u(e_n)$ be an eigenfunction corresponding to $\lambda_1(e_n)$:

$$a_{e_n}(u_n, v) = \lambda_1(e_n) b_{e_n}(u_n, v) \quad \forall v \in V. \tag{7.11}$$

We may assume that $\|u_n\|_{2,I} = 1 \; \forall n \in \mathbb{N}$ so that there is a subsequence of $\{u_n\}$ such that

$$u_n \rightharpoonup u \quad \text{in } V, \quad n \to \infty. \tag{7.12}$$

Letting $n \to \infty$ in (7.11), using (7.10), (7.12), and uniform convergence of $\{e_n\}$ to e in I we arrive at

$$a_e(u, v) = \lambda b_e(u, v) \quad \forall v \in V.$$

We now prove that $\{u_n\}$ tends strongly to u. We use the same approach as in Lemma 2.1. Let $|||\cdot|||$ be the norm defined by (2.9). Then

$$\begin{aligned} |||u_n|||^2 &= \int_I \beta e^3 (u_n'')^2 \, dx = \int_I \beta (e^3 - e_n^3)(u_n'')^2 \, dx + \int_I \beta e_n^3 (u_n'')^2 \, dx \\ &= \int_I \beta (e^3 - e_n^3)(u_n'')^2 \, dx + \lambda_1(e_n) \int_I \rho e_n u_n^2 \, dx \to \lambda \int_I \rho e u^2 \, dx = |||u|||^2. \end{aligned}$$

Since $|||\cdot|||$ and $\|\cdot\|_{2,I}$ are equivalent we see that $\|u\|_{2,I} = 1$, implying that λ is an eigenvalue and u is an eigenfunction corresponding to λ.

It remains to show that $\lambda := \lambda_1(e)$ is the smallest eigenvalue in $(\mathcal{P}_2(e))$. Since $\lambda_1(e_n)$ is the smallest eigenvalue in $(\mathcal{P}_2(e_n))$ for any $n \in \mathbb{N}$ it holds that

$$\lambda_1(e_n) = \mathcal{R}(e_n, u_n) := \frac{a_{e_n}(u_n, u_n)}{b_{e_n}(u_n, u_n)} \le \mathcal{R}(e_n, v) \quad \forall v \in V. \tag{7.13}$$

It is left as an easy exercise (see Problem 7.1) to show that the Rayleigh quotient is continuous in $U^{ad} \times V$. Passing to the limit with $n \to \infty$ in (7.13) we obtain

$$\lambda = \mathcal{R}(e, u) \le \mathcal{R}(e, v) \quad \forall v \in V,$$

making use of (7.10) and strong convergence of $\{u_n\}$ to u. From this it follows that $\lambda := \lambda_1(e)$ is the smallest eigenvalue in $(\mathcal{P}_2(e))$. Since any accumulation point of $\{\lambda_1(e_n)\}$ has this characterization the whole sequence $\{\lambda_1(e_n)\}$ tends to $\lambda_1(e)$. That J_3 is continuous can be proved in the same way. $\square$

From Lemma 7.1 and the compactness of U^{ad} we arrive at the following result.

Theorem 7.1. *Problem* $(\mathbb{P}_w)$ *has a solution.*

7.1.2 Approximation and numerical realization of $(\mathbb{P}_w)$

We proceed as in Section 2.1. The space V will be discretized by means of piecewise cubic polynomials over an equidistant partition $\Delta_h = \{a_i\}_{i=0}^d$ of the interval I:

$$\begin{aligned} V_h = \{v_h \in C^1(I) \mid v_{h|[a_{i-1}, a_i]} \in P_3([a_{i-1}, a_i]) \; \forall i = 1, \dots, d, \\ v_h(0) = v_h'(0) = v_h(\ell) = 0\}, \end{aligned}$$

while thicknesses are discretized by piecewise constant functions belonging to $\widetilde{U}_h^{ad}$ given by (2.32).

Let $e_h \in \widetilde{U}_h^{ad}$ be fixed. The discrete state problems are defined as follows:

$$\begin{cases} \text{Find } u_h := u_h(e_h) \in V_h \quad \text{such that} \\ a_{e_h}(u_h, v_h) = \int_I f v_h \, dx \quad \forall v \in V_h. \end{cases} \tag{$\mathcal{P}_{1h}(e_h)$}$$

$$\begin{cases} \text{Find } (u_h, \lambda_h) := (u_h(e_h), \lambda_h(e_h)) \in V_h \times \mathbb{R}, \ u_h \neq 0, \quad \text{such that} \\ a_{e_h}(u_h, v_h) = \lambda_h b_{e_h}(u_h, v_h) \quad \forall v_h \in V_h. \end{cases} \tag{$\mathcal{P}_{2h}(e_h)$}$$

$$\begin{cases} \text{Find } (u_h, \mu_h) := (u_h(e_h), \mu_h(e_h)) \in V_h \times \mathbb{R}, \ u_h \neq 0, \quad \text{such that} \\ a_{e_h}(u_h, v_h) = \mu_h d(u_h, v_h) \quad \forall v_h \in V_h. \end{cases} \tag{$\mathcal{P}_{3h}(e_h)$}$$

The discretization of $(\mathbb{P}_w)$ now reads as follows:

$$\begin{cases} \text{Find } e_h^* \in \widetilde{U}_h^{ad} \quad \text{such that} \\ J_0(e_h^*) \leq J_0(e_h) \quad \forall e_h \in \widetilde{U}_h^{ad}, \end{cases} \tag{$\mathbb{P}_{wh}$}$$

where $J_0(e_h) = \sum_{i=1}^3 w_i J_i(e_h)$, $J_1(e_h) = \int_I f u_h \, dx$, $J_2(e_h) = 1/\lambda_{1h}$, $J_3(e_h) = 1/\mu_{1h}$, with $u_h := u_h(e_h)$ being the solution of $(\mathcal{P}_{1h}(e_h))$ and $\lambda_{1h} := \lambda_{1h}(e_h)$, $\mu_{1h} := \mu_{1h}(e_h)$ being the smallest eigenvalues of $(\mathcal{P}_{2h}(e_h))$, $(\mathcal{P}_{3h}(e_h))$, respectively. Using exactly the same approach as in the continuous setting of the problem one can prove the following theorem.

Theorem 7.2. *For any $h > 0$ there exists a solution to $(\mathbb{P}_{wh})$.*

Next we shall investigate the relation between $(\mathbb{P}_w)$ and $(\mathbb{P}_{wh})$ for $h \to 0+$. To this end we need the following lemma.

Lemma 7.2. *Let $e_h \to e$ in $L^\infty(I)$, $e_h \in \widetilde{U}_h^{ad}$, $e \in U^{ad}$. Then*

$$\lambda_{1h}(e_h) \to \lambda_1(e), \quad \mu_{1h}(e_h) \to \mu_1(e) \quad \text{as } h \to 0+$$

and $\lambda_1(e)$, $\mu_1(e)$ are the smallest eigenvalues of $(\mathcal{P}_2(e))$, $(\mathcal{P}_3(e))$, respectively.

Proof. We first show that $\{\lambda_{1h}(e_h)\}$ is bounded. As in the previous subsection we obtain that

$$c_1 \frac{|v_h|_{1,I}^2}{\|v_h\|_{0,I}^2} \leq \mathcal{R}(e_h, v_h) = \frac{a_{e_h}(v_h, v_h)}{b_{e_h}(v_h, v_h)} \leq c_2 \frac{|v_h|_{2,I}^2}{\|v_h\|_{0,I}^2}$$

holds for any $v_h \in V_h$ and $h > 0$, where $c_1, c_2 > 0$ are constants independent of v_h and h. Therefore,

$$c_1 \overline{\lambda}_{1h} \leq \lambda_{1h}(e_h) \leq c_2 \overline{\lambda}_{1h} \quad \forall h > 0, \tag{7.14}$$

where $\overline{\lambda}_{1h}$ is the smallest eigenvalue of the auxiliary problem

$$(\overline{u}_h, \overline{\lambda}_h) \in V_h \times \mathbb{R},\ \overline{u}_h \neq 0: \quad \int_I \overline{u}_h'' v_h''\, dx = \overline{\lambda}_h \int_I \overline{u}_h v_h\, dx \quad \forall v_h \in V_h.$$

From the continuity of $\mathcal{R}$ and the density of $\{V_h\}$, $h \to 0+$, in V it follows that $\overline{\lambda}_h \to \overline{\lambda}$, $h \to 0+$, where $\overline{\lambda}$ is the smallest eigenvalue of (7.9). From this and (7.14) the boundedness of $\{\lambda_{1h}(e_h)\}$ immediately follows. We may assume again that the eigenfunctions u_h are normalized: $\|u_h\|_{2,I} = 1\ \forall h > 0$. The rest of the proof is straightforward: one can pass to convergent subsequences such that

$$\lambda_{1h}(e_h) \to \lambda_1, \quad u_h(e_h) \rightharpoonup u \quad \text{in } V \text{ as } h \to 0+. \tag{7.15}$$

Letting $h \to 0+$ in $(\mathcal{P}_{2h}(e_h))$, using (7.15) and the density of $\{V_h\}$ in V, we obtain

$$a_e(u, v) = \lambda_1 b_e(u, v) \quad \forall v \in V.$$

Strong convergence in (7.15) can be shown using exactly the same approach as in the proof of Lemma 7.1. Thus λ_1 is an eigenvalue and $u \neq 0$ is the corresponding eigenfunction in $(\mathcal{P}_2(e))$. The fact that $\lambda_1 := \lambda_1(e)$ is the smallest eigenvalue in $(\mathcal{P}_2(e))$ follows again from the continuity of the Rayleigh quotient and the density of $\{V_h\}$ in V. This implies that not only the subsequence, but also the whole sequence $\{\lambda_{1h}(e_h)\}$, tends to $\lambda_1(e)$. ◻

With this result in hand we arrive at the following convergence statement.

Theorem 7.3. *For any sequence $\{e_h^*\}$ of solutions to $(\mathbb{P}_{wh})$, $h \to 0+$, there exists a subsequence such that*

$$e_h^* \to e^* \quad \text{in } L^\infty(I), \quad h \to 0+, \tag{7.16}$$

and $e^* \in U^{ad}$ solves $(\mathbb{P}_w)$. In addition, any accumulation point of $\{e_h^*\}$ in the sense of (7.16) *possesses this characteristic.*

Proof. The proof is left as an easy exercise. ◻

Because any function $e_h \in \widetilde{U}_h^{ad}$ is constant in each interval $[a_i, a_{i+1}]$, $i = 0, \ldots, d-1$, of length h, it is uniquely determined by a vector $\boldsymbol{e} = (e_1, \ldots, e_d) \in \mathcal{U}$, $e_i = e_{h|[a_{i-1}, a_i]}$, where

$$\mathcal{U} = \Big\{ \boldsymbol{e} \in \mathbb{R}^d \mid e_{\min} \le e_i \le e_{\max},\ i = 1, \ldots, d,$$
$$|e_{i+1} - e_i| \le L_0 h,\ i = 1, \ldots, d-1,\ h \sum_{i=1}^d e_i = \gamma \Big\}.$$

The algebraic formulation of discrete state problems $(\mathcal{P}_{1h}(e_h))$–$(\mathcal{P}_{3h}(e_h))$ leads to one linear algebraic system and two generalized algebraic eigenvalue problems:

$$\boldsymbol{K}(\boldsymbol{e})\boldsymbol{q}(\boldsymbol{e}) = \boldsymbol{f}, \tag{7.17}$$
$$\boldsymbol{K}(\boldsymbol{e})\boldsymbol{z}(\boldsymbol{e}) = \lambda(\boldsymbol{e})\boldsymbol{B}(\boldsymbol{e})\boldsymbol{z}(\boldsymbol{e}), \tag{7.18}$$
$$\boldsymbol{K}(\boldsymbol{e})\boldsymbol{w}(\boldsymbol{e}) = \mu(\boldsymbol{e})\boldsymbol{D}\boldsymbol{w}(\boldsymbol{e}). \tag{7.19}$$

Here $\boldsymbol{K}(\boldsymbol{e})$, $\boldsymbol{B}(\boldsymbol{e})$, and $\boldsymbol{D}$ are the stiffness matrix, mass matrix, and geometric stiffness matrix, respectively, and $\boldsymbol{f}$ is the force vector. The elements of $\boldsymbol{K}(\boldsymbol{e})$, $\boldsymbol{B}(\boldsymbol{e})$, $\boldsymbol{D}$, and $\boldsymbol{f}$ are evaluated as follows:

$$\begin{aligned}
k_{ij}(\boldsymbol{e}) &= \int_I \beta e_h^3 \varphi_i'' \varphi_j''\, dx = \sum_{k=1}^{d} e_k^3 \int_{a_{k-1}}^{a_k} \beta \varphi_i'' \varphi_j''\, dx,\\
b_{ij}(\boldsymbol{e}) &= \int_I \rho e_h \varphi_i \varphi_j\, dx = \sum_{k=1}^{d} e_k \int_{a_{k-1}}^{a_k} \rho \varphi_i \varphi_j\, dx,\\
d_{ij} &= \int_I \varphi_i' \varphi_j'\, dx,\\
f_j &= \int_I f \varphi_j\, dx, \quad i, j = 1, \dots, n,
\end{aligned}$$

where $\{\varphi_i\}_{i=1}^n$ is a basis of V_h and $n = \dim V_h$.

Let $\mathcal{J}_i(\boldsymbol{e}) := J_i(\mathcal{T}_D^{-1}\boldsymbol{e})$, $i = 1, 2, 3$, where $\mathcal{T}_D$ is the isomorphism between $\widetilde{U}_h^{ad}$ and $\mathcal{U}$ (see also (2.21)). Then the algebraic form of $(\mathbb{P}_{wh})$ reads as follows:

$$\begin{cases} \text{Find } \boldsymbol{e}^* \in \mathcal{U} \quad \text{such that} \\ \mathcal{J}_0(\boldsymbol{e}^*) \le \mathcal{J}_0(\boldsymbol{e}) \quad \forall \boldsymbol{e} \in \mathcal{U}, \end{cases} \tag{$\mathbb{P}_{wd}$}$$

where $\mathcal{J}_0(\boldsymbol{e}) = \sum_{i=1}^3 w_i \mathcal{J}_i(\boldsymbol{e})$.

Let us briefly comment on the numerical solution of the algebraic eigenvalue problem. If we are only interested in the (simple) smallest eigenvalue and the respective eigenfunction for problems (7.18), (7.19), we can use one of the simplest methods, namely the classical inverse iteration method. The reduction of the generalized eigenproblem (7.18) to a standard eigenproblem can be done implicitly (for simplicity of notation the argument $\boldsymbol{e}$ is omitted): multiplying (7.18) by $\boldsymbol{B}^{-1}$ from the left we obtain the standard eigenproblem

$$\boldsymbol{B}^{-1}\boldsymbol{K}\boldsymbol{z} = \lambda \boldsymbol{z}.$$

The matrix $\boldsymbol{C} = \boldsymbol{B}^{-1}\boldsymbol{K}$ is neither symmetric nor sparse, but it is self-adjoint with respect to the B inner product $(\boldsymbol{x}, \boldsymbol{y})_B := \boldsymbol{x}^{\mathrm{T}}\boldsymbol{B}\boldsymbol{y}$. Thus one can solve the eigenvalue problem

$$\boldsymbol{C}\boldsymbol{z} = \lambda \boldsymbol{z}$$

using inverse iterations in which the inner products are represented by the B inner products (see [Par98]). If faster convergence is needed or one needs the k smallest eigenvalues ($k \ll n$), then the subspace iteration is an efficient method of calculating them. For more details see [Par98]. For example, subroutine F02FJF in the NAG subroutine library [NAG97] can be used in this way to solve problem (7.18).

The piecewise constant parametrization of the thickness will result in a large number of optimization parameters. Therefore we prefer gradient type methods for the numerical solution of $(\mathbb{P}_{wd})$. Sensitivity analysis for the cost function $\mathcal{J}_1$ can be done in the way described in Section 3.2. The cost functions $\mathcal{J}_2$ and $\mathcal{J}_3$ involve the smallest eigenvalues of (7.18) and (7.19). Thus one needs a formula for the directional derivatives of eigenvalues with respect to design parameters. Consider next the following generalized algebraic eigenvalue problem depending on a parameter $\boldsymbol{e} \in \mathcal{U}$:

$$\boldsymbol{A}(\boldsymbol{e})z(\boldsymbol{e}) = \lambda(\boldsymbol{e})\boldsymbol{M}(\boldsymbol{e})z(\boldsymbol{e}). \tag{7.20}$$

We assume that the eigenvectors are normalized in the following way:

$$z(\boldsymbol{e})^{\mathrm{T}}\boldsymbol{M}(\boldsymbol{e})z(\boldsymbol{e}) = 1. \tag{7.21}$$

Theorem 7.4. *Assume that $\boldsymbol{A}(\boldsymbol{e})$, $\boldsymbol{M}(\boldsymbol{e})$ are symmetric and positive definite for all $\boldsymbol{e} \in \mathcal{U}$. Let the kth eigenvalue $\lambda_k(\boldsymbol{e})$ in (7.20), $k = 1, \dots, n$, be simple at $\boldsymbol{e} = \overline{\boldsymbol{e}}$ and let $z_k(\overline{\boldsymbol{e}})$ be the corresponding eigenvector. If the matrix functions $\boldsymbol{A} : \boldsymbol{e} \mapsto \boldsymbol{A}(\boldsymbol{e})$ and $\boldsymbol{M} : \boldsymbol{e} \mapsto \boldsymbol{M}(\boldsymbol{e})$ are continuously differentiable in $\mathcal{U}$, so is the function $\lambda_k : \boldsymbol{e} \mapsto \lambda_k(\boldsymbol{e})$ at $\boldsymbol{e} = \overline{\boldsymbol{e}}$. In addition, the directional derivative of λ_k at $\overline{\boldsymbol{e}}$ and in any direction $\boldsymbol{e} \in \mathbb{R}^d$ is given by*

$$\lambda_k'(\overline{\boldsymbol{e}}; \boldsymbol{e}) = z_k(\overline{\boldsymbol{e}})^{\mathrm{T}}\boldsymbol{A}'(\overline{\boldsymbol{e}}; \boldsymbol{e})z_k(\overline{\boldsymbol{e}}) - \lambda_k(\overline{\boldsymbol{e}})z_k(\overline{\boldsymbol{e}})^{\mathrm{T}}\boldsymbol{M}'(\overline{\boldsymbol{e}}; \boldsymbol{e})z_k(\overline{\boldsymbol{e}}). \tag{7.22}$$

Proof. It is readily seen that if $\lambda_k(\overline{\boldsymbol{e}})$ is a simple eigenvalue, then $\lambda_k(\boldsymbol{e})$ remains simple for any $\boldsymbol{e}$ belonging to a sufficiently small neighborhood of $\overline{\boldsymbol{e}}$ and, in addition, the function $\lambda_k : \boldsymbol{e} \mapsto \lambda_k(\boldsymbol{e})$ is continuously differentiable in a vicinity of $\overline{\boldsymbol{e}}$ (see Problem 7.2). Let $\boldsymbol{e} \in \mathbb{R}^d$ be arbitrary and $|t| \le \delta$, δ small enough. Then differentiating the identity

$$\boldsymbol{A}(\overline{\boldsymbol{e}} + t\boldsymbol{e})z_k(\overline{\boldsymbol{e}} + t\boldsymbol{e}) = \lambda_k(\overline{\boldsymbol{e}} + t\boldsymbol{e})\boldsymbol{M}(\overline{\boldsymbol{e}} + t\boldsymbol{e})z_k(\overline{\boldsymbol{e}} + t\boldsymbol{e}), \quad |t| \le \delta, \tag{7.23}$$

with respect to t at $t = 0$ we obtain

$$\begin{aligned} &\boldsymbol{A}'(\overline{\boldsymbol{e}}; \boldsymbol{e})z_k(\overline{\boldsymbol{e}}) + \boldsymbol{A}(\overline{\boldsymbol{e}})z_k'(\overline{\boldsymbol{e}}; \boldsymbol{e}) \\ &\quad = \lambda'(\overline{\boldsymbol{e}}; \boldsymbol{e})\boldsymbol{M}(\overline{\boldsymbol{e}})z_k(\overline{\boldsymbol{e}}) + \lambda(\overline{\boldsymbol{e}})\boldsymbol{M}'(\overline{\boldsymbol{e}}; \boldsymbol{e})z_k(\overline{\boldsymbol{e}}) + \lambda(\overline{\boldsymbol{e}})\boldsymbol{M}(\overline{\boldsymbol{e}})z_k'(\overline{\boldsymbol{e}}; \boldsymbol{e}), \end{aligned} \tag{7.24}$$

using also that the mapping $z_k : \boldsymbol{e} \mapsto z_k(\boldsymbol{e})$ is continuously differentiable at $\boldsymbol{e} = \overline{\boldsymbol{e}}$ (see Problem 7.2). Multiplying (7.24) from the left by $z_k(\overline{\boldsymbol{e}})^{\mathrm{T}}$ and taking into account (7.20) and (7.21) we arrive at (7.22). $\square$

We use this theorem with $\boldsymbol{A}(\boldsymbol{e}) := \boldsymbol{K}(\boldsymbol{e})$ and $\boldsymbol{M}(\boldsymbol{e}) := \boldsymbol{B}(\boldsymbol{e})$ and $\boldsymbol{D}$.

From the previous theorem we see that if we differentiate functionals depending on eigenvalues only, no adjoint equation is needed. The computation of $\boldsymbol{K}'$ and $\boldsymbol{B}'$ is very simple due to the piecewise constant discretization of the thickness e. In particular the partial derivatives of their elements are given by

$$\frac{\partial k_{ij}(\boldsymbol{e})}{\partial e_k} = 3e_k^2 \int_{a_{k-1}}^{a_k} \beta\varphi_i''\varphi_j''\,dx, \tag{7.25}$$

$$\frac{\partial b_{ij}(\boldsymbol{e})}{\partial e_k} = \int_{a_{k-1}}^{a_k} \rho\varphi_i\varphi_j\,dx, \quad k = 1, \dots, d. \tag{7.26}$$

When a black box optimization routine using a gradient type method is used to minimize $\mathcal{J}_0$, the user must supply a subroutine that computes the gradients of the cost functionals $\mathcal{J}_i$, $i = 1, 2, 3$, at a given point $\boldsymbol{e}$. This can be done in the following way:

1. Form $\boldsymbol{K}(\boldsymbol{e})$, $\boldsymbol{B}(\boldsymbol{e})$, $\boldsymbol{D}$, and $\boldsymbol{f}$ corresponding to the current design $\boldsymbol{e}$.
2. Triangularize $\boldsymbol{K}(\boldsymbol{e}) = \boldsymbol{L}\boldsymbol{L}^{\mathrm{T}}$.
3. Solve the triangular systems $\boldsymbol{L}\boldsymbol{y} = \boldsymbol{f}$ and $\boldsymbol{L}^{\mathrm{T}}\boldsymbol{q} = \boldsymbol{y}$.
4. Compute $\mathcal{J}_1(\boldsymbol{e}) = \boldsymbol{f}^{\mathrm{T}}\boldsymbol{q}$.
5. Find iteratively the smallest eigenvalues λ_1, μ_1 and the corresponding eigenvectors $\boldsymbol{z}, \boldsymbol{w}$ for the problems $\boldsymbol{K}(\boldsymbol{e})\boldsymbol{z} = \lambda \boldsymbol{B}(\boldsymbol{e})\boldsymbol{z}$ and $\boldsymbol{K}(\boldsymbol{e})\boldsymbol{w} = \mu \boldsymbol{D}\boldsymbol{w}$ using the factorization of $\boldsymbol{K}(\boldsymbol{e})$.
6. Compute $\mathcal{J}_2(\boldsymbol{e}) = 1/\lambda_1$, $\mathcal{J}_3(\boldsymbol{e}) = 1/\mu_1$.
7. **do** $k = 1, \dots, n$.
   (i) Compute the partial derivatives of $\boldsymbol{K}(\boldsymbol{e})$ and $\boldsymbol{B}(\boldsymbol{e})$ with respect to e_k using (7.25), (7.26).
   (ii) Compute the partial derivatives of $\mathcal{J}_i$, $i = 1, \dots, 3$ by

$$\begin{cases} \dfrac{\partial \mathcal{J}_1}{\partial e_k}(\boldsymbol{e}) = -\boldsymbol{q}^{\mathrm{T}} \dfrac{\partial \boldsymbol{K}(\boldsymbol{e})}{\partial e_k} \boldsymbol{q}, \\ \dfrac{\partial \mathcal{J}_2}{\partial e_k}(\boldsymbol{e}) = -\lambda_1^{-2} \left(\boldsymbol{z}^{\mathrm{T}} \dfrac{\partial \boldsymbol{K}(\boldsymbol{e})}{\partial e_k} \boldsymbol{z} - \lambda_1 \boldsymbol{z}^{\mathrm{T}} \dfrac{\partial \boldsymbol{B}(\boldsymbol{e})}{\partial e_k} \boldsymbol{z} \right), \\ \dfrac{\partial \mathcal{J}_3}{\partial e_k}(\boldsymbol{e}) = -\mu_1^{-2} \boldsymbol{w}^{\mathrm{T}} \dfrac{\partial \boldsymbol{K}(\boldsymbol{e})}{\partial e_k} \boldsymbol{w}. \end{cases} \tag{7.27}$$

end do

Remark 7.1. Theorem 7.4 does not apply to the case of multiple eigenvalues. In fact it can be shown that multiple eigenvalues are generally only *directionally differentiable* with respect to $\boldsymbol{e}$. For details, see [HCK86].

It has been shown in [OR77] that if μ_1 is maximized for a clamped-clamped beam (i.e., $u(x) = u'(x) = 0$ at $x = 0$ and l) having a circular cross section, then the eigenvalue $\mu_1(e^*)$ is double at optimum e^*. Then gradient type methods may give unsatisfactory results and, e.g., methods of nonsmooth optimization should be used. Unfortunately, these methods are not widely available in standard subroutine libraries.

If eigenvalues are optimized using gradient type methods for smooth functions, it is necessary to monitor the multiplicity of the eigenvalues during the optimization process. If the first and second eigenvalues $\lambda_1(\boldsymbol{e}_m)$ and $\lambda_2(\boldsymbol{e}_m)$ at all iterations $\boldsymbol{e}_m$, $m = 1, 2, \dots,$ generated by gradient methods are clearly separated, then the results are trustworthy.

Example 7.1. Let us consider a beam of unit length ($\ell = 1$). The load f and the parameters related to the material properties and the cross-sectional shape of the beam have constant values: $f = -1$, $\beta = \rho = 1$. The set U^{ad} is defined by the following parameters: $e_{\min} = 0.01$, $e_{\max} = 0.1$, $L_0 = 0.5$, $\gamma = 0.05$. The beam is discretized by using 32 cubic Hermite elements; i.e., $h = 1/32$ and $d = 32$.

In optimization we used the sequential quadratic programming (SQP) algorithm E04UCF from the NAG subroutine library [NAG97]. The state problems were solved by the band Cholesky method. Instead of the simple inverse iteration method, the eigenvalue problems were solved using the subspace iteration algorithm F02FJF from the NAG

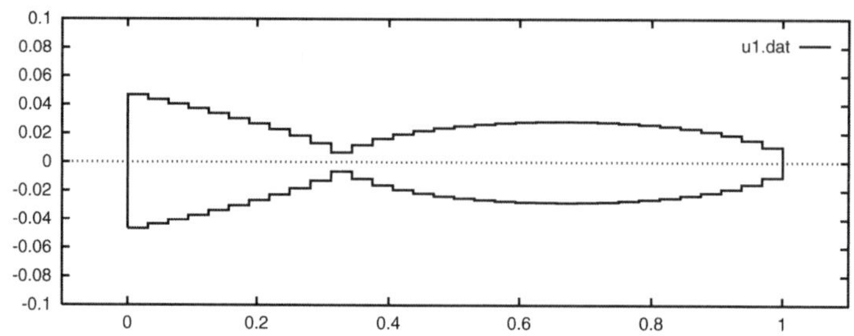

Figure 7.1. *A beam minimizing the cost functional $\mathcal{J}_1$. Compliance* $= 16.03$, $\lambda_1 = 0.9202$, $\mu_1 = 0.02963$.

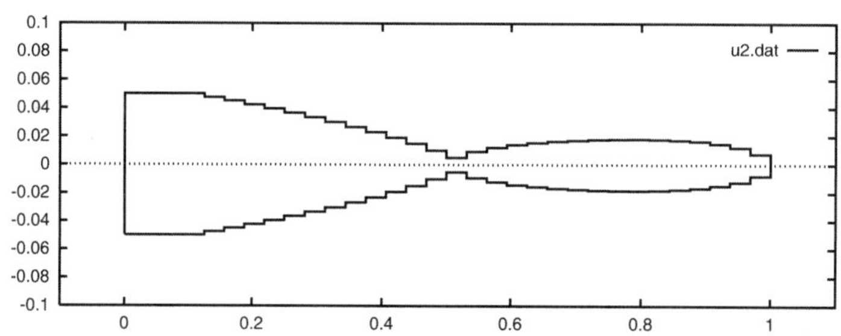

Figure 7.2. *A beam minimizing the cost functional $\mathcal{J}_2$. Compliance* $= 20.67$, $\lambda_1 = 1.059$, $\mu_1 = 0.01572$.

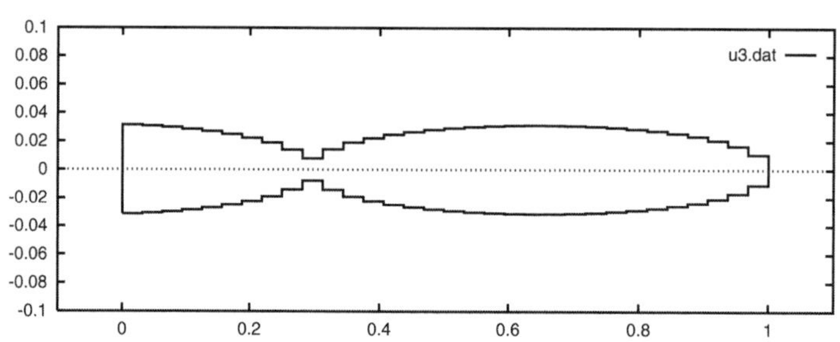

Figure 7.3. *A beam minimizing the cost functional $\mathcal{J}_3$. Compliance* $= 18.23$, $\lambda_1 = 0.7962$, $\mu_1 = 0.03530$.

library. The two smallest eigenvalues and their corresponding eigenvectors were calculated in order to check whether the smallest eigenvalue is simple or not and thus to ensure that the assumptions of Theorem 7.4 are satisfied. In all examples a constant thickness $e_h(x) = 0.05$ was used as an initial guess. The number of SQP iterations in different examples varied between 31 and 40.

We first minimize each of the cost functionals $\mathcal{J}_i$, $i = 1, 2, 3$, separately. In Figures 7.1, 7.2, and 7.3 the solutions of the individual optimization problems are shown. The

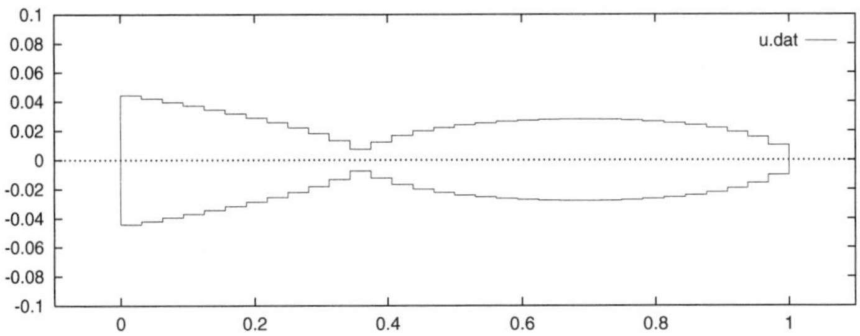

Figure 7.4. *A beam minimizing the cost functional* $\widehat{\mathcal{J}}_0$ *with* $w_1 = w_3 = 0.25,\ w_2 = 0.5$. *Compliance* $= 16.26$, $\lambda_1 = 0.9489$, $\mu_1 = 0.03131$.

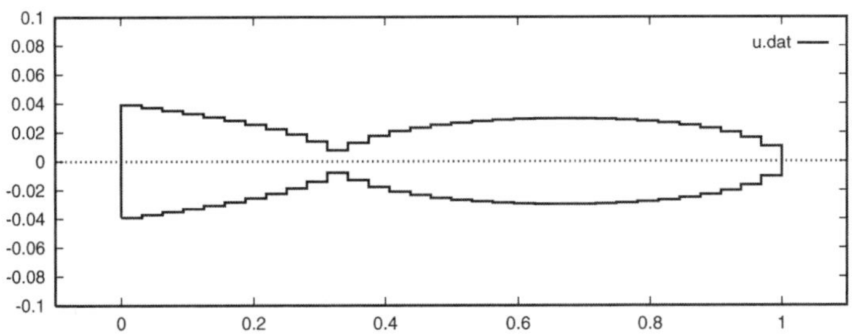

Figure 7.5. *A beam minimizing the cost functional* $\widehat{\mathcal{J}}_0$ *with* $w_1 = 0.1,\ w_2 = 0.4,\ w_3 = 0.5$. *Compliance* $= 16.71$, $\lambda_1 = 0.8881$, $\mu_1 = 0.03387$.

maximum thickness and the position of the "hinge-like" section of the beam vary considerably in the results of different optimization problems.

Let $z_i^* = \min_{e \in \mathcal{U}} \mathcal{J}_i(e),\ i = 1, 2, 3$, denote the components of the ideal criterion vector for the respective discrete multiobjective optimization problem. For proper scaling of the functions in $(\mathbb{P}_{wd})$ we set

$$\widehat{\mathcal{J}}_0(e) = w_1 \widehat{\mathcal{J}}_1(e) + w_2 \widehat{\mathcal{J}}_2(e) + w_3 \widehat{\mathcal{J}}_3(e),$$

where $\widehat{\mathcal{J}}_i(e) := \mathcal{J}_i(e)/z_i^*$. Then each $\widehat{\mathcal{J}}_i$ has minimum value equal to one if optimized separately.

Next we compute two (locally) Pareto optimal solutions by using the scalarization function $\widehat{\mathcal{J}}_0$. The results are shown in Figures 7.4 and 7.5 for two different choices of the weights $w_i,\ i = 1, 2, 3$.

7.2 Shape optimization of elasto-plastic bodies in contact

7.2.1 Setting of the problem

This section deals with the realization of one contact shape optimization problem. The setting will be exactly the same as in Subsection 2.5.5 except for one modification. Instead of a linearly elastic material, characterized by a linear Hooke's law (2.148), we shall now consider a more general class of materials obeying the theory of *small elasto-plastic deformations*. For the mechanical justification of such a model we refer to [Was74] and [NH81]. The respective stress-strain relation is now *nonlinear* and in the case of plane strain it reads as follows:

$$\tau_{ij} = \kappa\varepsilon_{ll}\delta_{ij} + 2\mu(\sqrt{\iota})\left(\varepsilon_{ij} - \frac{1}{3}\delta_{ij}\varepsilon_{ll}\right), \tag{7.28}$$

where κ and μ stand for the bulk and shear modulus, respectively; δ_{ij} is the Kronecker symbol; and ε_{ll} is the trace of ε (as in Subsection 2.5.5, the summation convention is used). The shear modulus μ is assumed to be a function of the invariant $\iota := \iota(\varepsilon_{ij})$ defined by

$$\iota = \frac{1}{3}\left[(\varepsilon_{11} - \varepsilon_{22})^2 + \varepsilon_{11}^2 + \varepsilon_{22}^2 + 6\varepsilon_{12}^2\right]. \tag{7.29}$$

We shall suppose that the functions $\kappa := \kappa(x)$, $\mu := \mu(t, x)$, $x \in \Omega$, $t \geq 0$, depend continuously on their arguments and μ is continuously differentiable with respect to t. In addition, the following assumptions on κ, μ are made:

$$0 < \kappa_0 \leq \kappa(x) \leq \kappa_1 \quad \forall x \in \Omega; \tag{7.30}$$

$$0 < \mu_0 \leq \mu(t, x) \leq \frac{3}{2}\kappa(x) \quad \forall x \in \Omega,\ \forall t > 0; \tag{7.31}$$

$$0 < \theta_0 \leq \mu(t, x) + \frac{\partial \mu(t, x)}{\partial t} t \leq \theta_1 \quad \forall x \in \Omega,\ \forall t > 0, \tag{7.32}$$

where κ_0, κ_1, μ_0, θ_0, and θ_1 are given positive constants.

The shapes of admissible domains are the same as in Figure 2.6 with $\alpha \in U^{ad}$ defined by (2.174). The boundary and unilateral conditions are expressed by (2.147), (2.151), and (2.170)–(2.172). To give the variational formulation of this problem we introduce the *total potential energy* Φ_α reflecting the nonlinear Hooke's law (7.28). It can be shown (see [NH81]) that

$$\Phi_\alpha(v) = \frac{1}{2}\int_{\Omega(\alpha)}\left(\kappa\, \varepsilon_{ll}^2(v) + \int_0^{\Sigma^2(v)} \mu(t)\, dt\right) dx - L_\alpha(v), \tag{7.33}$$

where L_α is given by (2.158) and $\Sigma^2(v) := \Psi(v, v)$. The symbol Ψ stands for the bilinear form defined by

$$\Psi(u, v) = -\frac{2}{3}\varepsilon_{ii}(u)\varepsilon_{jj}(v) + 2\varepsilon_{ij}(u)\varepsilon_{ij}(v).$$

By a *variational solution* to the Signorini problem obeying (7.28) we mean any function $u(\alpha) \in \mathbb{K}(\alpha)$ such that

$$\Phi_\alpha(u(\alpha)) \leq \Phi_\alpha(v) \quad \forall v \in \mathbb{K}(\alpha), \tag{$\mathcal{P}(\alpha)$}$$

where $\mathbb{K}(\alpha)$ is defined by (2.173). It can be shown that the functional Φ_α is *weakly lower semicontinuous*, *strictly convex*, and *coercive* in $\mathbb{K}(\alpha)$ in view of the assumptions imposed on κ and μ. Therefore $(\mathcal{P}(\alpha))$ has a unique solution $u(\alpha)$ for any $\alpha \in U^{ad}$ (for details we refer to [NH81]).

Remark 7.2. It is easy to verify that Φ_α is also Fréchet differentiable at any $u \in \mathbb{V}(\alpha)$ and the Fréchet derivative $D\Phi_\alpha(u) \in \mathbb{V}'(\alpha)$ is given by

$$D\Phi_\alpha(u, v) = a_\alpha(u, v) - L_\alpha(v) \quad \forall u, v \in \mathbb{V}(\alpha),$$

where a_α is defined by (2.157) with τ_{ij} expressed by (7.28) (see [NH81]). Problem $(\mathcal{P}(\alpha))$ is equivalent to the following variational inequality:

$$\begin{cases} \text{Find } u := u(\alpha) \in \mathbb{K}(\alpha) \quad \text{such that} \\ a_\alpha(u, v - u) \geq L_\alpha(v - u) \quad \forall v \in \mathbb{K}(\alpha). \end{cases} \qquad (\mathcal{P}(\alpha)')$$

As we have already mentioned, a typical goal in contact shape optimization is to avoid stress concentrations on the contact part $\Gamma_C(\alpha)$. Motivated by the results presented at the end of Subsection 3.3.1 it turns out that this aim can be achieved by choosing

$$J(\alpha, u(\alpha)) := \Phi_\alpha(u(\alpha)) \quad \forall \alpha \in U^{ad} \qquad (7.34)$$

as the cost; i.e., J is equal to the total potential energy evaluated at the equilibrium state $u(\alpha)$. Let (7.30)–(7.32) be satisfied for any $(t, x) \in (0, \infty) \times \widehat{\Omega}$, where $\Omega(\alpha) \subset \widehat{\Omega}\ \forall \alpha \in U^{ad}$. Using the abstract theory of Section 2.4 one can prove that the following optimal shape design problem:

$$\begin{cases} \text{Find } \alpha^* \in U^{ad} \quad \text{such that} \\ J(\alpha^*, u(\alpha^*)) \leq J(\alpha, u(\alpha)) \quad \forall \alpha \in U^{ad}, \end{cases} \qquad (\mathbb{P})$$

has at least one solution (see [HN96]).

7.2.2 Approximation and numerical realization of $(\mathbb{P})$

We shall proceed exactly as in Subsection 2.5.5 keeping the meaning of all notation introduced there. Let $s_\varkappa \in U^{ad}_\varkappa$ be given and define $\mathbb{V}_h(s_\varkappa)$, $\mathbb{K}_h(s_\varkappa)$ by (2.167), (2.179), respectively. On any $\Omega_h(s_\varkappa)$ the discretization of the state problem is defined as follows:

$$\begin{cases} \text{Find } u_h(s_\varkappa) \in \mathbb{K}_h(s_\varkappa) \quad \text{such that} \\ \Phi_{r_h s_\varkappa}(u_h(s_\varkappa)) \leq \Phi_{r_h s_\varkappa}(v_h) \quad \forall v_h \in \mathbb{K}_h(s_\varkappa), \end{cases} \qquad (\mathcal{P}_h(s_\varkappa))$$

where $\Phi_{r_h s_\varkappa}$ denotes the total potential energy functional (7.33) evaluated over $\Omega_h(s_\varkappa)$.

Remark 7.3. The approximate solution $u_h(s_\varkappa)$ can again be equivalently characterized as the unique solution to the following discrete inequality:

$$\begin{cases} \text{Find } u_h := u_h(s_\varkappa) \in \mathbb{K}_h(s_\varkappa) \quad \text{such that} \\ a_{r_h s_\varkappa}(u_h, v_h - u_h) \geq L_{r_h s_\varkappa}(v_h - u_h) \quad \forall v_h \in \mathbb{K}_h(s_\varkappa). \end{cases} \qquad (\mathcal{P}_h(s_\varkappa)')$$

The discretization of $(\mathbb{P})$ is now stated as follows:

$$\begin{cases} \text{Find } s^*_\varkappa \in U^{ad}_\varkappa \quad \text{such that} \\ J(s^*_\varkappa, u_h(s^*_\varkappa)) \le J(s_\varkappa, u_h(s_\varkappa)) \quad \forall s_\varkappa \in U^{ad}_\varkappa, \end{cases} \tag{$\mathbb{P}_{h\varkappa}$}$$

where $u_h(s_\varkappa)$ solves $(\mathcal{P}_h(s_\varkappa))$ and

$$J(s_\varkappa, u_h(s_\varkappa)) := \Phi_{r_h s_\varkappa}(u_h(s_\varkappa)).$$

Applying the abstract convergence results of Section 2.4 we obtain the following results.

THEOREM 7.5. *For any sequence $\{(s^*_\varkappa, u_h(s^*_\varkappa))\}$ of optimal pairs of $(\mathbb{P}_{h\varkappa})$, $h \to 0+$, there exists its subsequence such that*

$$\begin{cases} s^*_\varkappa \rightrightarrows \alpha^* \quad \text{in } [0,1], \\ \widetilde{u}_h(s^*_\varkappa) \rightharpoonup u^* \quad \text{in } (H^1(\widehat{\Omega}))^2, \quad h, \varkappa \to 0+, \end{cases} \tag{7.35}$$

where $\widetilde{u}_h(s^*_\varkappa)$ is the uniform extension of $u_h(s^*_\varkappa)$ from $\Omega_h(s^*_\varkappa)$ to $\widehat{\Omega}$. In addition, $(\alpha^*, u^*{}_{|\Omega(\alpha^*)})$ is an optimal pair of $(\mathbb{P})$. Furthermore, any accumulation point of $\{(s^*_\varkappa, u_h(s^*_\varkappa))\}$ in the sense of (7.35) possesses this property.

We now turn to the numerical realization of $(\mathbb{P}_{h\varkappa})$. Unlike the problem presented in Subsection 2.5.5, the total potential energy to be minimized is not quadratic any more. Therefore the algebraic form of $(\mathcal{P}_h(s_\varkappa))$ leads to a general *convex programming problem*. A possible way to realize this is to use the following *SQP type approach* known in the literature as the *secant modulus* or *Kachanov method*:

$$\begin{cases} \text{For } u_h^{(k)} \in \mathbb{K}_h(s_\varkappa), \ k \in \mathbb{N} \text{ known, define} \\ u_h^{(k+1)} = \underset{v_h \in \mathbb{K}_h(s_\varkappa)}{\operatorname{argmin}} \left\{ \frac{1}{2} B_{r_h s_\varkappa}(u_h^{(k)}, v_h, v_h) - L_{r_h s_\varkappa}(v_h) \right\}, \end{cases} \tag{7.36}$$

where $L_{r_h s_\varkappa}$ is the linear term evaluated on $\Omega_h(s_\varkappa)$ and

$$B_{r_h s_\varkappa}(u; v, w) = \int_{\Omega_h(s_\varkappa)} \left[\left(\kappa - \frac{2}{3}\mu(\Sigma^2(u)) \right) \varepsilon_{ii}(v)\varepsilon_{jj}(w) + 2\mu(\Sigma^2(u))\varepsilon_{ij}(v)\varepsilon_{ij}(w) \right] dx.$$

Each iterative step in (7.36) is already a quadratic programming problem that can be realized by standard methods. One can show (see [NH81]) that the previous algorithm converges for any choice of $u_h^{(0)} \in \mathbb{K}_h(s_\varkappa)$ provided that (7.30)–(7.32) hold; i.e., $u_h^{(k)} \to u_h$ as $k \to \infty$ and $u_h \in \mathbb{K}_h(s_\varkappa)$ solves $(\mathcal{P}_h(s_\varkappa))$.

Denote by $\boldsymbol{q}(\boldsymbol{\alpha}) = (q_1(\boldsymbol{\alpha}), \ldots, q_n(\boldsymbol{\alpha}))$, $\boldsymbol{\alpha} \in \mathcal{U}$, the nodal displacement vector corresponding to the solution $u_h(s_\varkappa)$. From Remark 7.3 it follows that $\boldsymbol{q}(\boldsymbol{\alpha})$ solves the following algebraic inequality:

$$\begin{cases} \text{Find } \boldsymbol{q}(\boldsymbol{\alpha}) \in \mathcal{K}(\boldsymbol{\alpha}) \quad \text{such that} \\ (\boldsymbol{v} - \boldsymbol{q}(\boldsymbol{\alpha}))^{\mathrm{T}} \boldsymbol{K}(\boldsymbol{\alpha}, \boldsymbol{q}(\boldsymbol{\alpha}))\boldsymbol{q}(\boldsymbol{\alpha}) \ge \boldsymbol{f}(\boldsymbol{\alpha})^{\mathrm{T}}(\boldsymbol{v} - \boldsymbol{q}(\boldsymbol{\alpha})) \quad \forall \boldsymbol{v} \in \mathcal{K}(\boldsymbol{\alpha}), \end{cases} \tag{$\mathcal{P}(\boldsymbol{\alpha})$}$$

where $\mathcal{K}(\boldsymbol{\alpha})$ is defined by (2.183), $\mathcal{U}$ is isometrically isomorphic with $U^{ad}_{\varkappa}$, and $\boldsymbol{K} : \mathcal{U} \times \mathbb{R}^n \to \mathbb{R}^{n\times n}$ is a nonlinear matrix function representing the inner energy $a_{r_h s_\varkappa}$. Next we shall show how to construct $\boldsymbol{K}$ by using the isoparametric technique of Subsection 3.3.2.

Let $\boldsymbol{\alpha} \in \mathcal{U}$ be given. *To simplify notation $\boldsymbol{\alpha}$ will be omitted in the arguments of functions.* We first express the stress-strain relation (7.28) in the following matrix form:

$$\boldsymbol{\tau} = \boldsymbol{D}(\iota)\,\boldsymbol{\varepsilon}, \tag{7.37}$$

$$\iota = \boldsymbol{\varepsilon}^{\mathrm{T}} \boldsymbol{S} \boldsymbol{\varepsilon}. \tag{7.38}$$

Here

$$\boldsymbol{\tau} = \begin{pmatrix} \tau_{11} \\ \tau_{22} \\ \tau_{12} \end{pmatrix}, \quad \boldsymbol{\varepsilon} = \begin{pmatrix} \varepsilon_{11} \\ \varepsilon_{22} \\ 2\varepsilon_{12} \end{pmatrix},$$

$\boldsymbol{S}$ is a symmetric matrix realizing (7.29), and $\boldsymbol{D}(\iota)$ is a symmetric matrix whose elements depend on ι. The matrix $\boldsymbol{D}(\iota)$ will be split as follows:

$$\boldsymbol{D}(\iota) = \boldsymbol{D}^0 + \mu(\sqrt{\iota})\boldsymbol{D}^1,$$

where

$$\boldsymbol{D}^0 = \begin{pmatrix} \kappa & \kappa & 0 \\ \kappa & \kappa & 0 \\ 0 & 0 & 0 \end{pmatrix} \quad \text{and} \quad \boldsymbol{D}^1 = \begin{pmatrix} 4/3 & -2/3 & 0 \\ -2/3 & 4/3 & 0 \\ 0 & 0 & 1 \end{pmatrix}.$$

Next we shall suppose that κ and μ *do not depend* on x. The matrix $\boldsymbol{K}(\boldsymbol{q})$ can be assembled element by element from the local contributions on each $T_e \in \mathcal{T}_h$ as we have done in Subsection 3.3.2. Let $\widehat{T}$ be a parent element and $F_e : \widehat{T} \to T_e$ be a one-to-one mapping constructed in (3.78). Then the local stiffness matrix $\boldsymbol{K}^e(\boldsymbol{q}^e)$ is given by

$$\boldsymbol{K}^e(\boldsymbol{q}^e) = \int_{\widehat{T}} \boldsymbol{B}^{\mathrm{T}}\, \boldsymbol{D}(\boldsymbol{q}^e)\, \boldsymbol{B} |\boldsymbol{J}|\, d\xi, \tag{7.39}$$

where $\boldsymbol{q}^e$, $\boldsymbol{B}$ are defined by (3.70), (3.92), respectively; $|\boldsymbol{J}|$ is the determinant of the Jacobian of F_e; and

$$\boldsymbol{D}(\boldsymbol{q}^e) := \boldsymbol{D}^0 + \mu(\boldsymbol{q}^e)\boldsymbol{D}^1 \quad \text{with}$$

$$\mu(\boldsymbol{q}^e) := \mu\left(\sqrt{(\boldsymbol{B}\boldsymbol{q}^e)^{\mathrm{T}} \boldsymbol{S} \boldsymbol{B} \boldsymbol{q}^e}\right).$$

Here we used that $\iota_{|T_e} := \iota(\boldsymbol{q}^e) = (\boldsymbol{B}\boldsymbol{q}^e)^{\mathrm{T}} \boldsymbol{S}\boldsymbol{B}\boldsymbol{q}^e$. In the rest of this subsection we shall suppose that the mappings Φ_j from (2.52) are continuously differentiable in $\mathcal{U}$. Then by virtue of (7.29) and (7.39) the mapping $\boldsymbol{K} : \mathcal{U} \times \mathbb{R}^n \to \mathbb{R}^{n\times n}$ is *continuously differentiable* with respect to both parameters of $(\boldsymbol{\alpha}, \boldsymbol{q}) \in \mathcal{U} \times \mathbb{R}^n$.

Remark 7.4. From (7.30)–(7.32) we know that the mapping $\boldsymbol{q} \mapsto \boldsymbol{K}(\cdot, \boldsymbol{q})$, $\boldsymbol{q} \in \mathbb{R}^n$, is strictly monotone in $\mathbb{R}^n$ uniformly with respect to $\boldsymbol{\alpha} \in \mathcal{U}$. From Problem 3.5 it follows that the mapping $\boldsymbol{q} : \boldsymbol{\alpha} \mapsto \boldsymbol{q}(\boldsymbol{\alpha})$, where $\boldsymbol{q}(\boldsymbol{\alpha})$ solves $(\mathcal{P}(\boldsymbol{\alpha}))$, is Lipschitz continuous in $\mathcal{U}$ and directionally differentiable at any point $\boldsymbol{\alpha} \in \mathcal{U}$ and in any direction $\boldsymbol{\beta}$.

Problem ($\mathbb{P}_{h\varkappa}$) is equivalent to the following *nonlinear programming problem*:

$$\begin{cases} \text{Find } \boldsymbol{\alpha}^* \in \mathcal{U} \text{ such that} \\ \widetilde{\mathcal{J}}(\boldsymbol{\alpha}^*) \le \widetilde{\mathcal{J}}(\boldsymbol{\alpha}) \quad \forall \boldsymbol{\alpha} \in \mathcal{U}, \end{cases} \tag{7.40}$$

where $\widetilde{\mathcal{J}}(\boldsymbol{\alpha}) := \mathcal{J}(\boldsymbol{\alpha}, \boldsymbol{q}(\boldsymbol{\alpha}))$ and $\mathcal{J}$ is the algebraic form of the cost functional $\Phi_{r_h s_\varkappa}$. Observe that $\nabla_{\boldsymbol{q}} \mathcal{J}(\boldsymbol{\alpha}, \boldsymbol{q}) = \boldsymbol{K}(\boldsymbol{\alpha}, \boldsymbol{q})\boldsymbol{q} - \boldsymbol{f}(\boldsymbol{\alpha})$ (see Remark 7.2).

We may assume that the function $\mathcal{J} : (\boldsymbol{\alpha}, \boldsymbol{q}) \mapsto \mathcal{J}(\boldsymbol{\alpha}, \boldsymbol{q})$, $\boldsymbol{\alpha} \in \mathcal{U}$, $\boldsymbol{q} \in \mathbb{R}^n$, is continuously differentiable in $\mathcal{U} \times \mathbb{R}^n$. On the other hand, the mapping $\boldsymbol{q} : \boldsymbol{\alpha} \mapsto \boldsymbol{q}(\boldsymbol{\alpha})$ is *only directionally differentiable*. Let us compute the directional derivative $\widetilde{\mathcal{J}}'(\boldsymbol{\alpha}; \boldsymbol{\beta})$:

$$\begin{aligned} \widetilde{\mathcal{J}}'(\boldsymbol{\alpha}; \boldsymbol{\beta}) &= \lim_{t \to 0+} \frac{\widetilde{\mathcal{J}}(\boldsymbol{\alpha} + t\boldsymbol{\beta}) - \widetilde{\mathcal{J}}(\boldsymbol{\alpha})}{t} \\ &= (\nabla_{\boldsymbol{\alpha}} \mathcal{J}(\boldsymbol{\alpha}, \boldsymbol{q}(\boldsymbol{\alpha})))^T \boldsymbol{\beta} + (\nabla_{\boldsymbol{q}} \mathcal{J}(\boldsymbol{\alpha}, \boldsymbol{q}(\boldsymbol{\alpha})))^T \boldsymbol{q}'(\boldsymbol{\alpha}), \end{aligned} \tag{7.41}$$

where

$$\boldsymbol{q}'(\boldsymbol{\alpha}) = \lim_{t \to 0+} \frac{\boldsymbol{q}(\boldsymbol{\alpha} + t\boldsymbol{\beta}) - \boldsymbol{q}(\boldsymbol{\alpha})}{t}.$$

Next we shall eliminate the directional derivative of $\boldsymbol{q}$ from (7.41). The x_2 components of the residual vector $\boldsymbol{r}(\boldsymbol{\alpha}) := \nabla_{\boldsymbol{q}} \mathcal{J}(\boldsymbol{\alpha}, \boldsymbol{q}(\boldsymbol{\alpha}))$ at $\boldsymbol{\alpha} \in \mathcal{U}$ associated with the contact nodes can be interpreted as the discrete counterpart of the x_2 component of the stress vector $T(u)$ along $\Gamma_C(\alpha)$. For the characterization of $\boldsymbol{r}(\boldsymbol{\alpha})$ see Problem 7.11. Since $\boldsymbol{q}$ depends continuously on $\boldsymbol{\alpha} \in \mathcal{U}$ and $\mathcal{J}$ is continuously differentiable with respect to $\boldsymbol{q}$, then $\boldsymbol{r}$ depends continuously on $\boldsymbol{\alpha}$ as well. Let $\mathcal{I}$ be the set containing indices of all constrained components of $\boldsymbol{q} \in \mathcal{K}(\boldsymbol{\alpha})$. Then for any $i \notin \mathcal{I}$ it holds that $r_i(\boldsymbol{\alpha}) = 0 \ \forall \boldsymbol{\alpha} \in \mathcal{U}$. On the other hand, if $r_{j_i}(\boldsymbol{\alpha}) > 0$ for some $j_i \in \mathcal{I}$, then $r_{j_i}(\boldsymbol{\alpha} + t\boldsymbol{\beta}) > 0$ for any $t > 0$ sufficiently small. This means that the corresponding contact node remains in contact regardless of small perturbations of $\Omega_h(s_\varkappa)$ (see (2.183) and Problem 7.11):

$$q_{j_i}(\boldsymbol{\alpha} + t\boldsymbol{\beta}) = -s_\varkappa(b_i, \boldsymbol{\alpha} + t\boldsymbol{\beta}). \tag{7.42}$$

Let us suppose that $s_\varkappa$ is a linear combination of $(d+1)$ linearly independent functions $\{\xi_l\}_{l=0}^d$:

$$s_\varkappa(x_1, \boldsymbol{\alpha}) = \sum_{l=0}^{d} \alpha_l \xi_l(x_1), \quad \boldsymbol{\alpha} = (\alpha_0, \dots, \alpha_d), \quad x_1 \in [0, 1].$$

Then (7.42) reads as follows:

$$q_{j_i}(\boldsymbol{\alpha} + t\boldsymbol{\beta}) = -\sum_{l=0}^{d} (\alpha_l + t\beta_l)\xi_l(b_i)$$

so that

$$\frac{\partial q_{j_i}(\boldsymbol{\alpha})}{\partial \alpha_k} = -\xi_k(b_i), \quad k = 0, \dots, d. \tag{7.43}$$

From (7.41) and (7.43) we see that the function $\widetilde{\mathcal{J}}$ is once continuously differentiable in $\mathcal{U}$ despite the fact that the inner mapping $\boldsymbol{q}$ is not. We summarize these results as follows.

THEOREM 7.6. *The function $\widetilde{\mathcal{J}}$ is once continuously differentiable in $\mathcal{U}$ and*

$$\frac{\partial \widetilde{\mathcal{J}}(\boldsymbol{\alpha})}{\partial \alpha_k} = \frac{\partial \mathcal{J}(\boldsymbol{\alpha}, \boldsymbol{q}(\boldsymbol{\alpha}))}{\partial \alpha_k} - \sum_{j_i \in \mathcal{I}} r_{j_i}(\boldsymbol{\alpha}) \xi_k(b_i), \quad k = 0, \dots, d.$$

EXAMPLE 7.2. The family of admissible domains consists of $\Omega(\alpha)$ defined by

$$\Omega(\alpha) = \{(x_1, x_2) \in \mathbb{R}^2 \mid 0 < x_1 < 4,\ \alpha(x_1) < x_2 < 1\}, \quad \alpha \in U^{ad},$$

where

$$U^{ad} = \left\{ \alpha \in C^{0,1}([0,4]) \mid 0 \le \alpha \le 0.2,\ |\alpha'| \le 1 \text{ in } [0,4],\ \int_0^4 \alpha \, dx_1 = 0.2 \right\}.$$

We consider the following nonlinear Hooke's law:

$$\tau_{ij} = \kappa \varepsilon_{ll} \delta_{ij} + 2\mu(\sqrt{\iota}) \left(\varepsilon_{ij} - \frac{1}{3} \delta_{ij} \varepsilon_{ll} \right), \tag{7.44}$$

where $\kappa = 0.83333$ and

$$\mu(t) = \frac{1}{4}(1 + \exp(-t)), \quad t \ge 0. \tag{7.45}$$

The partition of the boundary is done as follows: $\Gamma_u(\alpha) = \{0\} \times]\alpha(0), 1[$, $\Gamma_C(\alpha) = \{(x_1, x_2) \in \mathbb{R}^2 \mid x_1 \in]0, 4[,\ x_2 = \alpha(x_1)\}$, $\Gamma_P(\alpha) = \partial\Omega(\alpha) \setminus \overline{(\Gamma_u(\alpha) \cup \Gamma_C(\alpha))}$. The body force f is assumed to be equal to the zero vector and the external load P is of the form

$$P = \begin{cases} (0, -0.0009(x_1 - 1)), & x_2 = 1, x_1 \in]1, 4[, \\ (0, 0) & \text{otherwise.} \end{cases}$$

The contact parts $\Gamma_C(\alpha)$, $\alpha \in U^{ad}$, are approximated by Bézier functions of degree $d \ge 2$. Thus the set $U^{ad}_{\varkappa}$ consists of Bézier functions of degree d defined by the control points

$$z^{(i)} = ((i-1)\varkappa, 0.2\alpha_i), \quad i = 1, \dots, d,\ \varkappa = 4/(d-1),$$

where $\alpha_i \in \mathbb{R}$, $i = 1, \dots, d$, are the discrete design variables belonging to the set

$$\mathcal{U} = \left\{ \boldsymbol{\beta} \in \mathbb{R}^d \mid 0 \le \beta_i \le 1,\ i = 1, \dots, d; \right.$$
$$\left. \frac{|(\beta_{i+1} - \beta_i)0.2|}{\varkappa} \le 1,\ i = 1, \dots, d-1,\quad \int_0^4 r_h s_{\varkappa} \, dx_1 = 0.2 \right\}, \tag{7.46}$$

where $r_h s_{\varkappa}$ is the Lagrange interpolant of $s_{\varkappa}$ at the contact nodes. The integral in (7.46) is evaluated using the trapezoidal rule. As an initial guess, $s_{\varkappa}^{(0)} = 0.05$ was chosen. The

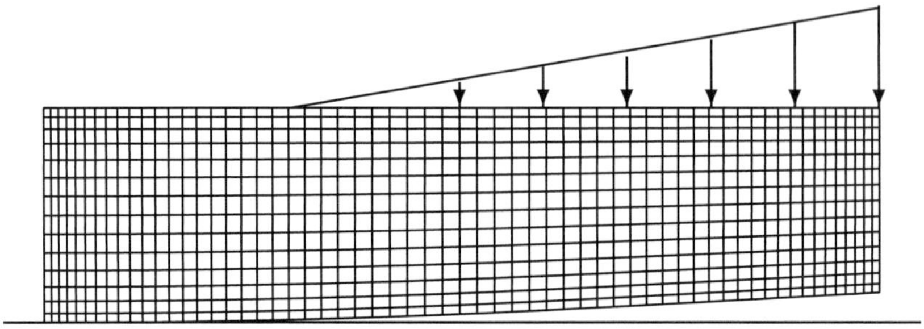

Figure 7.6. *Finite element mesh of the optimal structure.*

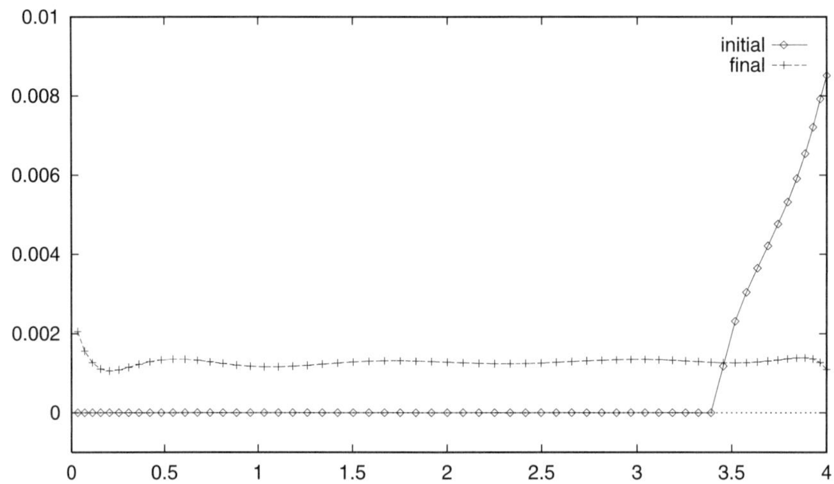

Figure 7.7. *Initial and final contact stress distributions.*

number of design variables was $d = 15$ and the finite element mesh consisted of 840 four-noded isoparametric elements. The initial cost was equal to -0.1581×10^{-3}. After 16 optimization iterations the cost was reduced to the value -0.3103×10^{-3}. The finite element mesh corresponding to the final (optimal) domain is shown in Figure 7.6. The initial and final contact stress distributions are plotted in Figure 7.7. The convergence history is shown in Figure 7.8.

In the optimization process the SQP subroutine E04UCF from the NAG library [NAG97] was used. The quadratic programming subproblem (7.36) was solved by the block successive overrelaxation (SOR) method with projection. The stopping criterion in the Kachanov method was $\|\boldsymbol{q}_{k+1} - \boldsymbol{q}_k\|_2 \leq 10^{-10}\|\boldsymbol{q}_{k+1}\|_2$, where $\boldsymbol{q}_k$ is the nodal displacement vector corresponding to $u_h^{(k)}$. It was usually fullfilled after five iterations. Computations were done in 64-bit floating-point arithmetic using an HP9000/J5600 workstation. The total CPU time needed was 42 s.

From the results of Subsection 3.3.1 it follows that the minimization of the total potential energy for a *linear elastic* body yields "almost" constant contact stress distributions along the optimal contact part. Similar behavior can be observed in the nonlinear material case.

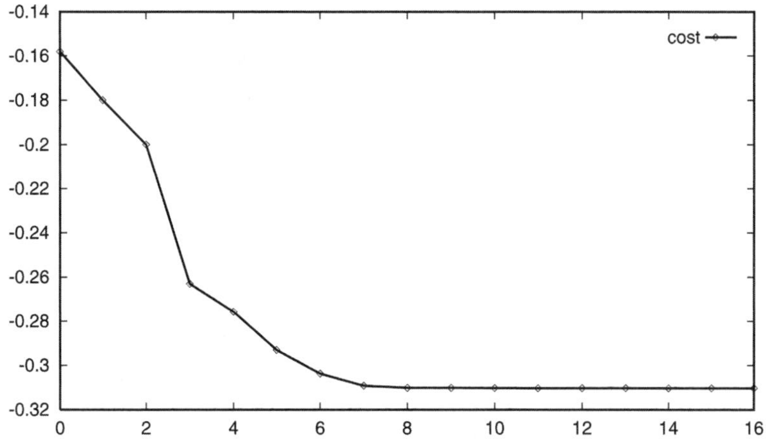

Figure 7.8. *Cost (multiplied by* 10^3*) vs. iteration.*

Problems

PROBLEM 7.1. Prove that the Rayleigh quotient

$$\mathcal{R}(e, v) = \frac{a_e(v, v)}{b_e(v, v)}, \quad e \in U^{ad}, \ v \in V, \ v \neq 0,$$

where a_e, b_e, and U^{ad} are defined by (7.5), (7.6), and (2.3), respectively, is continuous in $U^{ad} \times V$.

PROBLEM 7.2. Let $\lambda(\boldsymbol{e})$ be a simple eigenvalue of the generalized eigenvalue problem

$$\boldsymbol{A}(\boldsymbol{e})\boldsymbol{z}(\boldsymbol{e}) = \lambda(\boldsymbol{e})\boldsymbol{B}(\boldsymbol{e})\boldsymbol{z}(\boldsymbol{e}), \quad \boldsymbol{z}(\boldsymbol{e}) \neq \boldsymbol{0},$$

at $\boldsymbol{e} = \overline{\boldsymbol{e}}$, where $\boldsymbol{A}(\boldsymbol{e})$, $\boldsymbol{B}(\boldsymbol{e})$ are symmetric, positive definite matrices in an open set $\mathcal{U} \subset \mathbb{R}^d$ containing $\overline{\boldsymbol{e}}$. Suppose that the mappings $\boldsymbol{A} : \boldsymbol{e} \mapsto \boldsymbol{A}(\boldsymbol{e})$, $\boldsymbol{B} : \boldsymbol{e} \mapsto \boldsymbol{B}(\boldsymbol{e})$ are continuously differentiable in $\mathcal{U}$. Prove that

(i) $\lambda(\boldsymbol{e})$ is a simple eigenvalue for any $\boldsymbol{e} \in B_\delta(\overline{\boldsymbol{e}})$, $\delta > 0$, small enough;

(ii) the mappings $\lambda : \boldsymbol{e} \mapsto \lambda(\boldsymbol{e})$, $\boldsymbol{z} : \boldsymbol{e} \mapsto \boldsymbol{z}(\boldsymbol{e})$ are continuously differentiable in $B_\delta(\overline{\boldsymbol{e}})$.

PROBLEM 7.3. Prove (7.27) by using Theorem 7.4.

PROBLEM 7.4. Show that the Fréchet derivative of Φ_α defined by (7.33) is given by

$$D\Phi_\alpha(u, v) = a_\alpha(u, v) - L_\alpha(v) \quad \forall u, v \in \mathbb{V}(\alpha),$$

where

$$a_\alpha(u, v) = \int_{\Omega(\alpha)} \tau_{ij}(u)\varepsilon_{ij}(v)\, dx$$

with $\tau_{ij}(u)$ expressed by (7.28) and

$$L_\alpha(v) = \int_{\Omega(\alpha)} f_i v_i \, dx + \int_{\Gamma_P(\alpha)} P_i v_i \, ds.$$

Problem 7.5. Prove that, for any $u, v \in (H^1(\Omega(\alpha)))^2$,

$$D\Phi_\alpha(u + v, v) - D\Phi_\alpha(u, v) \geq c_1 \int_{\Omega(\alpha)} \varepsilon_{ij}(v)\varepsilon_{ij}(v)\, dx,$$

where $c_1 > 0$ is a constant depending solely on μ_0, θ_0 appearing in (7.31) and (7.32).
Hint:

(i) Express

$$D\Phi_\alpha(u + v, v) - D\Phi_\alpha(u, v) = \int_0^1 D^2\Phi_\alpha(u + tv, v, v)\, dt,$$

where $D^2\Phi_\alpha(u, v, w) := \left. \frac{d}{dt} D\Phi_\alpha(u + tw, v)\right|_{t=0}$.

(ii) Show that

$$D^2\Phi_\alpha(u + tv, v, v) \geq c_1 \int_{\Omega(\alpha)} \varepsilon_{ij}(v)\varepsilon_{ij}(v)\, dx,$$

where $c_1 := c_1(\mu_0, \theta_0) > 0$.

Problem 7.6. Show that

$$\Phi_\alpha(v) \geq \frac{1}{2} c_1 \int_{\Omega(\alpha)} \varepsilon_{ij}(v)\varepsilon_{ij}(v)\, dx - c_2 \|v\|_{0,\Omega(\alpha)} - c_3 \|v\|_{0,\Gamma_P(\alpha)},$$

where $c_1 > 0$ is the same as in Problem 7.5 and $c_2, c_3 > 0$ are constants depending on $\|F\|_{0,\Omega(\alpha)}$, $\|P\|_{0,\Gamma_P(\alpha)}$, respectively.

Problem 7.7. On the basis of the previous results show that $(\mathcal{P}(\alpha))$ from Subsection 7.2.1 has a unique solution $u(\alpha)$ such that

$$\|u(\alpha)\|_{1,\Omega(\alpha)} \leq c,$$

where $c > 0$ does not depend on $\alpha \in U^{ad}$.

Problem 7.8. Prove that $(\mathbb{P})$ defined in Subsection 7.2.1 has a solution.

Problem 7.9. Prove Theorem 7.5.

PROBLEM 7.10. Prove that the function μ defined by (7.45) satisfies (7.31) and (7.32) with $\kappa = 0.83333$.

PROBLEM 7.11. Let $\boldsymbol{r}(\boldsymbol{\alpha}) = \boldsymbol{K}(\boldsymbol{\alpha}, \boldsymbol{q}(\boldsymbol{\alpha})) - \boldsymbol{f}(\boldsymbol{\alpha})$ be the residual vector in $(\mathcal{P}(\boldsymbol{\alpha}))$, $\boldsymbol{\alpha} \in \mathcal{U}$. Show that

(i) $r_i(\boldsymbol{\alpha}) = 0 \; \forall i \notin \mathcal{I}, \; r_{j_i}(\boldsymbol{\alpha}) \geq 0 \quad \forall j_i \in \mathcal{I}$;

(ii) $r_{j_i}(\boldsymbol{\alpha})(q_{j_i}(\boldsymbol{\alpha}) + s_{\varkappa}(\overline{x}_i, \boldsymbol{\alpha})) = 0 \quad \forall j_i \in \mathcal{I}$ (no sum),

where $\mathcal{I}$ is the set containing indices of all constrained components of $\boldsymbol{q} \in \mathcal{K}(\boldsymbol{\alpha})$.

Chapter 8

Fluid Mechanical and Multidisciplinary Applications

Traditionally shape optimization has been restricted to one discipline, linear elasticity, only. Recently, there has been much interest in shape optimization of systems governed by equations of both fluid mechanics and electromagnetics. In this chapter we shall consider shape optimization with state problems related to fluid mechanics or combined fluid mechanics and electromagnetics problems.

Finite element and nonlinear programming methods in shape optimization of compressible subsonic flows have been used by Angrand [Ang83], and later by Beux and Dervieux [BD92]. Angrand computed optimal shapes of a nozzle and a lifting airfoil by using the full potential equation. Beux and Dervieux optimized the shape of a nozzle in the case of subsonic Euler flow. A slightly different approach has been used by Jameson [Jam88]. For further study on numerical methods in shape optimization problems governed by fluid flow or in multidisciplinary problems we refer to [MP01].

8.1 Shape optimization of a dividing tube

This section deals with a shape optimization problem governed by the Navier–Stokes equations for viscous incompressible fluids.

8.1.1 Introduction

The quality of paper produced is largely determined by phenomena taking place in the device of a paper machine called the "headbox." For example, the basis weight and the fiber orientation variations depend on the fluid dynamics in the headbox. The first flow passage in the headbox is a dividing tube (the "header"). It is designed to distribute a fiber suspension (wood fibers, filler clays, and chemicals mixed in water) in such a way that the paper produced will have a uniform thickness and an optimal fiber orientation across the width of the paper machine.

The fluid flowing in the headbox is a mixture of water and wood fibers and, therefore, a simulation of separation or mixing of different phases requires a multiphase model for

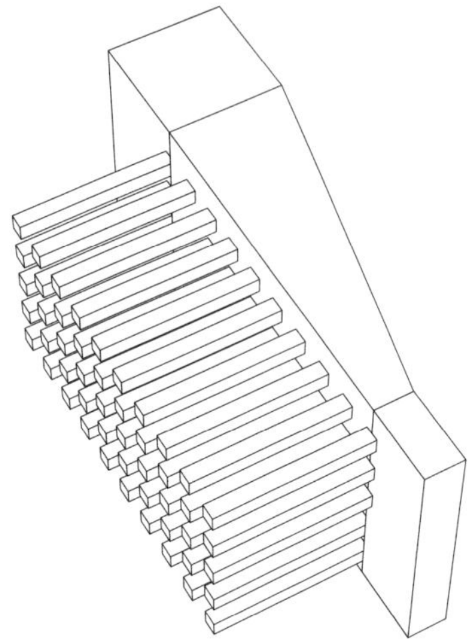

Figure 8.1. *Schematic picture of the header and the manifold tube bank.*

the water-fiber suspension. For a simulation of large-scale phenomena, however, one-phase modeling, where the fluid is pure water, is assumed to be sufficient. A typical Reynolds number is of order 10^6, requiring turbulence to be taken into account.

The geometry of the header presents additional difficulties in modeling the fluid flow because the fluid escapes from the header to the next chamber through a tube bank, which consists of hundreds of small identical tubes (see Figure 8.1). Using a homogenization approach the complicated geometry of this part will be replaced by a simple one represented by a part Γ_{out} but with a nonlinear boundary condition. A detailed description of the headbox flow modeling and the derivation of the homogenized outflow boundary condition can be found in [Häm93], [HT96].

We use a simplified flow model, namely the Reynolds-averaged Navier–Stokes (RANS) equations with an algebraic mixing length turbulence model. For details on turbulence modeling we refer to [Rod93]. The problem is also simplified by considering the two-dimensional geometry.

A more realistic simulation of header flows is based on the so-called k-ε turbulence model [Häm93]. The use of the k-ε model requires solving two additional nonlinear diffusion equations. The simplified model used in this section represents essentially the same aspects as the k-ε model from the standpoint of optimal shape design methods. The physical justification and mathematical analysis of different turbulence models go beyond the scope of this textbook, however.

8.1.2 Setting of the problem

We consider a two-dimensional fluid flow in a header $\Omega(\alpha)$ as given in Figure 8.2. The fluid flows in through the part Γ_{in} of the boundary and flows out through the small tubes on

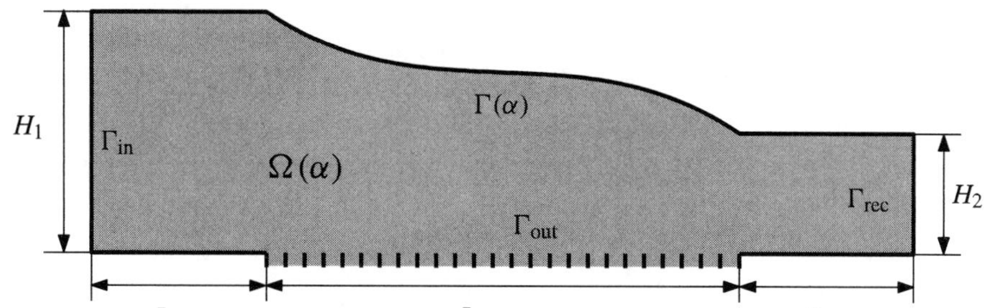

Figure 8.2. *Problem geometry.*

the boundary Γ_{out}. Also there is a small outflow on Γ_{rec}. The parameters H_1, H_2, L_1, L_2, and L_3 defining the geometry are *fixed*. Let the back wall $\Gamma(\alpha)$ of the tube (the only curved part of $\partial\Omega(\alpha)$) be defined by a function $\alpha \in U^{ad}$:

$$\Gamma(\alpha) = \{x = (x_1, x_2) \mid L_1 < x_1 < L_1 + L_2,\ x_2 = \alpha(x_1)\},$$

where

$$\begin{aligned} U^{ad} = \{\alpha \in C^{0,1}([L_1, L_1 + L_2]) \mid {} & 0 < \alpha_{\min} \le \alpha \le \alpha_{\max}, \\ & \alpha(L_1) = H_1,\ \alpha(L_1 + L_2) = H_2, \\ & |\alpha'| \le L_0 \text{ almost everywhere in } [L_1, L_1 + L_2]\} \end{aligned}$$

and $\alpha_{\min}$, $\alpha_{\max}$, L_0 are given parameters. For an incompressible fluid, conservation laws for momentum and mass in two dimensions read as follows (the summation convention is used):

$$-\frac{\partial \tau_{ij}(u, p)}{\partial x_j} + \varrho u_j \frac{\partial u_i}{\partial x_j} = 0 \quad \text{in } \Omega(\alpha), \quad i = 1, 2, \tag{8.1}$$

$$\operatorname{div} u = 0 \quad \text{in } \Omega(\alpha). \tag{8.2}$$

Here $u = (u_1, u_2)$ is the velocity vector, $\tau = (\tau_{ij})_{i,j=1}^2$ is the stress tensor, p is the static pressure, and ϱ is the density of the fluid. The components of the stress tensor are related to the components of the strain rate tensor $\varepsilon(u)$ and the pressure p by means of the constitutive law $\tau_{ij}(u) = 2\mu(u)\varepsilon_{ij}(u) - p\delta_{ij}$, $i, j = 1, 2$, where $\mu(u)$ is the viscosity and $\varepsilon(u) = (\varepsilon_{ij}(u))_{i,j=1}^2$ with $\varepsilon_{ij}(u) = \frac{1}{2}\left(\partial u_i/\partial x_j + \partial u_j/\partial x_i\right)$.

In addition, the following boundary conditions are prescribed (see Figure 8.2):

$$u = 0 \quad \text{on } \partial\Omega(\alpha) \setminus (\Gamma_{in} \cup \Gamma_{out} \cup \Gamma_{rec}), \tag{8.3}$$

$$u = u_{in} \quad \text{on } \Gamma_{in}, \tag{8.4}$$

$$u = u_{rec} \quad \text{on } \Gamma_{rec}, \tag{8.5}$$

$$u_1 = 0 \quad \text{on } \Gamma_{out}, \tag{8.6}$$

$$2\mu(u)\varepsilon_{2j}(u)\nu_j - p\nu_2 - c\,u_2^2 = 0 \quad \text{on } \Gamma_{out}. \tag{8.7}$$

The nonlinear boundary condition (8.7) corresponds to the homogenized outflow with a given constant c depending on the geometry of the outflow tube bank.

For turbulent flows, (8.1)–(8.2) represent the RANS equations with u and p being the *averaged* quantities. The viscosity for RANS equations is of the form $\mu(u) = \mu_0 + \mu_t(u)$, that is, the sum of a constant laminar viscosity $\mu_0 > 0$ and a turbulent viscosity $\mu_t(u)$.

In order to solve the RANS equations for turbulent flows, one has to select a model for the turbulent viscosity $\mu_t(u)$. Here, we use a simple algebraic model

$$\mu_t(u) = \varrho \ell_m^2 \left(\frac{1}{2} \varepsilon_{ij}(u) \varepsilon_{ij}(u) \right)^{1/2} \tag{8.8}$$

with $(\frac{1}{2}\varepsilon_{ij}\varepsilon_{ij})^{1/2}$ being the second invariant of the strain rate tensor and $\ell_m := \ell_m(x)$ denoting the mixing length defined by the so-called Nikuradse's formula [AB85], [Rod93]:

$$\ell_m(x) = \frac{1}{2} H(x) \left[0.14 - 0.08 \left(1 - \frac{2d(x)}{H(x)} \right)^2 - 0.06 \left(1 - \frac{2d(x)}{H(x)} \right)^4 \right], \tag{8.9}$$

where

$$H(x) = \begin{cases} H_1, & 0 \le x_1 \le L_1, \\ \alpha(x_1), & L_1 < x_1 \le L_1 + L_2, \\ H_2, & x_1 > L_1 + L_2, \end{cases}$$

and $d(x) = \text{dist}(x, \partial\Omega(\alpha) \setminus (\Gamma_{\text{in}} \cup \Gamma_{\text{out}} \cup \Gamma_{\text{rec}}))$.

To give the weak formulation of (8.1)–(8.7) we introduce the following sets of functions, defined in $\Omega(\alpha)$, $\alpha \in U^{ad}$:

$$\mathbb{W}_{\text{div}}^{1,3}(\alpha) = \{ v = (v_1, v_2) \in (W^{1,3}(\Omega(\alpha)))^2 \mid \text{div}\, v = 0 \text{ in } \Omega(\alpha) \}, \tag{8.10}$$

$$\mathbb{V}(\alpha) = \{ v \in (W^{1,3}(\Omega(\alpha)))^2 \mid v = 0 \text{ on } \partial\Omega(\alpha) \setminus \Gamma_{\text{out}},\ v_1 = 0 \text{ on } \Gamma_{\text{out}} \}, \tag{8.11}$$

$$\begin{aligned} \mathbb{V}_g(\alpha) = \{ v \in (W^{1,3}(\Omega(\alpha)))^2 \mid v = u_{\text{in}} \text{ on } \Gamma_{\text{in}},\ v = u_{\text{rec}} \text{ on } \Gamma_{\text{rec}}, \\ v_1 = 0 \text{ on } \Gamma_{\text{out}},\ v = 0 \text{ on } \partial\Omega(\alpha) \setminus (\Gamma_{\text{in}} \cup \Gamma_{\text{out}} \cup \Gamma_{\text{rec}}) \}, \end{aligned} \tag{8.12}$$

where for brevity of notation g denotes the function defining the nonhomogeneous Dirichlet boundary data on $\Gamma_{\text{in}} \cup \Gamma_{\text{rec}}$. The weak formulation of (8.1)–(8.7) reads as follows:

$$\begin{cases} \text{Find } u := u(\alpha) \in \mathbb{V}_g(\alpha) \cap \mathbb{W}_{\text{div}}^{1,3}(\alpha) \quad \text{such that} \\ \displaystyle\int_{\Omega(\alpha)} 2\mu(u)\varepsilon_{ij}(u)\varepsilon_{ij}(v)\, dx + \int_{\Omega(\alpha)} \varrho u_i \frac{\partial u_j}{\partial x_i} v_j\, dx \\ \qquad \displaystyle - \int_{\Gamma_{\text{out}}} c u_2^2 v_2\, dx_1 = 0 \quad \forall v \in \mathbb{V}(\alpha) \cap \mathbb{W}_{\text{div}}^{1,3}(\alpha), \end{cases} \tag{$\mathcal{P}(\alpha)$}$$

where $\mu(u) = \mu_0 + \varrho \ell_m^2 (\frac{1}{2}\varepsilon_{ij}(u)\varepsilon_{ij}(u))^{\frac{1}{2}}$.

Remark 8.1. The reason for introducing the space $(W^{1,3}(\Omega(\alpha)))^2$ in $(\mathcal{P}(\alpha))$ is the presence of the nonlinear term $\mu(u)$. It is left as an exercise to show that for $u \in \mathbb{V}_g(\alpha)$, $v \in \mathbb{V}(\alpha)$ all integrals in $(\mathcal{P}(\alpha))$ are finite.

For better performance of the header the outflow profile u_2 on Γ_{out} should be close to a given target profile $u_{ad} \in L^2(\Gamma_{\text{out}})$. Therefore we formulate the following optimization problem:

$$\begin{cases} \text{Find } \alpha^* \in U^{ad} \text{ such that} \\ J(u(\alpha^*)) \le J(u(\alpha)) \quad \forall \alpha \in U^{ad}, \end{cases} \tag{$\mathbb{P}$}$$

where

$$J(u(\alpha)) = \int_{\widetilde{\Gamma}} (u_2(\alpha) - u_{ad})^2 \, dx_1, \tag{8.13}$$

with $u(\alpha)$ being the solution to $(\mathcal{P}(\alpha))$ and $\widetilde{\Gamma} \subset \Gamma_{\text{out}}$ given. We use $\widetilde{\Gamma}$ instead of Γ_{out} because in practice the velocity profile must be close to the target profile only in the middle of the header. The edges are controlled by additional edge flow feeds.

To overcome difficulties arising from the incompressibility condition $\operatorname{div} u = 0$ in $\Omega(\alpha)$ we use a penalty approach (see also Subsection 2.5.6). The penalized form of $(\mathcal{P}(\alpha))$ is as follows:

$$\begin{cases} \text{Find } u_\varepsilon := u_\varepsilon(\alpha) \in \mathbb{V}_g(\alpha) \quad \text{such that} \\ \displaystyle\int_{\Omega(\alpha)} 2\mu(u_\varepsilon)\varepsilon_{ij}(u_\varepsilon)\varepsilon_{ij}(v)\, dx + \int_{\Omega(\alpha)} \varrho u_{\varepsilon i} \frac{\partial u_{\varepsilon j}}{\partial x_i} v_j \, dx \\ \displaystyle\qquad + \frac{1}{\varepsilon}\int_{\Omega(\alpha)} \operatorname{div} u_\varepsilon \operatorname{div} v \, dx - \int_{\Gamma_{\text{out}}} c u_{\varepsilon 2}^2 v_2 \, dx_1 = 0 \quad \forall v \in \mathbb{V}(\alpha), \end{cases} \tag{$\mathcal{P}_\varepsilon(\alpha)$}$$

where $\varepsilon > 0$ is a penalty parameter destined to tend to zero.

For any $\varepsilon > 0$ we define the new shape optimization problem using the penalized form $(\mathcal{P}_\varepsilon(\alpha))$ in place of the state problem:

$$\begin{cases} \text{Find } \alpha_\varepsilon^* \in U^{ad} \text{ such that} \\ J(u_\varepsilon(\alpha_\varepsilon^*)) \le J(u_\varepsilon(\alpha)) \quad \forall \alpha \in U^{ad}, \end{cases} \tag{$\mathbb{P}_\varepsilon$}$$

where

$$J(u_\varepsilon(\alpha)) = \int_{\widetilde{\Gamma}} (u_{\varepsilon 2}(\alpha) - u_{ad})^2 \, dx_1, \quad u_{ad} \in L^2(\Gamma_{\text{out}}), \tag{8.14}$$

and $u_\varepsilon(\alpha)$ solves $(\mathcal{P}_\varepsilon(\alpha))$.

8.1.3 Approximation and numerical realization of $(\mathbb{P}_\varepsilon)$

Let $U_\varkappa^{ad}$ be a discretization of U^{ad} by Bézier functions of a certain degree; i.e., the unknown back wall will be represented by the graph $\Gamma(s_\varkappa)$ of $s_\varkappa \in U_\varkappa^{ad}$. The computational domain will be realized by a polygonal approximation of $\Omega(s_\varkappa)$ determined by a piecewise linear Lagrange interpolant $r_h s_\varkappa$ of $s_\varkappa$ constructed on a partition Δ_h of $[L_1, L_1 + L_2]$ representing

Γ_{out}. Finally, the symbol $\Omega_h(s_\varkappa)$ stands for the computational domain with a given partition $\mathcal{R}(h, s_\varkappa)$ made of *quadrilaterals*. On any $\Omega_h(s_\varkappa)$ we define the following spaces:

$$\mathbb{V}_h(s_\varkappa) = \{v_h \in (C(\overline{\Omega_h(s_\varkappa)}))^2 \mid v_{h|R} \in (Q_1(R))^2 \ \forall R \in \mathcal{R}(h, s_\varkappa)\} \cap \mathbb{V}(r_h s_\varkappa), \quad (8.15)$$

$$\begin{aligned} \mathbb{V}_{gh}(s_\varkappa) = \{v_h \in (C(\overline{\Omega_h(s_\varkappa)}))^2 \mid v_{h|R} \in (Q_1(R))^2 \ \forall R \in \mathcal{R}(h, s_\varkappa), \\ v_h(a_i) = g(a_i) \ \forall a_i \in \mathcal{I}_1, \ v_{h1} = 0 \text{ on } \Gamma_{\text{out}}, \\ v_h = 0 \text{ on } \partial\Omega(r_h s_\varkappa) \setminus (\Gamma_{\text{in}} \cup \Gamma_{\text{out}} \cup \Gamma_{\text{rec}})\}, \end{aligned} \quad (8.16)$$

where $Q_1(R)$ is a four-noded isoparametric element on $R \in \mathcal{R}(h, s_\varkappa)$, $\mathbb{V}(r_h s_\varkappa)$ is the space defined by (8.11) on $\Omega(r_h s_\varkappa)$, and $\mathcal{I}_1$ contains all the nodes of $\mathcal{R}(h, s_\varkappa)$ on $\overline{\Gamma_{\text{in}}} \cup \overline{\Gamma_{\text{rec}}}$. In addition, g is supposed to belong to $(C(\overline{\Gamma_{\text{in}}} \cup \overline{\Gamma_{\text{rec}}}))^2$.

The nonlinear penalized state problem in $\Omega_h(s_\varkappa)$ will be numerically realized by using the following *linearization* approach ($s_\varkappa \in U_\varkappa^{ad}$ and $\varepsilon > 0$ are fixed):

$$\left\{ \begin{array}{l} \text{Choose } u_h^{(0)} \in \mathbb{V}_{gh}(s_\varkappa). \\ \text{For } u_h^{(k)} \in \mathbb{V}_{gh}(s_\varkappa), \ k \in \mathbb{N}, \ \text{known find } u_h^{(k+1)} := u_h^{(k+1)}(s_\varkappa) \in \mathbb{V}_{gh}(s_\varkappa): \\ B_{r_h s_\varkappa}(u_h^{(k)}, u_h^{(k+1)}, v_h) = 0 \quad \forall v_h \in \mathbb{V}_h(s_\varkappa), \end{array} \right. \quad (8.17)$$

where

$$\begin{aligned} B_{r_h s_\varkappa}(u_h, v_h, w_h) = \int_{\Omega_h(s_\varkappa)} 2\mu(u_h)\varepsilon_{ij}(v_h)\varepsilon_{ij}(w_h)\,dx + \int_{\Omega_h(s_\varkappa)} \varrho u_{hi} \frac{\partial v_{hj}}{\partial x_i} w_{hj}\,dx \\ - \int_{\Gamma_{\text{out}}} c u_{h2} v_{h2} w_{h2}\,dx_1 + \frac{1}{\varepsilon} \sum_{R \in \mathcal{R}(h, s_\varkappa)} \text{meas}(R) \text{div}\, v_h(Q_R)\, \text{div}\, w_h(Q_R). \end{aligned} \quad (8.18)$$

To avoid the locking effect the penalty term is evaluated by using the one-point numerical integration formula with Q_R being the center of gravity of $R \in \mathcal{R}(h, s_\varkappa)$.

Remark 8.2. Let us note that for the stationary Navier–Stokes equations with constant viscosity the previous linearization method converges provided the right-hand side of the system is small enough (see [GR79, p. 109]).

Let $\varepsilon, \varkappa > 0$ be fixed. Then the discretization of $(\mathbb{P}_\varepsilon)$ reads as follows:

$$\left\{ \begin{array}{l} \text{Find } s_\varkappa^* \in U_\varkappa^{ad} \text{ such that} \\ J(u_h^{(k)}(s_\varkappa^*)) \le J(u_h^{(k)}(s_\varkappa)) \quad \forall s_\varkappa \in U_\varkappa^{ad}, \end{array} \right. \quad (8.19)$$

where

$$J(u_h^{(k)}(s_\varkappa)) = \int_{\widetilde{\Gamma}} (u_{h2}^{(k)}(s_\varkappa) - u_{ad})^2\,dx_1, \quad u_{ad} \in L^2(\Gamma_{\text{out}}),$$

and $u_h^{(k)}(s_\varkappa) := (u_{h1}^{(k)}(s_\varkappa), u_{h2}^{(k)}(s_\varkappa))$ is the solution of (8.17) for $k \in \mathbb{N}$ large enough.

The back wall $\Gamma(\alpha)$ sought is represented by a Bézier function defined by the control points $z^{(0)}, \dots, z^{(d+1)} \in \mathbb{R}^2$ with $z^{(0)}$ and $z^{(d+1)}$ being fixed: $z^{(0)} = (L_1, H_1)$, $z^{(d+1)} =$

$(L_1 + L_2, H_2)$. The remaining control points are allowed to move in the x_2 direction between "move limits" determined by the parameters $\alpha_{\min}, \alpha_{\max}$ and satisfying the slope condition $|s'_\varkappa| \le L_0$ in $[L_1, L_1 + L_2]$. Let $z^{(i)} = (z_1^{(i)}, z_2^{(i)})$, $z_1^{(i)} = L_1 + i\varkappa$, $z_2^{(i)} = (1-\alpha_i)\alpha_{\min} + \alpha_i\alpha_{\max}$, $\varkappa = L_2/(d+1)$, $i = 1, \dots, d$, where $\boldsymbol{\alpha} = (\alpha_1, \dots, \alpha_d)$ is the vector of the discrete design variables belonging to the convex set $\mathcal{U}$ defined as follows:

$$\mathcal{U} = \Bigg\{ \boldsymbol{\beta} \in \mathbb{R}^d \mid 0 \le \beta_i \le 1,\ i = 1, \dots, d;$$
$$\frac{|(\beta_{i+1} - \beta_i)(\alpha_{\max} - \alpha_{\min})|}{\varkappa} \le L_0,\ i = 1, \dots, d-1;\ \frac{|\beta_1(\alpha_{\max} - \alpha_{\min}) + \alpha_{\min} - H_1|}{\varkappa} \le L_0;$$
$$\frac{|\beta_d(\alpha_{\max} - \alpha_{\min}) + \alpha_{\min} - H_2|}{\varkappa} \le L_0 \Bigg\}. \tag{8.20}$$

The matrix formulation of (8.17) for fixed $h, \boldsymbol{\alpha}$ leads to a sequence of linear algebraic equations

$$\boldsymbol{K}(\boldsymbol{\alpha}, \boldsymbol{q}_k)\boldsymbol{q}_{k+1} = \boldsymbol{f}(\boldsymbol{\alpha}), \quad k \in \mathbb{N}, \tag{8.21}$$

where $\boldsymbol{q}_k$ and $\boldsymbol{q}_{k+1}$ contain the nodal values of the velocity corresponding to the previous and current iterations, respectively, and $\boldsymbol{K}(\boldsymbol{\alpha}, \boldsymbol{q}_k)$ is the matrix representing the bilinear form $B_{r_h s_\varkappa}(u_h^{(k)}, \cdot, \cdot)$. The algebraic form of (8.19) is then represented by the following nonlinear programming problem:

$$\begin{cases} \text{Find } \boldsymbol{\alpha}^* \in \mathcal{U} \quad \text{such that} \\ \mathcal{J}(\boldsymbol{q}(\boldsymbol{\alpha}^*)) \le \mathcal{J}(\boldsymbol{q}(\boldsymbol{\alpha})) \quad \forall \boldsymbol{\alpha} \in \mathcal{U}, \end{cases} \tag{$\mathbb{P}_d$}$$

where

$$\mathcal{J}(\boldsymbol{q}(\boldsymbol{\alpha})) = \sum_{i \in I_0} \omega_i (q_i(\boldsymbol{\alpha}) - u_{ad,i})^2.$$

The vector $\boldsymbol{q}(\boldsymbol{\alpha})$ is the solution of (8.21) for k large enough, I_0 contains indices of the degrees of freedom corresponding to the nodal values of u_{h2} on $\overline{\widetilde{\Gamma}}$, $\{\omega_i\}$ are weights of a numerical integration formula used for the evaluation of the integral defining J, and $u_{ad,i}$ are the values of u_{ad} at the nodes of $\mathcal{R}(h, s_\varkappa)$ on $\overline{\widetilde{\Gamma}}$ (we assume that $u_{ad} \in C(\overline{\widetilde{\Gamma}})$).

Assume now that the exact solution of the discrete state problem is at our disposal, i.e., the vector $\boldsymbol{q}(\boldsymbol{\alpha})$ that satisfies the equation

$$\boldsymbol{K}(\boldsymbol{\alpha}, \boldsymbol{q}(\boldsymbol{\alpha}))\boldsymbol{q}(\boldsymbol{\alpha}) = \boldsymbol{f}(\boldsymbol{\alpha}), \tag{8.22}$$

and denote $\boldsymbol{r}(\boldsymbol{\alpha}, \boldsymbol{y}) := \boldsymbol{K}(\boldsymbol{\alpha}, \boldsymbol{y})\boldsymbol{y} - \boldsymbol{f}(\boldsymbol{\alpha})$. The directional derivative of $\mathcal{J}$ at $\boldsymbol{\alpha}$ and in the direction $\boldsymbol{\beta}$ is given by (see (5.19), (5.20))

$$\mathcal{J}'(\boldsymbol{q}(\boldsymbol{\alpha}); \boldsymbol{\beta}) = -\boldsymbol{p}(\boldsymbol{\alpha})^{\mathrm{T}} \nabla_\alpha \boldsymbol{r}(\boldsymbol{\alpha}, \boldsymbol{q}(\boldsymbol{\alpha}))\boldsymbol{\beta}, \tag{8.23}$$

where $\boldsymbol{p}(\boldsymbol{\alpha})$ solves the adjoint equation

$$\boldsymbol{J}(\boldsymbol{\alpha}, \boldsymbol{q}(\boldsymbol{\alpha}))^{\mathrm{T}} \boldsymbol{p}(\boldsymbol{\alpha}) = \nabla_q \mathcal{J}(\boldsymbol{q}(\boldsymbol{\alpha})) \tag{8.24}$$

and $\boldsymbol{J}(\boldsymbol{\alpha}, \boldsymbol{q}(\boldsymbol{\alpha}))$ denotes the partial Jacobian of $\boldsymbol{r}$ with respect to $\boldsymbol{y}$ at $(\boldsymbol{\alpha}, \boldsymbol{q}(\boldsymbol{\alpha}))$.

In the forthcoming numerical example we will use the automatic differentiation technique presented in Section 5.3 to calculate the partial derivatives of $\boldsymbol{r}$ with respect to $\boldsymbol{\alpha}$ and $\boldsymbol{y}$ only at the element level.

To apply (8.23) in computations, (8.22) should be solved almost up to floating-point precision, i.e., nearly exactly in terms of the computer arithmetic. This is not an entirely unrealistic requirement if we switch to the quadratically converging Newton method

$$\boldsymbol{J}(\boldsymbol{\alpha}, \boldsymbol{q}_k)(\boldsymbol{q}_{k+1} - \boldsymbol{q}_k) = \boldsymbol{f}(\boldsymbol{\alpha}) - \boldsymbol{K}(\boldsymbol{\alpha}, \boldsymbol{q}_k)\boldsymbol{q}_k, \quad k \in \mathbb{N}, \tag{8.25}$$

after a reasonable approximation is computed using the simple iterations (8.21), which are less sensitive to the choice of the initial guess $\boldsymbol{q}_0$.

8.1.4 Numerical example

We choose the following fixed-size parameters (in meters): $H_1 = 1.0$, $H_2 = 0.1$, $L_1 = 1.0$, $L_2 = 8.0$, and $L_3 = 0.5$. These parameters, however, do not correspond to any existing headbox design. The parameters defining U^{ad} and the cost function are $\alpha_{\min} = H_2$, $\alpha_{\max} = H_1$, $L_0 = 2$, and $\widetilde{\Gamma} =]1.5, 8.5[$.

The physical parameters are chosen as follows: the density $\varrho = 1000$, the viscosity $\mu_0 = 0.001$, the coefficient of the outflow boundary condition (see (8.7)), $c = 1000.0$, and the inflow and recirculation velocities (in meters per second) fixed to $u_{\text{in}} = (4(1 - (2x_2 - 1)^8), 0.0)$ and $u_{\text{rec}} = (2(1 - (20x_2 - 1)^8), 0)$, respectively. The target velocity u_{ad} was chosen to be equal to the constant value -0.425 m/s.

The state problem is discretized by using 5104 four-noded isoparametric elements. The number of degrees of freedom in the discrete state problem is then approximately 10,700 (including those having fixed values). The penalty parameter ε is equal to 10^{-6}. The number of design variables d was chosen to be equal to 15. The discrete state problem was solved using a combination of the simple iterations (8.21) and the Newton method (8.25). The iteration process was stopped when $\|\boldsymbol{q}_{k+1} - \boldsymbol{q}_k\|_2 \leq 10^{-10}\|\boldsymbol{q}_{k+1}\|_2$. The linearized state problem and the adjoint equation were solved by a direct method based on the LU factorization of the coefficient matrix. In optimization, the sequential quadratic programming (SQP) algorithm E04UCF from the NAG subroutine library [NAG97] was used. All computations were done in 64-bit floating-point arithmetic.

Optimization is started from the traditional design, i.e., the linearly tapering header. The value of the objective function corresponding to this design is 7.474×10^{-2}. After 17 SQP iterations requiring 22 function evaluations the value of the objective function was reduced to 2.950×10^{-2}. The total CPU time was 83 min on an HP9000/J5600 computer (550 MHz, 4 GB RAM). The initial and optimized designs of the header are shown in Figure 8.3. From Figure 8.4 one can see the distribution of the velocity profiles for the initial and optimal designs. The convergence history is shown in Figure 8.5.

8.2 Multidisciplinary optimization of an airfoil profile using genetic algorithms

In traditional optimal shape design problems in aerospace engineering only one objective function of one scientific discipline is minimized. However, one discipline, such as aerodynamics, electromagnetics, etc., is usually not enough to describe the essential properties of

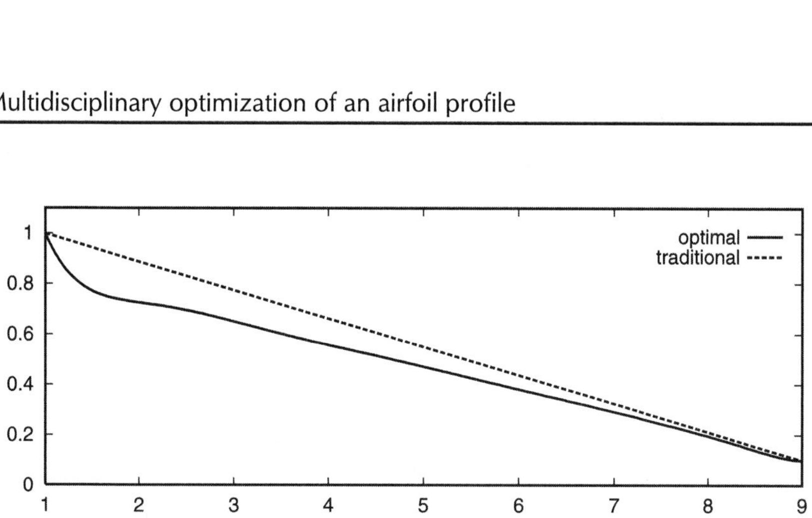

Figure 8.3. *The back wall of the header corresponding to a traditional linearly tapering design and the optimized design.*

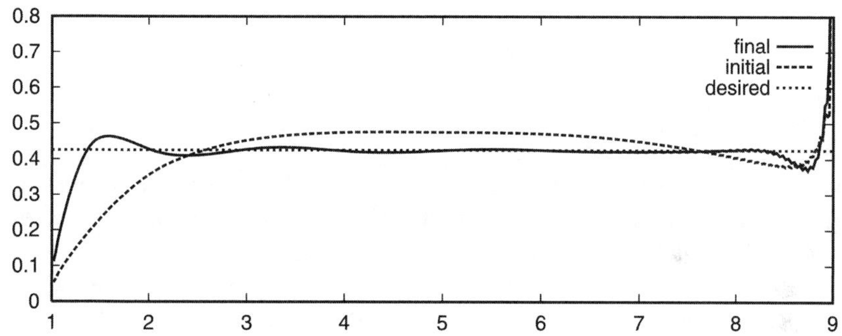

Figure 8.4. *Outlet velocity profiles.*

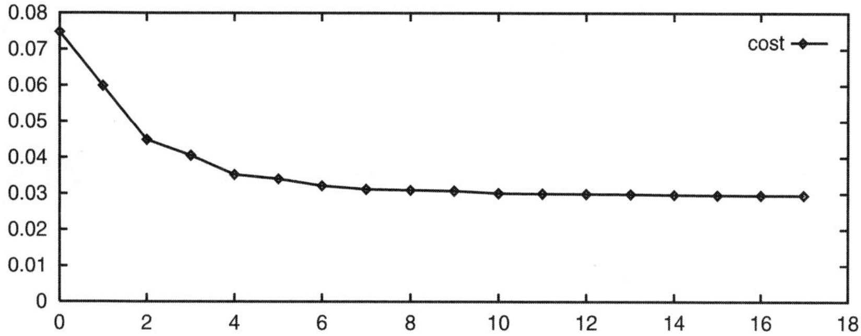

Figure 8.5. *Cost vs. iteration.*

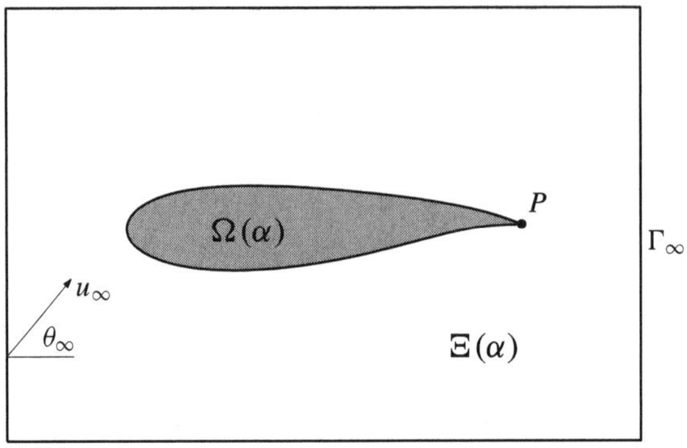

Figure 8.6. *Problem geometry for the scattering problem.*

the product to be optimized. Therefore it is necessary to consider *multidisciplinary* problems. For example, an airfoil should have certain aerodynamical properties while the radar visibility, i.e., the electromagnetic backscatter, should be minimized. In multidisciplinary optimization there are usually several conflicting criteria to be minimized.

8.2.1 Setting of the problem

Accurate numerical computation of a radar wave scattered by a flying obstacle (e.g., airfoil) and aerodynamical properties of the obstacle is a challenging problem because the solutions of the three-dimensional Maxwell and the compressible Navier–Stokes equations are needed. To reduce the huge computational burden several simplifications in the modeling of the physics are done. Assuming a two-dimensional design and the case of transverse magnetic polarization for the incoming wave the Maxwell equations reduce to the Helmholtz equation. One can also simplify the flow model by neglecting the viscous effects and solve the Euler equations instead of the Navier–Stokes equations. *To simplify our presentation we further assume that the flow is incompressible and irrotational. This model, of course, is valid only for flows with very small Mach numbers.*

We start with the shape parametrization of an airfoil. The Cartesian coordinates of the leading and trailing edges are $(0, 0)$ and $(1, 0)$, respectively. The profile is smooth except at the trailing edge P. The upper and lower surfaces of the airfoil are represented by two parametric curves $\alpha^+ \in U_+^{ad}$, $\alpha^- \in U_-^{ad}$, where

$$\begin{aligned} U_+^{ad} &= \{\gamma = (\gamma_1, \gamma_2) \in C^1([0, 1], \mathbb{R}^2) \mid \gamma(0) = (0, 0),\ \gamma(1) = (1, 0), \\ &\qquad 0 \le \gamma_2 \le t_{\max}/2,\ \gamma'(0) = (0, 1)\}, \\ U_-^{ad} &= \{\gamma = (\gamma_1, \gamma_2) \in C^1([0, 1], \mathbb{R}^2) \mid \gamma(0) = (0, 0),\ \gamma(1) = (1, 0), \\ &\qquad - t_{\max}/2 \le \gamma_2 \le 0,\ \gamma'(0) = (0, -1)\}, \end{aligned}$$

and $t_{\max}$ is the maximum thickness of the airfoil. In what follows we denote $\alpha = (\alpha^+, \alpha^-)$. The airfoil is then defined as a domain $\Omega(\alpha)$ surrounded by the curves α^+ and α^- (see Figure 8.6). Finally set $U^{ad} = U_+^{ad} \times U_-^{ad}$.

Two-dimensional modeling of a transverse magnetic polarized electromagnetic wave reflected by the obstacle $\Omega(\alpha)$ requires the solution of the wave equation

$$\frac{\partial^2 U}{\partial t^2} - \Delta U = 0 \quad \text{in } \mathbb{R}^2 \setminus \overline{\Omega(\alpha)}, \tag{8.26}$$

where $U := U(t, x)$ is the x_3 component of the electric field. Assuming the time harmonic case the solution of (8.26) is of the form

$$U(t, x) = \operatorname{Re}\left(w(x)e^{i\omega t}\right), \quad \omega \in \mathbb{R}, \tag{8.27}$$

where Re stands for the real part of complex numbers and $w(x)$ is the amplitude of the wave of frequency ω (we assume that the problem is rescaled so that the speed of light equals one). Substitution of (8.27) into (8.26) will result in the *Helmholtz equation*

$$\Delta w + \omega^2 w = 0 \quad \text{in } \mathbb{R}^2 \setminus \Omega(\alpha). \tag{8.28}$$

The total wave w occupying $\mathbb{R}^2 \setminus \overline{\Omega(\alpha)}$ consists of the incident wave u_∞ and the scattered wave u. We write $w = u_\infty + u$, where u_∞ is the plane wave propagating in the direction $(\cos\theta_\infty, \sin\theta_\infty)$, $\theta_\infty \in]-\pi/2, \pi/2[$; i.e., $u_\infty(x) = \exp(i\omega(x_1\cos\theta_\infty + x_2\sin\theta_\infty))$, $x = (x_1, x_2)$. The wave w vanishes on the conducting surface of the airfoil leading to the Dirichlet condition $u = -u_\infty$ on $\partial\Omega(\alpha)$. Since the incident wave u_∞ satisfies (8.28), the scattered wave $u : \mathbb{R}^2 \setminus \overline{\Omega(\alpha)} \to \mathbb{C}$ satisfies the exterior Helmholtz equation

$$\begin{cases} \Delta u + \omega^2 u = 0 & \text{in } \mathbb{R}^2 \setminus \overline{\Omega(\alpha)}, \\ u = -u_\infty & \text{on } \partial\Omega(\alpha). \end{cases} \tag{8.29}$$

To obtain a unique solution, problem (8.29) is completed by the additional *Sommerfeld radiation condition*

$$\lim_{r\to\infty} \sqrt{r}\left(\frac{\partial u}{\partial r} - i\omega u\right) = 0, \quad r = \|x\|_2, \tag{8.30}$$

describing the behavior of u at infinity, where the radial derivative is defined by $\partial/\partial r := (x_1/r)\partial/\partial x_1 + (x_2/r)\partial/\partial x_2$.

One is usually interested in the asymptotic behavior of the solution u. It can be shown (see [CK92]) that the scattered wave u (solving (8.29), (8.30)) has the following asymptotic expansion:

$$u(x) = \frac{e^{i(\omega r + \pi/4)}}{\sqrt{8\pi\omega r}}\widehat{u}_\omega(\theta) + O(r^{-3/2}), \quad r \to \infty, \tag{8.31}$$

where $x = re^{i\theta}$, $r > 0$, $\theta \in [0, 2\pi[$, and $\widehat{u}_\omega : [0, 2\pi[\to \mathbb{C}$ is the so-called far field pattern.

For the numerical treatment of (8.29), (8.30) we truncate the unbounded exterior domain by a rectangle Π large enough to contain $\overline{\Omega(\alpha)}$ in its interior. The complement of $\Omega(\alpha)$ in Π will be denoted by $\Xi(\alpha)$ (see Figure 8.6). On $\Gamma_\infty := \partial\Pi$ we pose a "nonreflecting" boundary condition approximating (8.30). The simplest one is

$$\frac{\partial u}{\partial \nu} - i\,\omega u = 0 \quad \text{on } \Gamma_\infty$$

and it will be used in what follows. For other choices of nonreflecting boundary conditions and for further details we refer to [BJR90], [EM79]. Thus we arrive at the following boundary value problem:

$$\begin{cases} \Delta u + \omega^2 u = 0 & \text{in } \Xi(\alpha), \\ u = g & \text{on } \partial\Omega(\alpha), \\ \dfrac{\partial u}{\partial \nu} - i\omega u = 0 & \text{on } \Gamma_\infty, \end{cases} \tag{8.32}$$

where $g = -u_\infty$ in $\mathbb{R}^2$.

In order to give a weak form of (8.32) we introduce the following spaces of *complex-valued functions* in $\Xi(\alpha)$, $\alpha \in U^{ad}$:

$$W(\alpha) = \{v \in \mathbb{H}^1(\Xi(\alpha)) \mid v = 0 \text{ on } \partial\Omega(\alpha)\}, \tag{8.33}$$

$$W_g(\alpha) = \{v \in \mathbb{H}^1(\Xi(\alpha)) \mid v = g \text{ on } \partial\Omega(\alpha)\}. \tag{8.34}$$

Here $\mathbb{H}^1(\Xi(\alpha)) = \{v \mid v = v_1 + iv_2 \mid v_1, v_2 \in H^1(\Xi(\alpha))\}$. The variational formulation of (8.32) involving the first order absorbing boundary condition reads as follows:

$$\begin{cases} \text{Find } u := u(\alpha) \in W_g(\alpha) \quad \text{such that} \\ \displaystyle\int_{\Xi(\alpha)} (\nabla u \cdot \nabla \overline{v} - \omega^2 u \overline{v})\, dx - i\omega \int_{\Gamma_\infty} u\overline{v}\, ds = 0 \quad \forall v \in W(\alpha), \end{cases} \tag{$\mathcal{P}_1(\alpha)$}$$

where $\overline{v}$ denotes the complex conjugate of v. We have [Hei97] the following result.

Theorem 8.1. *There exists a unique solution to* $(\mathcal{P}_1(\alpha))$ *for any* $\alpha \in U^{ad}$.

The fluid flow is modeled as a two-dimensional *incompressible* and *inviscid* flow. We introduce the stream function φ such that the velocity field v is given by $v = \operatorname{curl}\varphi$. On Γ_∞ the velocity is assumed to have a constant value $v_\infty = (v_{\infty 1}, v_{\infty 2})$. Inside the computational domain $\Xi(\alpha)$ the pressure is $p = p_{\text{ref}} - \frac{1}{2}|v|^2 = p_{\text{ref}} - \frac{1}{2}|\operatorname{curl}\varphi|^2$, where p_{ref} is a given constant reference pressure. The stream function formulation of this problem reads as follows (see [CO86]): Find a function $\varphi := \varphi(\alpha)$ and a constant $\beta := \beta(\alpha)$ satisfying the boundary value problem

$$\begin{cases} -\Delta\varphi = 0 & \text{in } \Xi(\alpha), \\ \varphi = \beta & \text{on } \partial\Omega(\alpha), \\ \varphi = \varphi_\infty & \text{on } \Gamma_\infty, \\ \nabla\varphi \text{ is regular in a vicinity of the trailing edge } P, \end{cases} \tag{8.35}$$

where $\varphi_\infty = v_{\infty 1} x_2 - v_{\infty 2} x_1$.

Remark 8.3. For any constant β there exists a unique solution to the nonhomogeneous Dirichlet boundary value problem $(8.35)_{1-3}$. Since the angle between two tangents to the profile at P is very sharp, the solution φ has a singularity at P for β given a priori. On the other hand it is known that there exists a (unique) particular value of β ensuring the regularity of φ, namely $\varphi \in H^2(\Xi(\alpha))$ (see [Gri85]). Among solutions to (8.35) we look for just the regular one. This is expressed by the last condition in (8.35). In computations this condition will be realized by requiring continuity of the flow at the trailing edge P:

$$\frac{\partial\varphi(P^+)}{\partial\nu} = -\frac{\partial\varphi(P^-)}{\partial\nu}, \tag{8.36}$$

where

$$\frac{\partial\varphi(P^\pm)}{\partial\nu} := \lim_{\substack{Q\to P\\ Q\in[\alpha^\pm]}} \frac{\partial\varphi(Q)}{\partial\nu}, \qquad [\alpha^\pm] := \text{graph of } \alpha^\pm.$$

Condition (8.36) is known as the *Kutta–Joukowski condition* and it will be used to fix the value of β in (8.35).

To give the weak form of (8.35) we need the spaces $V_r(\alpha)$, $r \in \mathbb{R}$, $\alpha \in U^{ad}$:

$$V_r(\alpha) = \left\{v \in H^1(\Xi(\alpha)) \mid v = \varphi_\infty \text{ on } \Gamma_\infty,\ v = r \text{ on } \partial\Omega(\alpha)\right\}. \tag{8.37}$$

The weak formulation of (8.35) reads as follows:

$$\left\{\begin{array}{l} \text{Find } \beta \in \mathbb{R} \text{ and } \varphi := \varphi(\alpha) \in V_\beta(\alpha) \quad \text{such that} \\ \displaystyle\int_{\Xi(\alpha)} \nabla\varphi\cdot\nabla v\, dx = 0 \quad \forall v \in H_0^1(\Xi(\alpha)), \end{array}\right. \tag{$\mathcal{P}_2(\alpha)$}$$

where β is such that $\nabla\varphi$ is regular in the vicinity of P.

Our aim is to find the shape of the profile $\Omega(\alpha)$ such that the intensity of the reflected electromagnetic wave into a given sector defined by two angles $\theta_* < \theta^*$, $\theta_*, \theta^* \in]0, 2\pi[$, is as small as possible while at the same time reasonable lift properties are maintained. To satisfy the latter condition the computed pressure distributions p^+, p^- on the upper and lower surfaces of the profile described by α^+, α^-, respectively, should not be too far from the given pressure distributions $p_d^+, p_d^- \in L^\infty([0, 1])$.

The choice of an appropriate single cost function is impossible and therefore the shape optimization problem is formulated as the following multicriteria optimization problem:

$$\left\{\begin{array}{l} \text{Find a Pareto optimal } \alpha^* \in U^{ad} \text{ for the problem} \\ \displaystyle\min_{\alpha\in U^{ad}} \{J_1(\alpha), J_2(\alpha)\}. \end{array}\right. \tag{$\mathbb{P}$}$$

The individual cost functions are defined by

$$J_1(\alpha) = \int_{\theta_*}^{\theta^*} |\widehat{u}_\omega(\theta)|^2 d\theta, \tag{8.38}$$

$$J_2(\alpha) = \int_0^1 (p^+(\alpha^+) - p_d^+)^2\, dx_1 + \int_0^1 (p^-(\alpha^-) - p_d^-)^2\, dx_1. \tag{8.39}$$

Here $\widehat{u}_\omega(\theta)$ is the far field pattern corresponding to the solution $u := u(\alpha)$ of $(\mathcal{P}_1(\alpha))$ (see (8.31)) and $p^\pm(\alpha^\pm)(x_1) := p(x_1, \alpha^\pm(x_1))$, $x_1 \in]0, 1[$. Recall that $p(\alpha) = p_{\text{ref}} -$

$\frac{1}{2}|\text{curl}\,\varphi(\alpha)|^2$, where $\varphi(\alpha)$ solves $(\mathcal{P}_2(\alpha))$. For other formulations of the problem see [BP93], [MTP99], and [MP01], for example.

Remark 8.4. For the sake of simplicity, we use the same rectangle Π in both problems $(\mathcal{P}_1(\alpha))$ and $(\mathcal{P}_2(\alpha))$. This, however, is neither necessary nor desired in computations. In practice the size of Π should be much larger for the flow problem than for the Helmholtz problem, especially if the second order absorbing boundary condition is used in the Helmholtz problem.

8.2.2 Approximation and numerical realization

Denote $U_\varkappa^{ad} = U_{+\varkappa}^{ad} \times U_{-\varkappa}^{ad}$, where $U_{\pm\varkappa}^{ad}$ is a subset of $U_\pm^{ad}$ realized by Bézier curves of degree d. The upper and lower surfaces of airfoils $\Omega(\alpha)$, $\alpha \in U^{ad}$, will be approximated by functions $s_\varkappa = (s_\varkappa^+, s_\varkappa^-) \in U_\varkappa^{ad}$. Let $\Omega(r_h s_\varkappa)$ be a polygonal approximation of $\Omega(s_\varkappa)$. Then $\Xi(s_\varkappa) := \Pi \setminus \overline{\Omega(r_h s_\varkappa)}$ is the computational domain for both state problems. As usual $\Xi_h(s_\varkappa)$ stands for the computational domain $\Xi(s_\varkappa)$ with a given triangulation $\mathcal{T}(h, s_\varkappa)$. Further, let $\mathcal{I}$ be the set of all the nodes of $\mathcal{T}(h, s_\varkappa)$ lying on $\partial\Omega(r_h s_\varkappa)$. For the discretization of state problems we use the following finite element spaces:

$$
\begin{aligned}
W_h(s_\varkappa) = \{v_h \mid v_h = v_{h1} + i v_{h2},\ v_{h1}, v_{h2} \in C(\overline{\Xi_h(s_\varkappa)}),\\
v_{hj\,|T} \in P_1(T),\ j = 1, 2,\ \forall T \in \mathcal{T}(h, s_\varkappa),\ v_h = 0 \text{ on } \partial\Omega(r_h s_\varkappa)\}, \qquad (8.40)\\
W_{gh}(s_\varkappa) = \{v_h \mid v_h = v_{h1} + i v_{h2},\ v_{h1}, v_{h2} \in C(\overline{\Xi_h(s_\varkappa)}),\\
v_{hj\,|T} \in P_1(T),\ j = 1, 2,\ \forall T \in \mathcal{T}(h, s_\varkappa),\ v_h(a) = g(a)\ \forall a \in \mathcal{I}\}, \qquad (8.41)\\
V_{rh}(s_\varkappa) = \{v_h \in C(\overline{\Xi_h(s_\varkappa)}) \mid v_{h|T} \in P_1(T)\ \forall T \in \mathcal{T}(h, s_\varkappa),\ v_h = \varphi_\infty \text{ on } \partial\Pi,\\
v_h = r \text{ on } \partial\Omega(r_h s_\varkappa)\},\ r \in \mathbb{R}. \qquad (8.42)
\end{aligned}
$$

Since the function φ_∞ is linear and r is constant, the conditions on $\partial\Pi$ and $\partial\Omega(r_h s_\varkappa)$ in the definition of $V_{rh}(s_\varkappa)$ can be realized exactly by piecewise linear functions.

The discrete state problems are defined as follows:

$$
\begin{cases}
\text{Find } u_h := u_h(s_\varkappa) \in W_{gh}(s_\varkappa) \quad \text{such that}\\
\displaystyle\int_{\Xi_h(s_\varkappa)} (\nabla u_h \cdot \nabla \bar{v}_h - \omega^2 u_h \bar{v}_h)\, dx - i\omega \int_{\Gamma_\infty} u_h \bar{v}_h\, ds = 0 \quad \forall v_h \in W_h(s_\varkappa)
\end{cases}
\qquad (\mathcal{P}_{1h}(s_\varkappa))
$$

and

$$
\begin{cases}
\text{Find } \beta := \beta(s_\varkappa) \in \mathbb{R},\ \varphi_h := \varphi_h(s_\varkappa) \in V_{\beta h}(s_\varkappa) \quad \text{such that}\\
\displaystyle\int_{\Xi_h(s_\varkappa)} \nabla\varphi_h \cdot \nabla v_h\, dx = 0 \quad \forall v_h \in V_{0h}(s_\varkappa)\\
\dfrac{\partial\varphi_h(P^+)}{\partial\nu} = -\dfrac{\partial\varphi_h(P^-)}{\partial\nu},
\end{cases}
\qquad (\mathcal{P}_{2h}(s_\varkappa))
$$

where $V_{0h}(s_\varkappa)$ is defined by (8.42) with $\varphi_\infty = r = 0$.

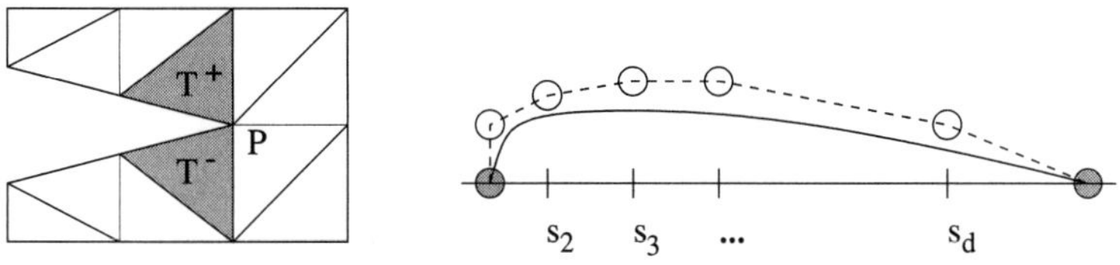

(a) (b)

Figure 8.7. (a) *Triangles involved in the Kutta–Joukowski condition.* (b) *Shape parametrization of the upper surface of the airfoil: control points marked* ● *are fixed whereas the ones marked* ○ *are allowed to move in the x_2 direction.*

Problem $(\mathcal{P}_{2h}(s_\varkappa))$ can be conveniently solved using the linearity in β. Let φ_{0h} and φ_{1h} be the solutions to the following:

$$\begin{cases} \text{Find } \varphi_{0h} := \varphi_{0h}(s_\varkappa) \in V_{0h}(s_\varkappa) \quad \text{such that} \\ \displaystyle\int_{\Xi_h(s_\varkappa)} \nabla\varphi_{0h} \cdot \nabla v_h \, dx = 0 \quad \forall v_h \in V_{0h}(s_\varkappa) \end{cases} \tag{8.43}$$

and

$$\begin{cases} \text{Find } \varphi_{1h} := \varphi_{1h}(s_\varkappa) \in V_{1h}(s_\varkappa) \quad \text{such that} \\ \displaystyle\int_{\Xi_h(s_\varkappa)} \nabla\varphi_{1h} \cdot \nabla v_h \, dx = 0 \quad \forall v_h \in V_{0h}(s_\varkappa), \end{cases} \tag{8.44}$$

where $V_{1h}(s_\varkappa)$ is defined by (8.42) with $r = 1$. Then the function

$$\varphi_h := \varphi_h(s_\varkappa) = \beta\varphi_{1h} + (1-\beta)\varphi_{0h},$$

where β is computed from the continuity flow condition (8.36):

$$\beta = -\frac{\nabla\varphi_{0h}(P^+)\cdot\nu + \nabla\varphi_{0h}(P^-)\cdot\nu}{\nabla(\varphi_{1h}-\varphi_{0h})(P^+)\cdot\nu + \nabla(\varphi_{1h}-\varphi_{0h})(P^-)\cdot\nu}, \tag{8.45}$$

and where the approximate (constant) values of the normal derivatives on the upper and lower surfaces of the airfoil are evaluated in the triangles T^+ and T^-, respectively (see Figure 8.7(a)), solves $(\mathcal{P}_{2h}(s_\varkappa))$.

The discretization of $(\mathbb{P})$ is defined as follows:

$$\begin{cases} \text{Find a Pareto optimal } s_\varkappa^* \in U_\varkappa^{ad} \text{ for the problem} \\ \displaystyle\min_{s_\varkappa \in U_\varkappa^{ad}} \{J_1(s_\varkappa), J_2(s_\varkappa)\}, \end{cases} \tag{$\mathbb{P}_h$}$$

where J_1 and J_2 defined by (8.38) and (8.39), respectively, are evaluated using the solutions to $(\mathcal{P}_{1h}(s_\varkappa))$ and $(\mathcal{P}_{2h}(s_\varkappa))$, respectively.

Next we derive the algebraic form of $(\mathbb{P}_h)$ for $h > 0$ fixed. The shape of the airfoil is parametrized using one Bézier curve for the upper part of the airfoil and another one for the lower part. The upper curve is defined by the control points

$$(0,0),\ (0,\alpha_1),\ (s_2,\alpha_2),\ (s_3,\alpha_3),\ldots,(s_d,\alpha_d),\ (1,0)$$

and the lower one by

$$(0,0),\ (0,\alpha_{d+1}),\ (s_2,\alpha_{d+2}),\ (s_3,\alpha_{d+3}),\ldots,(s_d,\alpha_{2d}),\ (1,0),$$

where $0 < s_2 < s_3 < \cdots < s_d < 1$ are given (see Figure 8.7(b)). The numbers α_i, $i = 1,\ldots,2d$, are the discrete design variables belonging to the set

$$\begin{aligned}\mathcal{U} = \{\boldsymbol{\gamma} \in \mathbb{R}^{2d} \mid 0 \le \gamma_i \le t_{\max}/2,\ i = 1,\ldots,d,\\ 0 \ge \gamma_i \ge -t_{\max}/2,\ i = d+1,\ldots,2d\}.\end{aligned} \tag{8.46}$$

REMARK 8.5. To satisfy the condition $(s_{\varkappa}^{\pm})'(0) = (0,\pm 1)$ we use two control points at the leading edge for each of the curves $s_{\varkappa}^{+}, s_{\varkappa}^{-}$.

The matrix formulations of $(\mathcal{P}_{1h}(s_{\varkappa}))$, (8.43), and (8.44) for fixed h and $\boldsymbol{\alpha} \in \mathcal{U}$ lead to a *complex* system of linear algebraic equations

$$\boldsymbol{C}(\boldsymbol{\alpha})\boldsymbol{u}(\boldsymbol{\alpha}) = \boldsymbol{b}(\boldsymbol{\alpha}) \tag{8.47}$$

and two real systems

$$\boldsymbol{A}(\boldsymbol{\alpha})\widetilde{\boldsymbol{\varphi}}_0(\boldsymbol{\alpha}) = \boldsymbol{f}_0(\boldsymbol{\alpha}), \tag{8.48}$$

$$\boldsymbol{A}(\boldsymbol{\alpha})\widetilde{\boldsymbol{\varphi}}_1(\boldsymbol{\alpha}) = \boldsymbol{f}_1(\boldsymbol{\alpha}). \tag{8.49}$$

The vector $\boldsymbol{u}(\boldsymbol{\alpha})$ contains the nodal values of the approximate scattered complex-valued wave $u_h(s_{\varkappa})$ at the nodes of $\mathcal{T}(h, s_{\varkappa})$ lying in $\overline{\Xi(s_{\varkappa})} \setminus \partial\Omega(r_h s_{\varkappa})$, while $\widetilde{\boldsymbol{\varphi}}_0(\boldsymbol{\alpha}), \widetilde{\boldsymbol{\varphi}}_1(\boldsymbol{\alpha})$ are the vectors of the nodal values of $\varphi_{0h}(s_{\varkappa}), \varphi_{1h}(s_{\varkappa})$, respectively, at all the inner nodes of $\mathcal{T}(h, s_{\varkappa})$.

REMARK 8.6. The nonzero right-hand sides in (8.47)–(8.49) are due to the nonhomogeneous Dirichlet boundary conditions in $(\mathcal{P}_{1h}(s_{\varkappa}))$, (8.43), and (8.44).

Extending the vectors $\widetilde{\boldsymbol{\varphi}}_0(\boldsymbol{\alpha}), \widetilde{\boldsymbol{\varphi}}_1(\boldsymbol{\alpha})$ by the known values of $\varphi_{0h}(s_{\varkappa}), \varphi_{1h}(s_{\varkappa})$, respectively, at the nodes of $\mathcal{T}(h, s_{\varkappa})$ on Γ_∞ and $\partial\Omega(r_h s_{\varkappa})$ we obtain the new vectors $\boldsymbol{\varphi}_0(\boldsymbol{\alpha}), \boldsymbol{\varphi}_1(\boldsymbol{\alpha})$. Then $\boldsymbol{\varphi}(\boldsymbol{\alpha}) = \beta\boldsymbol{\varphi}_1(\boldsymbol{\alpha}) + (1-\beta)\boldsymbol{\varphi}_0(\boldsymbol{\alpha})$ with β defined by (8.45) is the vector of the nodal values of the approximate velocity stream function $\varphi_h(s_{\varkappa})$.

The integral defining J_1 is evaluated by using the numerical integration

$$J_1(s_{\varkappa}) \approx \mathcal{J}_1(\boldsymbol{\alpha}) = \sum_i c_i |\widehat{u}_{\omega h}(\theta_i)|^2, \tag{8.50}$$

where c_i and θ_i are weights and integration points, respectively, of a quadrature formula; $\widehat{u}_{\omega h}$ is an approximation of the far field pattern; and $\boldsymbol{\alpha}$ is the vector of the control points of $s_{\varkappa} \in U_{\varkappa}^{ad}$.

Due to the piecewise linear approximation of the stream function φ there exists a set $\{t_i \mid i = 0,\ldots,m\}$, $0 = t_0 < t_1 < \cdots < t_m = 1$, such that the discrete pressure distributions $p_h^{\pm}(x_1) := p_h(x_1, r_h s_{\varkappa}^{\pm}(x_1))$ on the surface $\partial\Omega(r_h s_{\varkappa}^{\pm})$ of the airfoil $\Omega(r_h s_{\varkappa})$

are constant in each interval $]t_{i-1}, t_i[$, $i = 1, \ldots, m$. Recall that $p_h = p_{\text{ref}} - \frac{1}{2}|\text{curl}\, \varphi_h|^2$. Thus we obtain the following algebraic expression of J_2:

$$\mathcal{J}_2(\boldsymbol{\alpha}) = \sum_{i=1}^{m} (t_i - t_{i-1})(p_h^+(t_{i-\frac{1}{2}}) - p_d^+(t_{i-\frac{1}{2}}))^2 \\ + \sum_{i=1}^{m} (t_i - t_{i-1})(p_h^-(t_{i-\frac{1}{2}}) - p_d^-(t_{i-\frac{1}{2}}))^2, \quad (8.51)$$

where $t_{i-\frac{1}{2}} = (t_{i-1} + t_i)/2$ (we assume that $p_d^{\pm} \in C([0, 1])$).

The matrix form of $(\mathbb{P}_h)$ reads as follows:

$$\begin{cases} \text{Find a Pareto optimal } \boldsymbol{\alpha}^* \in \mathcal{U} \text{ for the problem} \\ \min\limits_{\boldsymbol{\alpha} \in \mathcal{U}} \{\mathcal{J}_1(\boldsymbol{\alpha}), \mathcal{J}_2(\boldsymbol{\alpha})\}. \end{cases} \quad (\mathbb{P}_d)$$

8.2.3 Numerical example

To be able to capture the oscillating nature of solutions to the Helmholtz problem for large ω, the mesh used for the discretization of $(\mathcal{P}_1(\alpha))$ has to be fine in the *whole* computational domain $\Xi_h(s_\varkappa)$. A traditional rule of thumb says that it should be at least 10 nodes per wavelength. This means that the size of system (8.47) is very large, making its direct solution infeasible. In the forthcoming numerical example the system is solved iteratively using the generalized minimum residual (GMRES) method with a special preconditioning technique using the so-called fast direct solvers. We briefly describe how to construct a good preconditioner. Let us consider the Helmholtz equation posed in the *whole* rectangle Π:

$$\begin{cases} \Delta\widetilde{u} + \omega^2\widetilde{u} = 0 & \text{in } \Pi, \\ \dfrac{\partial \widetilde{u}}{\partial \nu} - i\omega\widetilde{u} = 0 & \text{on } \Gamma_\infty. \end{cases} \quad (8.52)$$

Let $\mathcal{T}_\Pi(h_1, h_2)$ be a *uniform* triangulation of Π obtained by dividing Π into rectangles of lengths h_1, h_2 in the x_1, x_2 directions, respectively. Problem (8.52) is now approximated by piecewise linear functions over $\mathcal{T}_\Pi(h_1, h_2)$ resulting in the system of algebraic equations

$$\boldsymbol{B}\widetilde{\boldsymbol{u}} = \boldsymbol{0}. \quad (8.53)$$

Due to the uniformity of $\mathcal{T}_\Pi(h_1, h_2)$ the matrix $\boldsymbol{B}$ has a special structure, and any linear system of equations with $\boldsymbol{B}$ as a coefficient matrix can be solved very efficiently (i.e., the number of arithmetic operations needed is proportional to $N \log N$, where N is the number of unknowns) using a fast direct solver (the cyclic reduction method [Dor70], for example). Therefore the matrix $\boldsymbol{B}$ can be used as a preconditioner for solving (8.47). For this purpose $\boldsymbol{C}(\boldsymbol{\alpha})$ should be "close" to $\boldsymbol{B}$ in some sense or, in other words, the triangulation $\mathcal{T}(h, s_\varkappa)$ of $\overline{\Xi(r_h s_\varkappa)}$ should not differ too much from $\mathcal{T}_\Pi(h_1, h_2)_{|\Xi(r_h s_\varkappa)}$. To this end we use the so-called locally fitted meshes when the mesh $\mathcal{T}(h, s_\varkappa)$ coincides with $\mathcal{T}_\Pi(h_1, h_2)_{|\Xi(r_h s_\varkappa)}$ except on a narrow strip around the profile (see Figure 8.8). System (8.53) has, however,

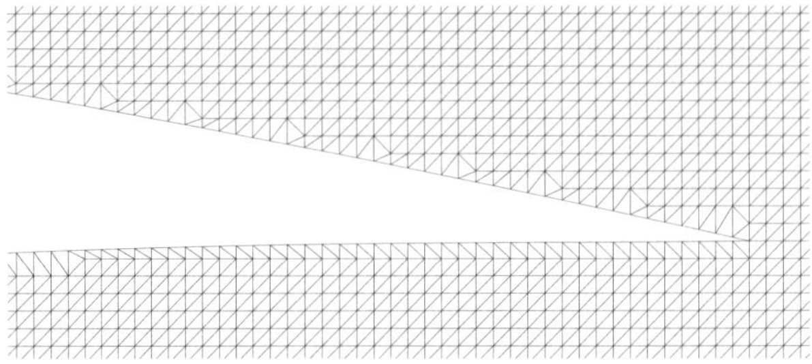

Figure 8.8. *A detail of the finite element mesh of* $(\mathcal{P}_{1h}(s_{\varkappa}))$ *near the trailing edge* P.

more unknowns than (8.47) so that the latter has to be extended. The simplest way of doing this is to use the zero extension

$$\mathring{\boldsymbol{C}}\mathring{\boldsymbol{u}} = \mathring{\boldsymbol{b}}, \tag{8.54}$$

where

$$\mathring{\boldsymbol{C}} = \begin{pmatrix} \boldsymbol{C} & \boldsymbol{0} \\ \boldsymbol{0} & \boldsymbol{0} \end{pmatrix}, \quad \mathring{\boldsymbol{u}} = \begin{pmatrix} \boldsymbol{u} \\ \boldsymbol{u}' \end{pmatrix}, \quad \mathring{\boldsymbol{b}} = \begin{pmatrix} \boldsymbol{b} \\ \boldsymbol{0} \end{pmatrix}, \quad \boldsymbol{C} := \boldsymbol{C}(\boldsymbol{\alpha}),\ \boldsymbol{b} := \boldsymbol{b}(\boldsymbol{\alpha}),\ \boldsymbol{u} := \boldsymbol{u}(\boldsymbol{\alpha}).$$

Then instead of the original system (8.47) we solve iteratively the following preconditioned system:

$$\boldsymbol{B}^{-1}\mathring{\boldsymbol{C}}\mathring{\boldsymbol{u}} = \boldsymbol{B}^{-1}\mathring{\boldsymbol{b}}. \tag{8.55}$$

Using more sophisticated extensions than (8.54), one can show that this type of preconditioning technique is almost optimal; i.e., the number of iterations in the GMRES method is almost independent of h. For details we refer to [Toi97], [Hei97]. The same preconditioning techniques can be used to solve flow problems (8.48) and (8.49).

Locally fitted meshes that are needed for constructing good preconditioners *are not* topologically equivalent in the sense of Section 2.2 (the number of triangles may vary after any change in the airfoil profile), resulting in the possible loss of the differentiability of the minimized functions. This and the desire to get several Pareto optimal solutions motivate us to use the multiobjective variant of the genetic algorithm (GA) presented in Subsection 4.4.4.

The state problems were solved with the following data: $\Pi = [-\frac{1}{2}, \frac{3}{2}] \times [-1, 1]$ for the Helmholtz problem and $\Pi = [-4, 5] \times [-5, 5]$ for the flow problem, $u_\infty = \exp(i20\pi(x_1 \cos 45^\circ + x_2 \sin 45^\circ))$ (i.e., $\omega = 20\pi$, $\theta_\infty = 45^\circ$), $\varphi_\infty = x_2 \cos 2^\circ - x_1 \sin 2^\circ$, $p_{\mathrm{ref}} = 1$. The cost functionals are specified as follows: $\theta_* = \pi$, $\theta^* = 3\pi/2$ in J_1, and the target pressure distributions $p_d^{\pm}$ defining J_2 are the computed pressure distributions on an NACA0012 airfoil. The mesh $\mathcal{T}_\Pi(h_1, h_2)$ is defined by a rectangular grid of 301×301 nodes. The far field pattern $\widehat{u}_\omega$ needed for the evaluation of J_1 is computed by using the following approximate expression:

$$\widehat{u}_\omega(\theta) \approx \sum_{k,l} b_{kl} e^{-i\omega(kh_1 \cos\theta + lh_2 \sin\theta)},$$

where b_{kl} are the nodal values of the so-called grid function b and h_1, h_2 are the parameters characterizing $\mathcal{T}_\Pi(h_1, h_2)$ (for details we refer to [Toi97]). Since the piecewise linear

Table 8.1. *The parameters in the multiobjective variant of the simple GA.*

Population size	64	Crossover probability	0.8
Number of generations	100	Mutation probability	0.2
Tournament size	3	Mutation exponent	4
Sharing distance	0.25		

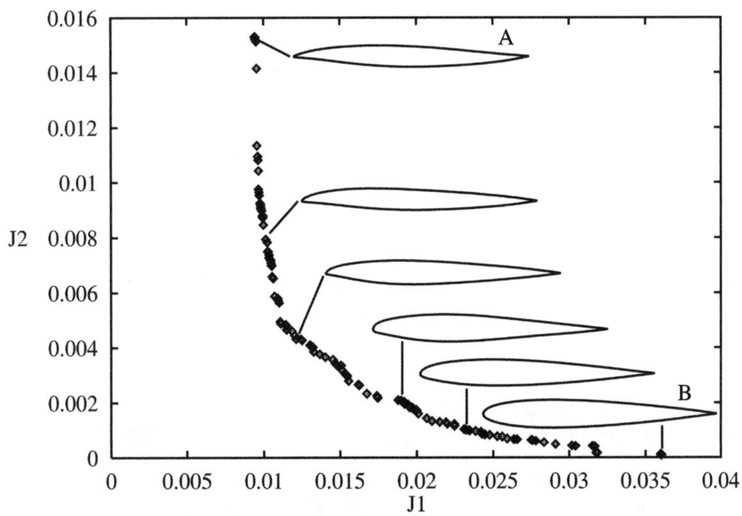

Figure 8.9. *Optimal airfoil profiles.*

elements produce a poor approximation of curl φ on the boundary (needed to calculate the pressure) a quite fine rectangular but not uniform grid with 361×401 nodes was used to solve the flow problem.

The computations were done on an HP9000/J5600 computer. The parameters used in the multiobjective variant of the simple GA are collected in Table 8.1. The total CPU time needed was approximately 15 h. The nondominated points of the final population in the J_1-J_2 space are shown in Figure 8.9. The airfoil profiles corresponding to the selected points are shown in the same figure.

The electromagnetic and aerodynamic properties of two extreme profiles A and B in Figure 8.9 are shown in Figures 8.10–8.12. Instead of $|\widehat{u}_\omega|$ and p, engineers prefer to illustrate the *radar cross section*

$$\mathrm{RCS}(\theta) = 10 \log_{10} \left(\frac{1}{8\pi} |\widehat{u}_\omega(\theta)|^2 \right)$$

and the *pressure coefficient*

$$C_p^{\pm}(x_1) = \frac{p^{\pm}(x_1) - p_{\mathrm{ref}}}{\varrho_{\mathrm{ref}} \frac{1}{2} |v_\infty|^2},$$

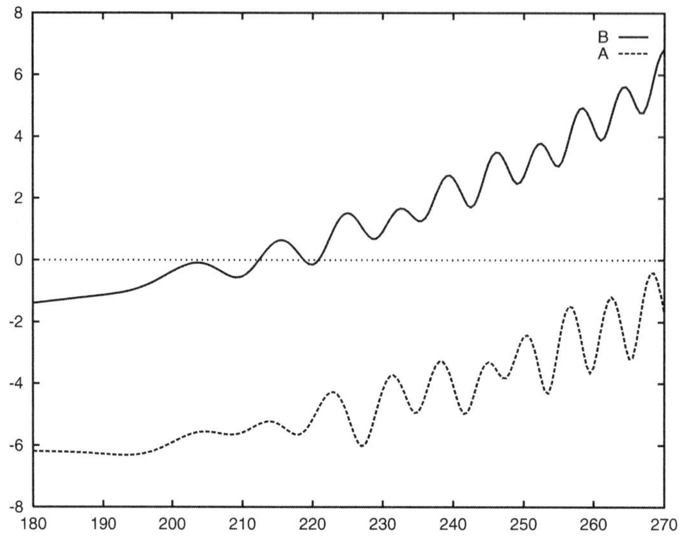

Figure 8.10. *Radar cross sections in the sector* [180°, 270°] *corresponding to airfoils A and B from Figure* 8.9.

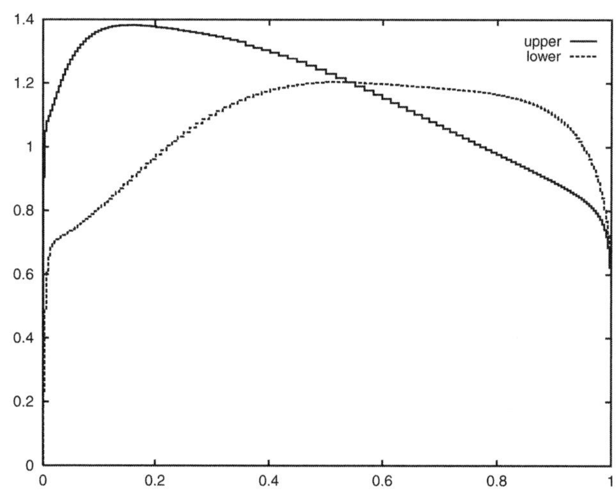

Figure 8.11. *Pressure coefficient on the upper and lower surfaces of airfoil A from Figure* 8.9.

where p_{ref}, ϱ_{ref} are reference values of the pressure and density in the undisturbed free stream (here we used $p_{\text{ref}} = \varrho_{\text{ref}} = 1$). The profile B with $J_2 \approx 0$ is essentially the NACA0012 profile. On the other hand, the radar visibility into the sector $[\theta_*, \theta^*]$ of the profile A (which has a peculiar shape) is very low.

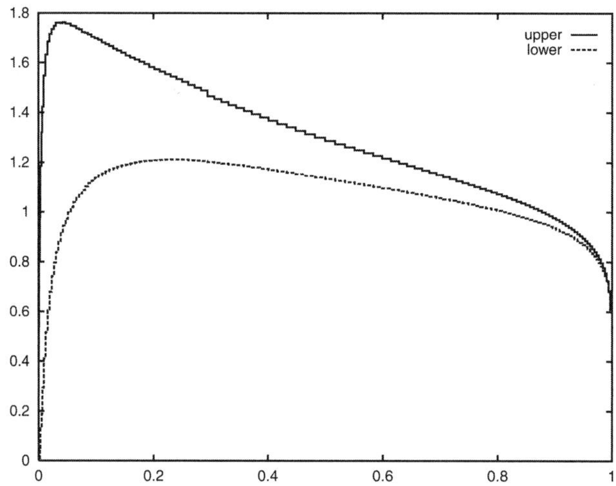

Figure 8.12. *Pressure coefficient on the upper and lower surfaces of airfoil B from Figure* 8.9.

Problems

Problem 8.1. Prove that for $u, v \in (W^{1,3}(\Omega(\alpha)))^2$, $\alpha \in U^{ad}$, all integrals in $(\mathcal{P}(\alpha))$ from Subsection 8.1.2 are finite.

Problem 8.2. Prove the formal equivalence between the classical formulation (8.1)–(8.7) and $(\mathcal{P}(\alpha))$.

Problem 8.3. Prove (8.23) and (8.24).

Problem 8.4. Prove that (8.32) and $(\mathcal{P}_1(\alpha))$ are formally equivalent.

Problem 8.5. Show that the solution φ_h of $(\mathcal{P}_{2h}(s_\varkappa))$ from Subsection 8.2.2 can be written in the form $\varphi_h = \beta\varphi_{0h} + (1-\beta)\varphi_{1h}$, where $\varphi_{0h}, \varphi_{1h}$ are solutions to (8.43), (8.44), respectively.

Appendix A

Weak Formulations and Approximations of Elliptic Equations and Inequalities

We recall in brief basic results of the theory of linear elliptic equations and inequalities, Sobolev spaces, and finite element approximations within the range needed in the previous text. For more details on weak formulations of elliptic equations we refer to [Neč67]. Finite element approximations of elliptic problems are discussed in [Cia02].

Let V be a real Hilbert space, K a *nonempty, closed, and convex* subset of V, and V' a dual of V. We denote by $\|\cdot\|$, $\|\cdot\|_*$ the norms in V, V', respectively. Further, let $\langle\cdot,\cdot\rangle$ stand for a duality pairing between V' and V and let $a : V \times V \to \mathbb{R}$ be a bilinear form satisfying the following assumptions:

(boundedness)

$$\exists M = \text{const.} > 0 : \quad |a(u, v)| \leq M\|u\|\,\|v\| \quad \forall u, v \in V; \tag{A.1}$$

(V-ellipticity)

$$\exists \alpha = \text{const.} > 0 : \quad a(v, v) \geq \alpha\|v\|^2 \quad \forall v \in V. \tag{A.2}$$

Definition A.1. *A triple $\{K, a, \ell\}$, where $\ell \in V'$ is given, defines* an abstract elliptic inequality. *An element $u \in K$ is said to be a solution of $\{K, a, \ell\}$ iff*

$$a(u, v - u) \geq \langle \ell, v - u\rangle \quad \forall v \in K. \tag{A.3}$$

The existence and uniqueness of a solution to $\{K, a, \ell\}$ follow from Lemma A.1.

Lemma A.1. *(Lax–Milgram.) Let* (A.1) *and* (A.2) *be satisfied. Then $\{K, a, \ell\}$ has a unique solution for any $\ell \in V'$ and*

$$\|u\| \leq \frac{1}{\alpha}\|\ell\|_*,$$

where α is the constant from (A.2).

In addition, if a is symmetric in V, i.e.,

$$a(u, v) = a(v, u) \quad \forall u, v \in V, \tag{A.4}$$

then $\{K, a, \ell\}$ is equivalent to the following minimization problem:

$$\text{Find } u \in K: \quad J(u) = \min_{v \in K} J(v), \tag{A.5}$$

where

$$J(v) = \frac{1}{2}a(v, v) - \langle \ell, v \rangle. \tag{A.6}$$

If $K = V$, then (A.3) takes the form

$$u \in V: \quad a(u, v) = \langle \ell, v \rangle \quad \forall v \in V. \tag{A.7}$$

In this case the respective triple defines an *abstract elliptic equation.*

Let H denote another Hilbert space with the scalar product $(\cdot, \cdot)$ and $\|\cdot\|_H$ be the corresponding norm. Let us suppose that $V \subset H$ with continuous embedding and let V be dense in H. An *abstract elliptic spectral problem* is defined as follows:

$$\begin{cases} \text{Find } \lambda \in \mathbb{R} \text{ and } u \in V,\ u \neq 0, \text{ such that} \\ a(u, v) = \lambda(u, v) \quad \forall v \in V. \end{cases} \tag{A.8}$$

If u exists it is called an *eigenfunction* corresponding to the *eigenvalue* λ. The basic result on the existence of solutions to (A.8) is given as follows.

Theorem A.1. *Let the embedding V into H be compact and $a : V \times V \to \mathbb{R}$ be bounded, V-elliptic, and symmetric in V. Then there exists an increasing sequence of positive eigenvalues tending to ∞:*

$$0 < \lambda_1 \leq \lambda_2 \leq \lambda_3 \leq \cdots, \quad \lim_{k \to \infty} \lambda_k = \infty,$$

and an orthonormal basis $\{w_n\}$ of H consisting of the normalized eigenfunctions corresponding to λ_n:

$$a(w_n, v) = \lambda_n(w_n, v) \quad \forall v \in V,\ \|w_n\|_H = 1.$$

Denote by

$$\mathcal{R}(v) = \frac{a(v, v)}{\|v\|_H^2}, \quad v \in V,\ v \neq 0,$$

the Rayleigh quotient. It is very easy to verify that

$$\lambda_1 = \min_{\substack{v \in V \\ v \neq 0}} \mathcal{R}(v) = \mathcal{R}(u_1),$$

where $u_1 \in V$ is an eigenfunction corresponding to λ_1.

The approximation of $\{K, a, \ell\}$ will be based on the classical *Ritz–Galerkin method.* Let $\{V_h\}$, $h \to 0+$, be a system of *finite dimensional subspaces* of V and h be a discretization parameter tending to zero and characterizing the dimension of V_h: dim $V_h = n(h) \to \infty$ as $h \to 0+$. Further, let K_h be a *nonempty, closed, and convex* subset of V_h $\forall h > 0$. The set

K_h will be considered to be an approximation of K *but not necessarily* its subset. By the Ritz–Galerkin approximation of $\{K, a, \ell\}$ on K_h we mean any solution u_h of $\{K_h, a, \ell\}$:

$$u_h \in K_h: \quad a(u_h, v_h - u_h) \geq \langle \ell, v_h - u_h \rangle \quad \forall v_h \in K_h. \tag{A.9}$$

If (A.4) is satisfied, then u_h solving $\{K_h, a, \ell\}$ can be equivalently characterized by

$$u_h \in K_h: \quad J(u_h) = \min_{v_h \in K_h} J(v_h), \tag{A.10}$$

where J is defined by (A.6). If $K_h = V_h \ \forall h > 0$, then $\{V_h, a, \ell\}$ is an approximation of $\{V, a, \ell\}$ and (A.9) becomes

$$u_h \in V_h: \quad a(u_h, v_h) = \langle \ell, v_h \rangle \quad \forall v_h \in V_h. \tag{A.11}$$

The existence and uniqueness of a solution to $\{K_h, a, \ell\}$ and $\{V_h, a, \ell\}$ follow again from the Lax–Milgram lemma.

Denote by $\varepsilon(h) := \|u - u_h\|$ the error between the exact solution u of $\{K, a, f\}$ and its approximation u_h. We say that the Ritz–Galerkin method is *convergent* if $\varepsilon(h) \to 0$ as $h \to 0+$. To ensure this property we need the following assumptions on $\{K_h\}$, $h \to 0+$:

$$\forall v \in K \ \exists \{v_h\},\ v_h \in K_h: \quad v_h \to v \text{ in } V,\ h \to 0+; \tag{A.12}$$

$$v_h \rightharpoonup v \text{ (weakly) in } V,\ h \to 0+,\ v_h \in K_h \implies v \in K. \tag{A.13}$$

The following convergence result holds.

Theorem A.2. *Let* (A.1), (A.2), (A.12), *and* (A.13) *be satisfied. Then the Ritz–Galerkin method for the approximation of* $\{K, a, f\}$ *is convergent.*

Remark A.1. Condition (A.12) is a density type assumption. If $K_h \subset K \ \forall h > 0$, then (A.13) is automatically satisfied. In particular it is satisfied when $K = V$ and $K_h = V_h \ \forall h > 0$. Assumption (A.13) has to be verified if the so-called external approximations of K are used, i.e., when $K_h \not\subset K$.

Let the discretization parameter $h > 0$ be fixed, $\dim V_h = n$, and $\{\varphi_i\}_{i=1}^n$ be a basis of V_h. Then one can define the isomorphism $\mathcal{T}$ between V_h and $\mathbb{R}^n$ by identifying $v_h \in V_h$ with a unique vector $\boldsymbol{x} = (x_1, \ldots, x_n)^{\mathrm{T}} \in \mathbb{R}^n$ such that $v_h = \sum_{j=1}^n x_j \varphi_j$. Denote by $\mathcal{K}$ a nonempty, closed, and convex subset of $\mathbb{R}^n$: $\mathcal{K} = \{\boldsymbol{x} \in \mathbb{R}^n \mid \mathcal{T}^{-1}\boldsymbol{x} \in K_h\}$, where $\mathcal{T}^{-1}: \mathbb{R}^n \to V_h$ is the inverse mapping to $\mathcal{T}$. Then the algebraic form of $\{K_h, a, \ell\}$ reads as follows:

$$\text{Find } \boldsymbol{x}^* \in \mathcal{K}: \quad (\boldsymbol{y} - \boldsymbol{x}^*)^{\mathrm{T}} \boldsymbol{A} \boldsymbol{x}^* \geq (\boldsymbol{y} - \boldsymbol{x}^*)^{\mathrm{T}} \boldsymbol{\ell} \quad \forall \boldsymbol{y} \in \mathcal{K}, \tag{A.14}$$

where $\boldsymbol{A} \in \mathbb{R}^{n \times n}$, $\boldsymbol{\ell} \in \mathbb{R}^n$ are defined by

$$\boldsymbol{A} = (a_{ij})_{i,j=1}^n, \quad a_{ij} = a(\varphi_j, \varphi_i),\ i, j = 1, \ldots, n;$$
$$\boldsymbol{\ell} = (\ell_i)_{i=1}^n, \quad \ell_i = \langle \ell, \varphi_i \rangle,\ i = 1, \ldots, n.$$

If a is symmetric in V, then $\boldsymbol{A}$ is symmetric and (A.14) is equivalent to the following *mathematical programming problem:*

$$\text{Find } \boldsymbol{x}^* \in \mathcal{K}: \quad \mathcal{J}(\boldsymbol{x}^*) = \min_{\boldsymbol{x}\in\mathcal{K}} \mathcal{J}(\boldsymbol{x}), \tag{A.15}$$

where

$$\mathcal{J}(\boldsymbol{x}) = \frac{1}{2}\boldsymbol{x}^{\mathrm{T}}\boldsymbol{A}\boldsymbol{x} - \boldsymbol{x}^{\mathrm{T}}\boldsymbol{\ell}.$$

Observe that if $K_h = V_h$, then $\mathcal{K} = \mathbb{R}^n$ and (A.14) leads to a linear system of algebraic equations

$$\boldsymbol{A}\boldsymbol{x}^* = \boldsymbol{\ell}. \tag{A.16}$$

In order to give a weak formulation of particular elliptic problems involving differential operators one needs appropriate function spaces. In what follows we shall introduce Sobolev spaces $H^k(\Omega)$, $k \geq 0$ integer, and summarize some of their basic properties. Before doing that let us specify a class of domains in which function spaces will be defined.

Definition A.2. *Let $\Omega \subset \mathbb{R}^n$ be a domain. We say that Ω has a Lipschitz boundary $\partial\Omega$ if there exist positive numbers α, β such that for each $x_0 \in \partial\Omega$ the Cartesian coordinate system can be rotated and shifted to x_0 in such a way that the following statement holds: Let*

$$\Delta_{n-1} = \{x' = (x_1, \dots, x_{n-1}),\ |x_i| \leq \alpha\ \forall i = 1, \dots, n-1\}$$

be an $(n-1)$-dimensional open cube. Then there exists a Lipschitz function $a : \Delta_{n-1} \to \mathbb{R}$ such that points $(x', a(x')) \in \partial\Omega$, $x' \in \Delta_{n-1}$. In addition, all the points (x', x_n) such that $x' \in \Delta_{n-1}$ and $a(x') < x_n < a(x') + \beta$ are supposed to lie inside Ω and all the points (x', x_n), $x' \in \Delta_{n-1}$, $a(x') - \beta < x_n < a(x')$, are supposed to lie outside $\overline{\Omega}$ (see Figure A.1).

The system of all domains in $\mathbb{R}^n$ with the Lipschitz boundary will be denoted by $\mathcal{N}^{0,1}$. *Throughout the whole book we shall use domains $\Omega \in \mathcal{N}^{0,1}$.* We start with spaces of *continuous and continuously differentiable* functions in $\Omega \in \mathcal{N}^{0,1}$.

By $C^k(\overline{\Omega})$, $k \geq 0$ integer, we denote the space of all functions whose partial derivatives up to order k are continuous in Ω and are continuously extendible up to the boundary $\partial\Omega$. The space $C^k(\overline{\Omega})$ endowed with the norm

$$\|v\|_{C^k(\overline{\Omega})} := \max_{|\alpha|\leq k} \max_{x\in\overline{\Omega}} |D^\alpha v(x)|$$

is a Banach space. Here we use the standard multi-index notation for partial derivatives: If $\alpha = (\alpha_1, \dots, \alpha_n)$, $\alpha_i \geq 0$ integer, is a multi-index, then $D^\alpha v(x) := \partial^{|\alpha|} v(x)/\partial x_1^{\alpha_1} \cdots x_n^{\alpha_n}$, where $|\alpha| = \sum_{i=1}^n \alpha_i$ is the length of α, with the following convention of notation: $D^0 v(x) := v(x)\ \forall x$, $C(\overline{\Omega}) := C^0(\overline{\Omega})$. Further we define

$$C^\infty(\overline{\Omega}) = \bigcap_{k=0}^{\infty} C^k(\overline{\Omega}),$$

$$C_0^\infty(\overline{\Omega}) = \{v \in C^\infty(\overline{\Omega}) \mid v \text{ vanishes in a neighborhood of } \partial\Omega\}.$$

By $C^{0,1}(\overline{\Omega})$ we denote the space of all Lipschitz continuous functions in $\overline{\Omega}$.

Next we introduce Sobolev spaces. Let $L^2(\Omega)$ stand for the Hilbert space of square integrable (in the Lebesgue sense) real functions in Ω endowed with the norm

$$\|v\|_{0,\Omega} := \sqrt{\int_\Omega v^2\,dx}, \quad v \in L^2(\Omega). \tag{A.17}$$

If $\Gamma \subseteq \partial\Omega$ is a nonempty open part in $\partial\Omega$, the symbol $L^2(\Gamma)$ stands for the space of square integrable functions (in the Lebesgue sense) on Γ equipped with the norm

$$\|v\|_{0,\Gamma} := \sqrt{\int_\Gamma v^2\,ds}. \tag{A.18}$$

The space of all measurable and bounded functions in Ω will be denoted by $L^\infty(\Omega)$. It is a Banach space with the norm

$$\|v\|_{0,\infty,\Omega} := \inf_{c\geq 0} c, \tag{A.19}$$

where inf is taken over all $c \geq 0$ satisfying $|v(x)| \leq c$ almost everywhere (a.e.) in Ω.

Let $f \in L^2(\Omega)$ and $\alpha = (\alpha_1, \ldots, \alpha_n)$ be a multi-index, $|\alpha| \geq 1$. A function $g_\alpha : \Omega \to \mathbb{R}$ is said to be the αth *generalized derivative* of f iff the integral identity

$$\int_\Omega f D^\alpha \varphi\,dx = (-1)^{|\alpha|} \int_\Omega g_\alpha \varphi\,dx \tag{A.20}$$

holds for any $\varphi \in C_0^\infty(\Omega)$. From (A.20) it easily follows that if such a g_α exists, then it is unique. This makes it possible to use the same symbol for generalized and classical derivatives, i.e., to write $D^\alpha f$ instead of g_α.

The Sobolev space $H^k(\Omega)$ of order k, $k \in \mathbb{N}$, is defined as follows:

$$H^k(\Omega) = \{v \in L^2(\Omega) \mid D^\alpha v \in L^2(\Omega)\ \forall |\alpha| \leq k\}; \tag{A.21}$$

i.e., $H^k(\Omega)$ contains all functions from $L^2(\Omega)$ whose generalized derivatives up to order k are square integrable in Ω. From the definition it easily follows that $H^k(\Omega)$ is a Hilbert space with the scalar product

$$(u, v)_{k,\Omega} := \sum_{|\alpha|\leq k} \int_\Omega D^\alpha u D^\alpha v\,dx \tag{A.22}$$

and the norm

$$\|v\|_{k,\Omega} := \sqrt{\sum_{|\alpha|\leq k} \int_\Omega (D^\alpha v)^2\,dx}. \tag{A.23}$$

The expression

$$|v|_{k,\Omega} := \sqrt{\sum_{|\alpha|=k} \int_\Omega (D^\alpha v)^2\,dx} \tag{A.24}$$

defines the seminorm in $H^k(\Omega)$.

By $H^{k,\infty}(\Omega)$, $k \in \mathbb{N}$, we denote the following subspace of $L^\infty(\Omega)$:

$$H^{k,\infty}(\Omega) = \{v \in L^\infty(\Omega) \mid D^\alpha v \in L^\infty(\Omega)\ \forall |\alpha| \le k\}$$

equipped with the norm

$$\|v\|_{k,\infty,\Omega} := \max_{|\alpha| \le k} \|D^\alpha v\|_{0,\infty,\Omega} \tag{A.25}$$

and the seminorm

$$|v|_{k,\infty,\Omega} := \max_{|\alpha| = k} \|D^\alpha v\|_{0,\infty,\Omega}. \tag{A.26}$$

In what follows we collect basic properties of the Sobolev spaces.

Theorem A.3. *(Density result.) The space $C^\infty(\overline{\Omega})$ is dense in $H^k(\Omega)$, $k \in \mathbb{N}$, $\Omega \in \mathcal{N}^{0,1}$, with respect to* (A.23).

Theorem A.4. *(Trace theorem.) There exists a unique linear compact mapping T of $H^1(\Omega)$ into $L^2(\partial\Omega)$, $\Omega \in \mathcal{N}^{0,1}$, such that $Tv = v_{|\partial\Omega}$ for any $v \in C^\infty(\overline{\Omega})$.*

Remark A.2. T is the *trace mapping* and its value Tv at $v \in H^1(\Omega)$ is called the *trace* of v on $\partial\Omega$. The trace mapping is an extension of the classical restriction mapping from $C(\overline{\Omega})$ into $C(\partial\Omega)$. Instead of Tv we write simply v.

Theorem A.5. *(Rellich.) The embedding of $H^k(\Omega)$ into $H^{k-1}(\Omega)$, $k \in \mathbb{N}$, $\Omega \in \mathcal{N}^{0,1}$, is compact with the following convention of notation: $H^0(\Omega) := L^2(\Omega)$.*

Corollary A.1. *From $v_n \rightharpoonup v$ (weakly) in $H^k(\Omega)$ it follows that $v_n \to v$ in $H^{k-1}(\Omega)$, $k \in \mathbb{N}$, $\Omega \in \mathcal{N}^{0,1}$.*

To formulate elliptic problems with Dirichlet boundary data one needs appropriate subspaces of $H^k(\Omega)$, $k \in \mathbb{N}$. Let $\Gamma \subset \partial\Omega$ be a *nonempty, open* part in $\partial\Omega$. We define

$$V = \{v \in H^k(\Omega) \mid D^\alpha v = 0 \text{ on } \Gamma\ \forall |\alpha| \le k-1\}; \tag{A.27}$$

i.e., V is a subspace of $H^k(\Omega)$ containing all functions whose derivatives up to order $(k-1)$ vanish on Γ in the sense of traces. In particular if $\Gamma = \partial\Omega$, then we use the symbol $H_0^k(\Omega)$ instead of V. The following density result holds for the space $H_0^k(\Omega)$, $k = 1, 2$.

Theorem A.6. *Let $\Omega \in \mathcal{N}^{0,1}$. Then*

$$H_0^k(\Omega) = \overline{C_0^\infty(\overline{\Omega})} \quad \textit{for } k = 1, 2,$$

where the closure is taken with respect to (A.23).

Remark A.3. The same result holds for any $k \in \mathbb{N}$, $k \ge 3$, provided that Ω has a sufficiently smooth boundary.

Functions belonging to $H^k(\Omega)$, $k \in \mathbb{N}$, are more regular compared with those from $L^2(\Omega)$. In particular one may ask whether for some k the space $H^k(\Omega)$ is embedded into a class of continuously differentiable functions. The answer is given as follows.

THEOREM A.7. *(Embedding theorem.) Let $\Omega \in \mathcal{N}^{0,1}$, $\Omega \subset \mathbb{R}^n$. Then $H^k(\Omega)$ is compactly embedded into $C^s(\overline{\Omega})$ provided that $k - s > n/2$, $k, s \in \mathbb{N}$. In addition, there exists a constant $c > 0$ such that*

$$\|v\|_{C^s(\overline{\Omega})} \leq c\|v\|_{k,\Omega} \quad \forall v \in H^k(\Omega). \tag{A.28}$$

In what follows we restrict ourselves to $k = 1$. Let $v \in H^1(\Omega)$ and $\widehat{\Omega} \in \mathcal{N}^{0,1}$ be a domain containing $\Omega \in \mathcal{N}^{0,1}$. Then one can construct a function $\widetilde{v} \in H^1(\widehat{\Omega})$ such that $\widetilde{v} = v$ in Ω and

$$\|\widetilde{v}\|_{1,\widehat{\Omega}} \leq c\|v\|_{1,\Omega} \quad \forall v \in H^1(\Omega) \tag{A.29}$$

with a constant c independent of v. A function $\widetilde{v}$ is called an *extension* of v to $\widehat{\Omega}$. We write $\widetilde{v} = p_\Omega v$, where p_Ω denotes the *linear extension mapping* from $H^1(\Omega)$ into $H^1(\widehat{\Omega})$ whose norm is bounded by the constant $c > 0$ in (A.29). This constant may, however, depend on a particular choice of $\Omega \in \mathcal{N}^{0,1}$. Below we define a class $\mathcal{M}$ of bounded domains in $\mathbb{R}^n$ such that the norm of p_Ω can be bounded independently of $\Omega \in \mathcal{M}$.

Let $h > 0$, $\theta \in]0, \frac{\pi}{2}[$, and $\xi \in \mathbb{R}^n$, $\|\xi\| = 1$, be given. The set

$$C(\xi, \theta, h) = \{x \in \mathbb{R}^n \mid (x, \xi) > \|x\| \cos\theta, \ \|x\| < h\}$$

is called the cone of angle θ, height h, and axis ξ.

DEFINITION A.3. *A domain $\Omega \subset \mathbb{R}^n$ is said to satisfy the cone property iff there exist numbers $\theta \in]0, \frac{\pi}{2}[$, $h > 0$, $r \in]0, h/2[$ with the property that $\forall\, x \in \partial\Omega\, \exists C_x := C(\xi_x, \theta, h)$ such that $\forall\, y \in B_r(x) \cap \Omega$ the set $y + C_x \subset \Omega$ (see Figure A.1).*

It can be shown that Ω possesses the cone property iff $\Omega \in \mathcal{N}^{0,1}$ (see [Che75]).

DEFINITION A.4. *Let $D \subset \mathbb{R}^n$ be a bounded domain and $\overline{\theta} \in]0, \pi/2[$, $\overline{h} > 0$, $\overline{r} \in]0, \overline{h}/2[$ be given. The set of all domains contained in D and satisfying the cone property with the numbers $\overline{\theta}$, $\overline{h}$, $\overline{r}$ will be denoted by $\mathcal{M}(\overline{\theta}, \overline{h}, \overline{r})$. We say that the system $\mathcal{M}(\overline{\theta}, \overline{h}, \overline{r})$ contains domains satisfying the uniform cone property.*

Let $\widehat{\Omega} \in \mathcal{N}^{0,1}$ be such that $\widehat{\Omega} \supset \overline{\Omega}\ \forall \Omega \in \mathcal{M}(\overline{\theta}, \overline{h}, \overline{r})$. The domains $\Omega \in \mathcal{M}(\overline{\theta}, \overline{h}, \overline{r})$ possess the uniform extension property as follows.

THEOREM A.8. *There exists an extension operator $p_\Omega : H^m(\Omega) \to H^m(\widehat{\Omega})$, $m \in \mathbb{N}$, and a constant $c > 0$ such that*

$$\|p_\Omega\|_{\mathcal{L}(H^m(\Omega), H^m(\widehat{\Omega}))} \leq c \quad \forall \Omega \in \mathcal{M}(\overline{\theta}, \overline{h}, \overline{r}).$$

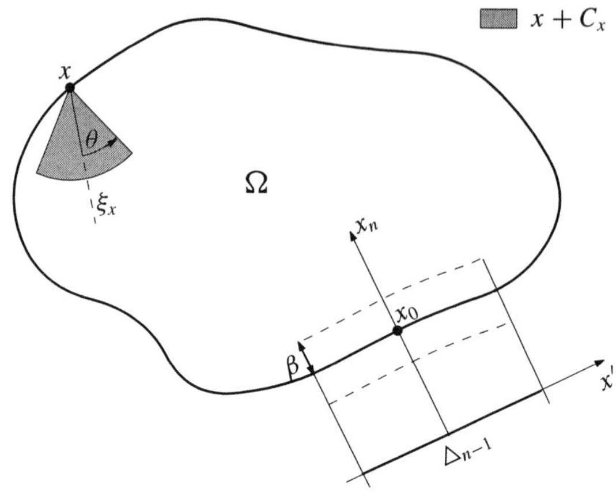

Figure A.1. *Cone property.*

For the rather technical proof of this result we refer to [Che75].

Let $\Gamma \subset \partial\Omega$ be a nonempty, open set in $\partial\Omega$ and denote

$$|||v|||_{1,\Omega} := \sqrt{\int_\Omega |\nabla v|^2\,dx + \int_\Gamma v^2\,ds}. \tag{A.30}$$

It is readily seen that $|||\cdot|||_{1,\Omega}$ defines a norm in $H^1(\Omega)$. A less trivial result says that $\|\cdot\|_{1,\Omega}$ and $|||\cdot|||_{1,\Omega}$ are equivalent.

Theorem A.9. *There exist constants* $c_1, c_2 > 0$ *such that*

$$c_1\|v\|_{1,\Omega} \le |||v|||_{1,\Omega} \le c_2\|v\|_{1,\Omega} \tag{A.31}$$

holds for any $v \in H^1(\Omega)$.

A direct consequence of (A.31) is as follows.

Theorem A.10. *(Generalized Friedrichs inequality.) Let* V *be defined by* (A.27) *with* $k = 1$. *Then there is a constant* $c > 0$ *such that*

$$c\|v\|_{1,\Omega} \le |v|_{1,\Omega}, \qquad v \in V. \tag{A.32}$$

Remark A.4. If $\Gamma = \partial\Omega$, then (A.32) is called the *Friedrichs inequality*, which is valid in $H_0^k(\Omega)$ for any $k \in \mathbb{N}$: there exists a constant $c > 0$ such that

$$c\|v\|_{k,\Omega} \le |v|_{k,\Omega}$$

holds for any $v \in H_0^k(\Omega)$.

We now turn to weak formulations of linear elliptic equations in $\Omega \in \mathcal{N}^{0,1}$ involving the second order elliptic operators. We use the summation convention again.

Let

$$\mathcal{A}u := -\frac{\partial}{\partial x_i}\left(a_{ij}\frac{\partial u}{\partial x_j}\right) + a_0 u$$

be such that the coefficients $a_{ij},\ a_0$ of $\mathcal{A}$ satisfy the following assumptions:

$$a_{ij}, a_0 \in L^\infty(\Omega)\ \forall\, i, j; \quad a_0 \geq 0 \text{ a.e. in } \Omega; \tag{A.33}$$

$$\exists\, \alpha = \text{const.} > 0: \quad a_{ij}(x)\xi_i\xi_j \geq \alpha\xi_i\xi_i \text{ a.e. in } \Omega \quad \forall \xi \in \mathbb{R}^n. \tag{A.34}$$

With $\mathcal{A}$ we associate the bilinear form $a : H^1(\Omega) \times H^1(\Omega) \to \mathbb{R}$ defined by

$$a(u, v) = \int_\Omega a_{ij}\frac{\partial u}{\partial x_j}\frac{\partial v}{\partial x_i}\,dx + \int_\Omega a_0 uv\,dx. \tag{A.35}$$

From (A.33) it follows that a is bounded in $H^1(\Omega)$. Let us check (A.2):

$$a(v, v) = \int_\Omega a_{ij}\frac{\partial v}{\partial x_j}\frac{\partial v}{\partial x_i}\,dx + \int_\Omega a_0 v^2\,dx \geq \alpha|v|^2_{1,\Omega} \quad \forall v \in H^1(\Omega), \tag{A.36}$$

taking into account (A.34).

Let V be defined by (A.27) with $k = 1$. Then from the generalized Friedrichs inequality and (A.36) we see that a is $H^1(\Omega)$-elliptic in V. For the same property in the whole space $H^1(\Omega)$ one has to suppose that there is a positive constant $\overline{a}$ such that

$$a(x) \geq \overline{a} > 0 \text{ a.e. in } \Omega. \tag{A.37}$$

If this is so, then $a(v, v) \geq \min(\alpha, \overline{a})\|v\|^2_{1,\Omega}\ \forall v \in H^1(\Omega)$.

Let $\Gamma \subset \partial\Omega$ be a nonempty open part in $\partial\Omega$; $\Gamma_1 = \partial\Omega \setminus \overline{\Gamma}$; and $f \in L^2(\Omega)$, $g \in L^2(\Gamma_1)$ be given functions. We set

$$\langle \ell, v\rangle := \int_\Omega fv\,dx + \int_{\Gamma_1} gv\,ds \tag{A.38}$$

(note that if $\Gamma = \partial\Omega$, then the integral over Γ_1 is not present). From the trace theorem it easily follows that $\ell \in V'$. We now define the linear elliptic problem $\{V, a, \ell\}$ with V, a, and ℓ defined by (A.27), (A.35), and (A.38), respectively.

If $\Gamma = \partial\Omega$, then $V = H^1_0(\Omega)$ and we obtain the homogeneous Dirichlet boundary value problem in Ω. If $\Gamma \subsetneq \partial\Omega$, $\Gamma \neq \emptyset$, then $\{V, a, \ell\}$ defines the mixed Dirichlet–Neumann boundary value problem in Ω. Finally, if $\Gamma = \emptyset$ then $V = H^1(\Omega)$, corresponding to the Neumann boundary condition prescribed on the whole boundary $\partial\Omega$. To ensure the existence and uniqueness of a solution in the latter case we suppose that (A.37) is satisfied (observe that this condition is not needed when $\Gamma \neq \emptyset$).

In the remainder of this appendix we briefly describe the simplest finite element spaces, which have been used throughout the book to approximate second order elliptic problems. We restrict ourselves to a plane case.

Let $\Omega \subset \mathbb{R}^2$ be a *polygonal* domain and $\mathcal{T}_h$ be its triangulation; i.e., $\overline{\Omega}$ is the union of a finite number of closed triangles T with nonoverlapping interiors whose diameters are less than or equal to h. If $T_1, T_2 \in \mathcal{T}_h$ are two different triangles with a nonempty intersection

T', then T' is either a common vertex or a common edge. In addition $\mathcal{T}_h$ has to be *consistent* with the decomposition of $\partial\Omega$ into Γ and $\Gamma_1 = \partial\Omega \setminus \overline{\Gamma}$; i.e., the boundary nodes of Γ in $\partial\Omega$ belong to $\mathcal{T}_h$. Let $\theta_0(h)$ be the minimal inner angle of all triangles $T \in \mathcal{T}_h$. We say that a family of triangulations $\{\mathcal{T}_h\}$, $h \to 0+$, is *regular* iff there exists a positive number θ_0 such that $\theta_0(h) \geq \theta_0$ $\forall h > 0$ (the so-called minimal angle condition for $\{\mathcal{T}_h\}$, $h \to 0+$). With any $\mathcal{T}_h$ we associate the following space of continuous, piecewise linear functions over $\mathcal{T}_h$:

$$X_h = \{v_h \in C(\overline{\Omega}) \mid v_{h|T} \in P_1(T)\ \forall T \in \mathcal{T}_h\}.$$

It is easy to prove that $X_h \subset H^1(\Omega)$. Denote by $\mathcal{N}_h$ the system of all the nodes of $\mathcal{T}_h$ in $\overline{\Omega}$. If $v \in C(\overline{\Omega})$, then there exists a unique function $r_h v \in X_h$ called the *piecewise linear Lagrange interpolant* of v such that

$$r_h v(A_i) = v(A_i) \quad \forall A_i \in \mathcal{N}_h.$$

This function enjoys important approximation properties as follows.

Theorem A.11. *Let $\{\mathcal{T}_h\}$, $h \to 0+$, be a regular system of triangulations of $\overline{\Omega}$. Then there exists a constant $c > 0$ that does not depend on h such that*

$$\|v - r_h v\|_{1,\Omega} \leq ch|v|_{2,\Omega} \tag{A.39}$$

holds for any $v \in H^2(\Omega)$. If $v \in H^1(\Omega)$, then there exists a sequence $\{v_h\}$, $v_h \in X_h$, such that

$$\|v - v_h\|_{1,\Omega} \to 0 \quad \text{as } h \to 0+\,. \tag{A.40}$$

Let V be a subspace of $H^1(\Omega)$ defined by (A.27) with $k = 1$. Then

$$V_h = \{v_h \in X_h \mid v_h = 0 \text{ on } \Gamma\} = X_h \cap V$$

is a finite element space discretizing V. If $v \in H^2(\Omega) \cap V$ and $\{\mathcal{T}_h\}$, $h \to 0+$, is a regular system of triangulations consistent with the decomposition of $\partial\Omega$ into Γ and $\Gamma_1 = \partial\Omega \setminus \overline{\Gamma}$, then $r_h v \in V_h$ and the error estimate (A.39) holds true. Also (A.40) remains valid in problems we meet in industrial applications.

As we reach the end of this appendix we briefly describe finite element spaces made of quadrilateral elements used in our computations.

Let $\mathcal{R}_h$ be a partition of $\overline{\Omega}$ into a finite number of closed *convex quadrilaterals R* satisfying the same conditions on their mutual position as triangles of $\mathcal{T}_h$ and such that $\operatorname{diam} R \leq h\ \forall R \in \mathcal{R}_h$.

Let $\widehat{K} = [0, 1]^2$ be the unit square and $Q_1(\widehat{K})$ be the space of all *bilinear functions* defined in $\widehat{K}$. Under appropriate assumptions on the geometry of R one can prove (see [Cia02]) that $R = F_R(\widehat{K})$, where F_R is a one-to-one mapping of $\widehat{K}$ onto R and $F_R \in (Q_1(\widehat{K}))^2$. This makes it possible to introduce the following set of functions defined in any $R \in \mathcal{R}_h$:

$$Q_1(R) = \left\{q \in C(R) \mid \exists \widehat{q} \in Q_1(\widehat{K}) : \ q = \widehat{q} \circ F_R^{-1}\right\},$$

where $\widehat{K} = F_R^{-1}(R)$ (observe that $Q_1(R)$ no longer contains bilinear functions unless R is a rectangle). With any $\mathcal{R}_h$ the following finite dimensional space X_h will be associated:

$$X_h = \left\{v_h \in C(\overline{\Omega}) \mid v_{h|R} \in Q_1(R)\ \forall R \in \mathcal{R}_h\right\}.$$

Functions from X_h are uniquely defined by their values at the nodes of $\mathcal{R}_h$. Under appropriate assumptions on the system $\{\mathcal{R}_h\}$, $h \to 0+$, one can prove that the respective Lagrange X_h interpolant satisfies (A.39) (for details see [Cia02]).

Appendix B

On Parametrizations of Shapes and Mesh Generation

B.1 Parametrization of shapes

In order to realize numerically discrete shape and sizing optimization problems one has to first find a suitable parametrization of admissible shapes using a finite number of parameters. In shape optimization with a discretized state problem it would seem that the most obvious choice is to use the positions of boundary nodes of a partition into finite elements as design parameters. For linear triangular elements this would mean that the boundary is represented by a piecewise linear curve. This choice, however, has many drawbacks, such as the large number of design variables and the need for additional constraints to keep the boundary regular enough.

Let the domain $\Omega(\alpha)$ be as in Figure 2.2. The number of design variables can be reduced by representing the unknown part of the boundary as the polynomial

$$\alpha(t) = \sum_{i=1}^{d} \alpha_i t^{i-1}. \tag{B.1}$$

The advantage of this parametrization is its simplicity and the smoothness of the resulting curve. Unfortunately the coefficients in the power basis form (B.1) contain practically no geometric insight into the shape of $\Omega(\alpha)$. In addition the implementation of constraints such as $\alpha_{\min} \leq \alpha \leq \alpha_{\max}$ is very cumbersome. To overcome these difficulties we adopt a technique used in computer graphics and computer-aided design for the parametrization of shapes.

Bézier curves were introduced in the 1960s by P. de Casteljau and P. Bézier while working in French automobile companies to design complicated curves and surfaces defining car parts. In what follows we briefly present the properties of parametric Bézier curves in $\mathbb{R}^2$.

Let $B_i^{(n)}$, $i = 0, \ldots, n$, denote the *Bernstein polynomials* in $[0, 1]$ defined by

$$B_i^{(n)}(t) = \binom{n}{i} t^i (1-t)^{n-i}, \quad t \in [0, 1]. \tag{B.2}$$

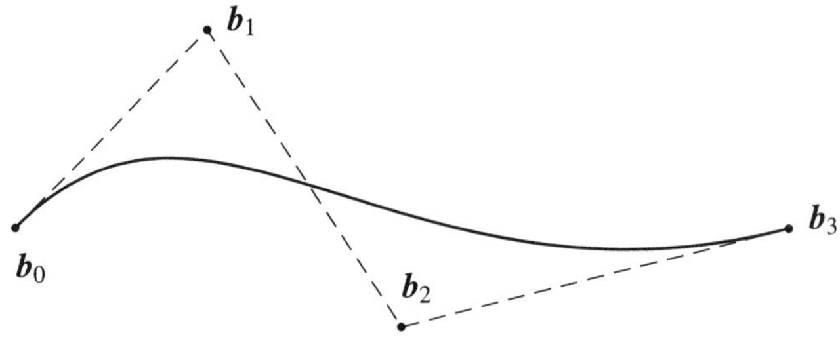

Figure B.1. *A Bézier curve (solid) and its control polygon (dashed).*

Bernstein polynomials enjoy the following useful properties:

$$B_i^{(n)}(t) \geq 0 \quad \forall t \in [0, 1]; \tag{B.3}$$

$$B_i^{(n)}(0) = \begin{cases} 0, & i \neq 0, \\ 1, & i = 0, \end{cases} \qquad B_i^{(n)}(1) = \begin{cases} 0, & i \neq n, \\ 1, & i = n; \end{cases} \tag{B.4}$$

$$\sum_{j=0}^{n} B_j^{(n)}(t) = 1 \quad \forall t \in [0, 1]; \tag{B.5}$$

$$\frac{d}{dt} B_i^{(n)}(t) = n(B_{i-1}^{(n-1)}(t) - B_i^{(n-1)}(t)) \quad \forall t \in [0, 1]. \tag{B.6}$$

Let $\boldsymbol{b}_0, \boldsymbol{b}_1, \dots, \boldsymbol{b}_n \in \mathbb{R}^2$ be a set of given *control points*. The broken line defined by the control points will be called a *control polygon* (see Figure B.1). We define a *parametric Bézier curve* $b : [0, 1] \to \mathbb{R}^2$ *of degree* n as a linear combination

$$b(t) = \sum_{j=0}^{n} \boldsymbol{b}_j B_j^{(n)}(t). \tag{B.7}$$

Due to (B.3)–(B.6) Bézier curves possess several interesting properties:

(i) b is *axes independent*; i.e., it is independent of the coordinate system defining the location of the control points;

(ii) b *interpolates* the endpoints of the control polygon: $b(0) = \boldsymbol{b}_0$, $b(1) = \boldsymbol{b}_n$;

(iii) b has the *convex hull property*: $b(t) \in \operatorname{conv}\{\boldsymbol{b}_0, \boldsymbol{b}_1, \dots, \boldsymbol{b}_n\}\ \forall t \in [0, 1]$;

(iv) b is *variation diminishing*: the curve b has fewer intersections with any straight line than the control polygon;

(v) derivatives of b at the endpoints are given by

$$b'(0) = n(\boldsymbol{b}_1 - \boldsymbol{b}_0), \qquad b'(1) = n(\boldsymbol{b}_n - \boldsymbol{b}_{n-1}); \tag{B.8}$$

i.e., the curve b is tangential to the vectors $\boldsymbol{b}_1 - \boldsymbol{b}_0$ and $\boldsymbol{b}_n - \boldsymbol{b}_{n-1}$ at the initial and endpoints of the curve, respectively.

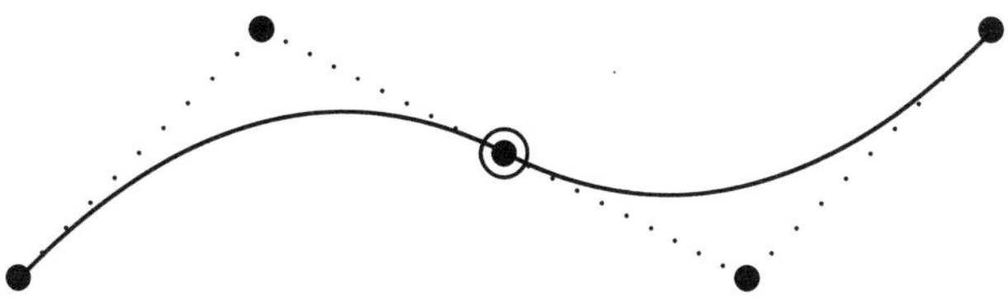

Figure B.2. *Piecewise quadratic Bézier curve possessing C^1-continuity at the joint.*

Let us consider the simplest case in which the curve is represented by the graph of a function $\alpha : [0, 1] \to \mathbb{R}$ (see Figure 2.2), where α is a linear combination of the Bernstein polynomials

$$\alpha(t) = \sum_{j=0}^{n} \alpha_j B_j^{(n)}(t), \quad \alpha_j \in \mathbb{R}. \tag{B.9}$$

Then α is termed a *Bézier function of degree n*. In this case one can consider $(\alpha_j, j/n) \in \mathbb{R}^2$, $j = 0, \dots, n$, as the control points of α, and $\{\alpha_j\}_{j=0}^n$ as the discrete design variables characterizing the shape of $\Omega(\alpha)$.

From Chapter 2 we know that admissible functions α defining the part of the boundary subject to optimization are supposed to satisfy at least the following constraints: $\alpha_{\min} \le \alpha(t) \le \alpha_{\max}$ $\forall t \in [0, 1]$ and $|\alpha'| \le L_0$ almost everywhere in $]0, 1[$. Let us find out how to restrict the coefficients α_j in (B.9) in order to satisfy these constraints. If $\alpha_{\min} \le \alpha_j \le \alpha_{\max}$, $j = 0, \dots, n$, then the convex hull of the points $(\alpha_j, j/n)$, $j = 0, \dots, n$, is contained in $[\alpha_{\min}, \alpha_{\max}] \times [0, 1]$ and thus $(\alpha(t), t) \in [\alpha_{\min}, \alpha_{\max}] \times [0, 1]$ for any $t \in [0, 1]$. From (B.6) it follows that the derivative of α is given by

$$\alpha'(t) = \sum_{j=0}^{n} \alpha_j B_j^{(n)'}(t) = n \sum_{j=0}^{n-1} (\alpha_{j+1} - \alpha_j) B_j^{(n-1)}(t).$$

If $|\alpha_{j+1} - \alpha_j| \le L_0/n$, $j = 0, \dots, n-1$, then

$$|\alpha'(t)| \le n \sum_{j=0}^{n-1} |\alpha_{j+1} - \alpha_j| B_j^{(n-1)}(t) \le n \sum_{j=0}^{n-1} \frac{1}{n} L_0 B_j^{(n-1)}(t) = L_0 \quad \forall t \in [0, 1]$$

using (B.5). Thus in order to satisfy the previous two constraints with functions α for all $t \in [0, 1]$ it is sufficient to satisfy them with the discrete design variables $\{\alpha_i\}_{i=0}^n$.

The degree of the polynomial defining a Bézier curve increases as the number of control points increases. To avoid excessive high order polynomials one can piece together several low order Bézier curves. Using simple geometric rules for the positions of control points, continuity at joints can be guaranteed. To achieve C^0-continuity at a joint, it is sufficient that the endpoints of the control polygons of neighboring pieces coincide. For curves defined by quadratic or higher order polynomials one can achieve C^1-continuity by requiring that the edges of two control polygons adjacent to the common endpoint lie on a straight line (see Figure B.2).

Piecewise kth order Bézier curves with C^{k-1}-continuity, $k \in \mathbb{N}$, across joints are called *B-spline curves*. For different ways of constructing them we refer to [Far88], [dB78]. B-spline curves share the good properties of single Bézier curves. In addition there is a very important property called the *local control*. By changing the position of one control node of a single Bézier curve we affect the *whole* curve. On the other hand by changing one control point of, e.g., a quadratic B-spline curve, we affect at most three curve segments.

B.2 Mesh generation in shape optimization

A finite element mesh is called *structured* if all interior nodes of the mesh have the same number of adjacent elements. In an *unstructured* mesh any number of elements may meet at a single node. Triangular (tetrahedral in 3D) meshes are mostly thought of when referring to unstructured meshes, although quadrilateral (hexahedral) meshes can also be unstructured.

Structured meshes are usually simpler and faster to generate than unstructured ones. On the other hand, it can be very difficult to construct a structured mesh for a domain having a complicated shape. Furthermore, a structured mesh may need many more elements than an unstructured one for the same problem because the element size cannot vary very rapidly. The advantage of structured meshes in the framework of optimal shape design consists of the fact that the nodal positions are smooth functions of design parameters.

Unstructured meshes are usually produced using Delaneu, advancing front, or quadtree techniques. Structured mesh generations are based either on algebraic (interpolation) or on partial differential equation (PDE) methods. The PDE approach solves a set of PDEs in order to map the domain Ω onto another domain with a simpler shape (such as a rectangle). For further details on mesh generation, see [Geo91]. In what follows we shall describe very briefly a simple way to generate two-dimensional structured meshes using the algebraic approach.

Assume that the domain $\Omega(\alpha)$ is given by

$$\Omega(\alpha) = \left\{(x_1, x_2) \in \mathbb{R}^2 \mid 0 < x_1 < \alpha(x_2),\ x_2 \in]0, 1[\right\}.$$

One can easily generate a structured mesh with $2m_x m_y$ triangular (or $m_x m_y$ quadrilateral) elements by defining the positions of the nodal points X_{ij} of the mesh by

$$X_{ij} = \left(\frac{i}{m_x}\alpha\left(\frac{j}{m_y}\right), \frac{j}{m_y}\right), \quad i = 0, \dots, m_x, \quad j = 0, \dots, m_y.$$

If α is parametrized using (B.9), then X_{ij} is given as a function of the discrete design variables α_s, $s = 0, \dots, n$, by the following explicit formula:

$$X_{ij} = \left(\frac{i}{m_x}\sum_{k=0}^{n} \alpha_k B_k^{(n)}\left(\frac{j}{m_y}\right), \frac{j}{m_y}\right).$$

Thus the calculation of the matrix X' needed in the sensitivity formulas (3.84)–(3.87) is very simple:

$$\frac{\partial X_{ij}}{\partial \alpha_s} = \left(\frac{i}{m_x}B_s^{(n)}\left(\frac{j}{m_y}\right), 0\right). \tag{B.10}$$

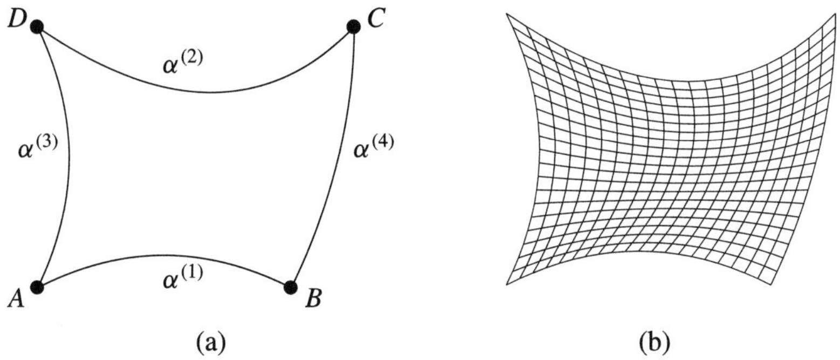

Figure B.3. *Domain defined by four parametric curves and meshed using transfinite interpolation.*

The previous technique can be generalized to the case when $\Omega(\alpha)$ is bounded by four parametric curves $\alpha^{(i)} : [0, 1] \to \mathbb{R}^2$, $i = 1, \dots, 4$, with $\alpha^{(1)}(0) = A$, $\alpha^{(2)}(0) = D$, $\alpha^{(3)}(0) = A$, $\alpha^{(4)}(0) = B$, as shown in Figure B.3. The positions of the nodes of the mesh can now be computed using the *transfinite interpolation:*

$$\begin{aligned} X_{ij} = {} & w_1(\xi_i)\alpha^{(4)}(\eta_j) + (1 - w_1(\xi_i))\alpha^{(3)}(\eta_j) + w_2(\eta_j)\alpha^{(2)}(\xi_i) \\ & + (1 - w_2(\eta_j))\alpha^{(1)}(\xi_i) - w_1(\xi_i)w_2(\eta_j)C - w_1(\xi_i)(1 - w_2(\eta_j))B \\ & - (1 - w_1(\xi_i))w_2(\eta_j)D - (1 - w_1(\xi_i))(1 - w_2(\eta_j))A, \end{aligned} \tag{B.11}$$

where $0 = \xi_0 < \xi_1 < \cdots < \xi_{m_x} = 1$, $0 = \eta_0 < \eta_1 < \cdots < \eta_{m_y} = 1$ are given and $w_1, w_2 : [0, 1] \to [0, 1]$ are the so-called blending functions. A natural choice of blending functions is to take $w_1(\xi) = \xi$, $w_2(\eta) = \eta$.

In practical shape optimization problems, however, it is not usually possible to represent admissible shapes by a simply connected domain bounded by a few parametric curves. This difficulty can be circumvented by a sort of multiblock approach: the domain is divided into a set of *design elements* bounded by, e.g., parametric Bézier curves. Some of the control points defining the curves are allowed to move in order to deform the shape of the domain. Each design element is meshed separately using the algebraic technique. It is necessary to ensure that the meshes in neighboring design elements are compatible on the common boundary; i.e., when the separate meshes are patched together an adequate mesh of the whole structure is produced.

Bibliography

[AB85] H. Andersson and C. Benocci, *An improved pressure iteration technique for the SOLA algorithm*, in Numerical Methods in Laminar and Turbulent Flow, vol. 1, C. Taylor, M. Olson, P. Gresho, and W. Habashi, eds., Pineridge Press, Swansea, U.K., 1985, 364–375.

[ABB+99] E. Anderson, Z. Bai, C. Bischof, S. Blackford, J. Demmel, J. Dongarra, J. Du Croz, A. Greenbaum, S. Hammarling, A. McKenney, and D. Sorensen, *LAPACK Users' Guide*, 3rd ed., SIAM, Philadelphia, 1999.

[Ack81] A. Acker, *An extremal problem involving current flow through distributed resistance*, SIAM J. Math. Anal. **12** (1981), 169–172.

[Ang83] F. Angrand, *Optimum design for potential flows*, Int. J. Numer. Methods Fluids **3** (1983), 265–282.

[Aro89] J. S. Arora, *Introduction to Optimum Design*, McGraw–Hill Book Company, New York, 1989.

[ATV97] M. M. Ali, A. Törn, and S. Viitanen, *A numerical comparison of some controlled random search algorithms*, J. Global Optim. **11** (1997), 377–385.

[Bäc96] T. Bäck, *Evolutionary Algorithms in Theory and Practice*, Oxford University Press, New York, 1996.

[BCG+92] C. H. Bischof, G. F. Corliss, L. Green, A. Griewank, K. Haigler, and P. Newman, *Automatic differentiation of advanced CFD codes for multidisciplinary design*, J. Comput. Systems Engrg. **3** (1992), 625–638.

[BCKM96] C. Bischof, A. Carle, P. Khademi, and A. Mauer, *ADIFOR 2.0: Automatic differentiation of Fortran 77 programs*, IEEE Comput. Sci. Engrg. **3** (1996), 18–32.

[BD92] F. Beux and A. Dervieux, *Exact-gradient shape optimization of a 2-D Euler flow*, Finite Elem. Anal. Des. **12** (1992), 281–302.

[BF91] F. Brezzi and M. Fortin, *Mixed and Hybrid Finite Element Methods*, Springer Series in Computational Mathematics, vol. 15, Springer-Verlag, New York, 1991.

[BG75] D. Begis and R. Glowinski, *Application de la méthode des éléments finis à l'approximation d'un problème de domaine optimal*, Appl. Math. Optim. **2** (1975), 130–169.

[BJR90] A. Bamberger, P. Joly, and J. E. Roberts, *Second-order absorbing boundary conditions for the wave equation: A solution for the corner problem*, SIAM J. Numer. Anal. **27** (1990), 323–352.

[BP93] F. J. Baron and O. Pironneau, *Multidisciplinary optimal design of a wing profile*, in Structural Optimization 93, vol. 2, The World Congress on Optimal Design of Structural Systems (Rio de Janeiro), J. Herskovits, ed., COPPE/Federal University of Rio de Janeiro, 1993, pp. 61–68.

[BS93] T. Bäck and H. P. Schwefel, *An overview of evolutionary algorithms for parameter optimization*, Evolutionary Computation **1** (1993), 1–23.

[BSS93] M. S. Bazaraa, H. D. Sherali, and C. M. Shetty, *Nonlinear Programming*, 2nd ed., John Wiley, New York, 1993.

[Céa71] J. Céa, *Optimisation: Théorie et algorithmes*, Dunod, Paris, 1971.

[Céa81] J. Céa, *Problems of shape optimal design*, in Optimization of Distributed Parameter Structures, E. J. Haug and J. Céa, eds., Sijthoff and Noordhoff, Alphen aan den Rijn, The Netherlands, 1981, 1005–1048.

[Che75] D. Chenais, *On the existence of a solution in a domain identification problem*, J. Math. Anal. Appl. **52** (1975), 189–219.

[Cia02] P. G. Ciarlet, *The Finite Element Method for Elliptic Problems*, Classics in Applied Mathematics 40, SIAM, Philadelphia, 2002.

[CK92] D. Colton and R. Kress, *Inverse Acoustic and Electromagnetic Scattering Theory*, Springer-Verlag, New York, 1992.

[CO86] G. F. Carey and J. T. Oden, *Finite Elements. Vol. VI. Fluid Mechanics*, Prentice–Hall, Englewood Cliffs, NJ, 1986.

[dB78] C. de Boor, *A Practical Guide to Splines*, Springer-Verlag, New York, Berlin, 1978.

[Dor70] F. W. Dorr, *The direct solution of the discrete Poisson equation on a rectangle*, SIAM Rev. **12** (1970), 248–263.

[DS96] J. E. Dennis, Jr. and R. B. Schnabel, *Numerical Methods for Unconstrained Optimization and Nonlinear Equations*, Classics in Applied Mathematics 16, SIAM, Philadelphia, 1996.

[DZ01] M. C. Delfour and J.-P. Zolésio, *Shapes and Geometries: Analysis, Differential Calculus, and Optimization*, Advances in Design and Control 4, SIAM, Philadelphia, 2001.

[EM79] B. Engquist and A. Majda, *Radiation boundary conditions for acoustic and elastic wave calculations*, Comm. Pure Appl. Math. **32** (1979), 313–357.

[Far88] G. Farin, *Curves and Surfaces for Computer Aided Geometric Design*, Academic Press, Boston, 1988.

[Fas92] A. Fasano, *Some free boundary problems with industrial applications*, in Shape Optimization and Free Boundaries, M. C. Delfour and G. Sabidussi, eds., Kluwer Academic Publishers, Dordrecht, The Netherlands, 1992, pp. 113–142.

[Fle87] R. Fletcher, *Practical Methods of Optimization*, 2nd ed., John Wiley & Sons, Chichester, U.K., 1987.

[FR97] M. Flucher and M. Rumpf, *Bernoulli's free-boundary problem, qualitative theory and numerical approximation*, J. Reine Angew. Math. **486** (1997), 165–204.

[Geo91] P. L. George, *Automatic Mesh Generation: Application to Finite Element Methods*, John Wiley & Sons, New York, 1991.

[GG95] V. Girault and R. Glowinski, *Error analysis of a fictitious domain method applied to a Dirichlet problem*, Japan J. Indust. Appl. Math. **12** (1995), 487–514.

[GMSW89] P. E. Gill, W. Murray, M. A. Saunders, and M. H. Wright, *Constrained nonlinear programming*, in Handbooks in Operations Research and Management Science, vol. 1, G. L. Nemhauser and A. H. G. Rinnooy Kan, eds., North-Holland, Amsterdam, 1989, pp. 171–210.

[GMW81] P. E. Gill, W. Murray, and M. H. Wright, *Practical Optimization*, Academic Press, New York, 1981.

[Gol89] D. E. Goldberg, *Genetic Algorithms in Search, Optimization and Machine Learning*, Addison–Wesley, Reading, MA, 1989.

[GR79] V. Girault and P.-A. Raviart, *Finite Element Approximation of the Navier–Stokes Equations*, Lecture Notes in Mathematics, vol. 749, Springer-Verlag, Berlin, Heidelberg, New York, 1979.

[Gri85] P. Grisvard, *Elliptic Problems in Nonsmooth Domains*, Monographs and Studies in Mathematics, 24, Pitman, Boston, 1985.

[Gri89] A. Griewank, *On automatic differentiation*, in Mathematical Programming: Recent Developments and Applications, M. Iri and K. Tanabe, eds., Kluwer Academic Publishers, Amsterdam, 1989, pp. 83–108.

[Gri00] A. Griewank, *Evaluating Derivatives. Principles and Techniques of Algorithmic Differentiation*, SIAM, Philadelphia, 2000.

[HA79] E. J. Haug and J. S. Arora, *Applied Optimal Design*, John Wiley & Sons, New York, 1979.

[Häm93] J. Hämäläinen, *Mathematical Modelling and Simulation of Fluid Flows in Headbox of Paper Machines*, Ph.D. thesis, University of Jyväskylä, Department of Mathematics, Jyväskylä, Finland, 1993.

[Has02] J. Haslinger, *A note on contact shape optimization with semicoercive state problems*, Appl. Math. **47** (2002), 397–410.

[HCK86] E. J. Haug, K. K. Choi, and V. Komkov, *Design Sensitivity Analysis of Structural Systems*, Academic Press, Orlando, FL, 1986.

[Hei97] E. Heikkola, *Domain Decomposition Method with Nonmatching Grids for Acoustic Scattering Problems*, Ph.D. thesis, University of Jyväskylä, Department of Mathematics, Jyväskylä, Finland, 1997.

[HGK90] R. T. Haftka, Z. Gürdal, and M. P. Kamat, *Elements of Structural Optimization*, 2nd rev. ed., Kluwer Academic Publishers, Dordrecht, The Netherlands, 1990.

[HHN96] J. Haslinger, I. Hlaváček, and J. Nečas, *Numerical methods for unilateral problems in solid mechanics*, in Handbook of Numerical Analysis, vol. IV, P. G. Ciarlet and J. L. Lions, eds., North–Holland, Amsterdam, 1996, 313–485.

[HJKT00] J. Haslinger, D. Jedelský, T. Kozubek, and J. Tvrdík, *Genetic and random search methods in optimal shape design problems*, J. Global Optim. **16** (2000), 109–131.

[HK00] J. Haslinger and T. Kozubek, *A fictitious domain approach for a class of Neumann boundary value problems with applications in shape optimization*, East-West J. Numer. Math. **8** (2000), 1–23.

[HM92] J. Haslinger and R. Mäkinen, *Shape optimization of elasto-plastic bodies under plane strains: Sensitivity analysis and numerical implementation*, Structural Optim. **4** (1992), 133–141.

[HMT01] J. Haslinger, J.-F. Maitre, and L. Tomas, *Fictitious domain methods with distributed Lagrange multipliers Part I: Application to the solution of elliptic state problems*, Math. Models Methods Appl. Sci. **11** (2001), 521–547.

[HN96] J. Haslinger and P. Neittaanmäki, *Finite Element Approximation for Optimal Shape, Material and Topology Design*, 2nd ed., John Wiley & Sons, Chichester, U.K., 1996.

[HNT86] J. Haslinger, P. Neittaanmäki, and T. Tiihonen, *Shape optimization in contact problems based on penalization of the state inequality*, Appl. Mat. **31** (1986), 54–77.

[HO77] E. Hinton and D. R. J. Owen, *Finite Element Programming*, Academic Press, London, 1977.

[Hol75] J. H. Holland, *Adaptation in Neural and Artificial Systems*, University of Michigan Press, Ann Arbor, MI, 1975.

[HT96] J. Hämäläinen and T. Tiihonen, *Modelling and simulation of fluid flows in a paper machine headbox*, in ICIAM-95, Issue 4: Applied Sciences, Especially Mechanics (Minisymposia), E. Kreuzer and O. Mahrenholtz, eds., Akademie Verlag, Berlin, 1996, pp. 62–66.

[IG00] I. Charpentier and M. Ghemires, *Efficient adjoint derivatives: Application to the atmospheric model Meso-NH*, Optim. Methods Software **13** (2000), 35–63.

[IKP00] K. Ito, K. Kunisch, and G. Peichl, *On the regularization and the numerical treatment of the inf-sup condition for saddle point problems*, Comput. Appl. Math. **21** (2002), 245–274.

[IMS94] IMSL, *The IMSL FORTRAN77 Mathematics and Statistics Libraries*, Visual Numerics Inc., Houston, TX, 1994.

[Ing93] L. Ingber, *Simulated annealing: Practice versus theory*, Math. Comput. Modelling **18** (1993), 29–57.

[Jam88] A. Jameson, *Aerodynamic design via control theory*, J. Sci. Comput. **3** (1988), 233–260.

[KL91] D. Kalman and R. Lindell, *Automatic differentiation in astrodynamical modeling*, in Automatic Differentiation of Algorithms: Theory, Implementation, and Application, A. Griewank and G. F. Corliss, eds., SIAM, Philadelphia, 1991, pp. 228–243.

[KT95] I. Křivý and J. Tvrdík, *The controlled random search algorithm in optimizing regression models*, Comput. Statist. Data Anal. **20** (1995), 229–234.

[KTK00] I. Křivý, J. Tvrdík, and R. Krpec, *Stochastic algorithms in nonlinear regression*, Comput. Statist. Data Anal. **33** (2000), 277–290.

[Lio69] J.-L. Lions, *Quelques méthodes de résolution des problèmes aux limites non linéaires*, Dunod, Gauthier–Villars, Paris, 1969.

[LM68] J.-L. Lions and E. Magenes, *Problèmes aux limites non homogènes et applications*, vol. 1, Travaux et Recherches Mathèmatiques, no. 17, Dunod, Paris, 1968.

[Mic92] Z. Michalewicz, *Genetic Algorithms + Data Structures = Evolution Programs*, Springer-Verlag, Berlin, New York, 1992.

[Mie94] K. Miettinen, *On the Methodology of Multiobjective Optimization with Applications*, Ph.D. thesis, University of Jyväskylä, Department of Mathematics, Jyväskylä, Finland, 1994.

[Mie99] K. M. Miettinen, *Nonlinear Multiobjective Optimization*, Kluwer Academic Publishers, Boston, 1999.

[MM95] K. Miettinen and M. M. Mäkelä, *Interactive bundle-based method for nondifferentiable multiobjective optimization: Nimbus*, Optimization **34** (1995), 231–246.

[MM00] K. Miettinen and M. M. Mäkelä, *Interactive multiobjective optimization system WWW-NIMBUS on the Internet*, Comput. Oper. Res. **27** (2000), 709–723.

[MMM96] K. Miettinen, M. M. Mäkelä, and R. A. E. Mäkinen, *Interactive multiobjective optimization system NIMBUS applied to nonsmooth structural design problems*, in System Modelling and Optimization, J. Dolezal and J. Fidler, eds., Chapman & Hall, London, 1996, pp. 379–385.

[MMM98] K. Miettinen, M. M. Mäkelä, and T. Männikkö, *Optimal control of continuous casting by nondifferentiable multiobjective optimization*, Comput. Optim. Appl. **11** (1998), 177–194.

[MMRS96] B. Mohammadi, J.-M. Malé, and N. Rostaing-Schmidt, *Automatic differentiation in direct and reverse modes: Application to optimum shapes design in fluid mechanics*, in Computational Differentiation: Techniques, Applications, and Tools, M. Berz, C. Bischof, G. Corliss, and A. Griewank, eds., SIAM, Philadelphia, 1996, pp. 309–318.

[MN92] M. M. Mäkelä and P. Neittaanmäki, *Nonsmooth Optimization: Analysis and Algorithms with Applications to Optimal Control*, World Scientific, Singapore, 1992.

[MNP+97] R. A. E. Mäkinen, P. Neittaanmäki, J. Periaux, M. Sefrioui, and J. Toivanen, *Parallel genetic solution for multiobjective MDO*, in Algorithms and Results Using Advanced Computers, Proceedings of the Parallel CFD'96 conference, A. Schiano, A. Ecer, J. Periaux, and N. Satofuka, eds., Elsevier, New York, 1997, 352–359.

[MP01] B. Mohammadi and O. Pironneau, *Applied Shape Optimization for Fluids*, Oxford University Press, Oxford, U.K., 2001.

[MR96] M. Metcalf and J. Reid, *Fortran 90/95 Explained*, Oxford University Press, Oxford, U.K., 1996.

[MRR+53] N. Metropolis, A. W. Rosenbluth, M. N. Rosenbluth, A. H. Teller, and E. Teller, *Equation of state calculations for fast computing machines*, J. Chem. Phys. **21** (1953), 1087–1092.

[MT94] R. A. E. Mäkinen and J. Toivanen, *Optimal shape design for Helmholtz/potential flow problem using fictitious domain method*, in Proceedings of the 5th AIAA/USAF/NASA/ISSMO Symposium on Multidisciplinary Analysis and Optimization, 1994, AIAA Paper 94-4307-CP, 529–536.

[MTP99] R. A. E. Mäkinen, J. Toivanen, and J. Périaux, *Multidisciplinary shape optimization in aerodynamics and electromagnetics using genetic algorithms*, Int. J. Numer. Methods Fluids **30** (1999), 149–159.

[NAG97] NAG, *The NAG Fortran Library Manual: Mark 18*, NAG Ltd., Oxford, U.K., 1997.

[Neč67] J. Nečas, *Les méthodes directes en théorie des équations elliptiques*, Academia, Prague, 1967.

[NH81] J. Nečas and I. Hlaváček, *Mathematical Theory of Elastic and Elastico-plastic Bodies: An Introduction*, Elsevier, Amsterdam, 1981.

[Nit81] J. A. Nitsche, *On Korn's second inequality*, RAIRO Anal. Numér. **15** (1981), 237–248.

[NM64] J. A. Nelder and R. Mead, *A simplex method for function minimization*, Comput. J. **7** (1964), 308–313.

[OR77] N. Olhoff and S. H. Rasmussen, *On single and bimodal optimum buckling loads of clamped columns*, Int. J. Solids Structures **13** (1977), 605–614.

[Par98] B. N. Parlett, *The Symmetric Eigenvalue Problem*, Classics in Applied Mathematics 20, SIAM, Philadelphia, 1998.

[Pri76] W. L. Price, *A controlled random search procedure for global optimization*, Comput. J. **20** (1976), 367–370.

[Rod93] W. Rodi, *Turbulence Models and Their Application in Hydraulics – a State of the Art Review*, A. A. Balkema, Rotterdam, 1993.

[RS93] N. Rostaing-Schmidt, *Différentiation automatique: Application à un problème d'optimisation en météorologie*, Ph.D. thesis, Université de Nice, Sophia Antipolis, 1993.

[Sch82] K. Schittkowski, *The nonlinear programming method of Wilson, Han, and Powell with an augmented Lagrangian type line search function. I. Convergence analysis*, Numer. Math. **38** (1981/82), 83–114.

[SD95] N. Srinivas and K. Deb, *Multiobjective optimization using nondominated sorting in genetic algorithms*, Evolutionary Comput. **3** (1995), 221–248.

[SZ92] J. Sokolowski and J.-P. Zolesio, *Introduction to Shape Optimization: Shape Sensitivity Analysis*, Springer-Verlag, Berlin, 1992.

[SZZ94] K. Schittkowski, C. Zillober, and R. Zotemantel, *Numerical comparison of nonlinear programming algorithms for structural optimization*, Structural Optim. **7** (1994), 1–28.

[Tem77] R. Temam, *Navier–Stokes Equations: Theory and Numerical Analysis*, Studies in Mathematics and Its Applications, vol. 2, North-Holland, Amsterdam, New York, Oxford, 1977.

[TMH98] P. Tarvainen, R. A. E. Mäkinen, and J. Hämäläinen, *Shape optimization for laminar and turbulent flows with applications to geometry design of paper machine headboxes*, in Proceedings of the Tenth International Conference on Finite Elements in Fluids, M. Hafez and J. C. Heinrich, eds., The University of Arizona, Tucson, AZ, 1998, 536–541.

[Toi97] J. Toivanen, *Fictitious Domain Method Applied to Shape Optimization*, Ph.D. thesis, Department of Mathematics, University of Jyväskylä, Jyväskylä, Finland, 1997.

[TŽ89] A. Törn and A. Žilinskas, *Global Optimization*, Lecture Notes in Computer Science, vol. 350, Springer-Verlag, Berlin, New York, 1989.

[Was74] K. Washizu, *Variational Methods in Elasticity and Plasticity*, 2nd ed., Pergamon Press, Oxford, U.K., 1974.

[Wol99] S. Wolfram, *The Mathematica® Book*, 4th ed., Cambridge University Press, Cambridge, U.K., 1999.

Index